Evolutionary Economics and Social Complexity Science

Volume 30

The Japan Association for Evolutionary Economics (JAFEE) always has adhered to its original aim of taking an explicit "integrated" approach. This path has been followed steadfastly since the Association's establishment in 1997 and, as well, since the inauguration of our international journal in 2004. We have deployed an agenda encompassing a contemporary array of subjects including but not limited to: foundations of institutional and evolutionary economics, criticism of mainstream views in the social sciences, knowledge and learning in socio-economic life, development and innovation of technologies, transformation of industrial organizations and economic systems, experimental studies in economics, agent-based modeling of socio-economic systems, evolution of the governance structure of firms and other organizations, comparison of dynamically changing institutions of the world, and policy proposals in the transformational process of economic life. In short, our starting point is an "integrative science" of evolutionary and institutional views. Furthermore, we always endeavor to stay abreast of newly established methods such as agent-based modeling, socio/econo-physics, and network analysis as part of our integrative links. More fundamentally, "evolution" in social science is interpreted as an essential key word, i.e., an integrative and/or communicative link to understand and re-domain various preceding dichotomies in the sciences: ontological or epistemological, subjective or objective, homogeneous or heterogeneous, natural or artificial, selfish or altruistic, individualistic or collective, rational or irrational, axiomatic or psychological-based, causal nexus or cyclic networked, optimal or adaptive, micro- or macroscopic, deterministic or stochastic, historical or theoretical, mathematical or computational, experimental or empirical, agent-based or socio/econo-physical, institutional or evolutionary, regional or global, and so on. The conventional meanings adhering to various traditional dichotomies may be more or less obsolete, to be replaced with more current ones vis-à-vis contemporary academic trends. Thus we are strongly encouraged to integrate some of the conventional dichotomies. These attempts are not limited to the field of economic sciences, including management sciences, but also include social science in general. In that way, understanding the social profiles of complex science may then be within our reach. In the meantime, contemporary society appears to be evolving into a newly emerging phase, chiefly characterized by an information and communication technology (ICT) mode of production and a service network system replacing the earlier established factory system with a new one that is suited to actual observations. In the face of these changes we are urgently compelled to explore a set of new properties for a new socio/economic system by implementing new ideas. We thus are keen to look for "integrated principles" common to the above-mentioned dichotomies throughout our serial compilation of publications. We are also encouraged to create a new, broader spectrum for establishing a specific method positively integrated in our own original way.

Yuichi Ikeda

Data Science of Renewable Energy Integration

The Nexus of Energy, Environment, and Economic Growth

Springer

Yuichi Ikeda
Adv. Integ. Stu. in Human Survivability
Kyoto University
Kyoto, Kyoto, Japan

ISSN 2198-4204 ISSN 2198-4212 (electronic)
Evolutionary Economics and Social Complexity Science
ISBN 978-981-99-8778-8 ISBN 978-981-99-8779-5 (eBook)
https://doi.org/10.1007/978-981-99-8779-5

This Springer imprint is published by the registered company Springer Nature Singapore Pte Ltd.
The registered company address is: 152 Beach Road, #21-01/04 Gateway East, Singapore 189721, Singapore

Paper in this product is recyclable.

Foreword

Dr. Fatih Birol of the International Energy Agency (IEA) says that renewable energy becomes a mainstream and it is getting more important in the future as a main tool to achieve carbon neutral. As everyone notices, energy issue is very complex and it is difficult for any single discipline to understand the whole nature of the energy issue. For example, electrical engineers analyze power system design and study technological elements of various devices. Economists concentrate on the economics of power systems and the institutional design of electricity markets. Environmentalists point out the danger of rising temperatures and sea level rise due to increased greenhouse gases. Undoubtedly, these individual studies are essential, but the actual implementation of their recommendations may often contradict each other. When such contradictions occur, a methodology is needed to resolve the various contradictions and obtain a higher-level solution. The authors of this book argue that such a methodology can be possible by a trans-disciplinary research approach using data science, computational science, network science, and economics. The book attempts to explain the contents as simply as possible without assuming a wide range of knowledge in the natural sciences, economics, and informatics. Practitioners and industrialists may not have the time to read this book in its entirety from beginning to end. In that case, I recommend reading through the chapters that interest you. This book will give you an unprecedented insight. I recommend this book to all those who want to think about and understand the energy issues about renewable energy and its significant positive impact on society.

Tokyo, Japan
2023

Nobuo Tanaka
Chair of the Steering Committee of Innovation for Cool Earth Forum,
Executive Director Emeritus of International Energy Agency, and CEO, Tanaka Global, Inc.

Preface

This book explains various data scientific approaches to analyze the grid integration of renewable energy, for which grid flexibility is the key to coping with its intermittency or fluctuation. It provides the readers with the scope to view renewable energy integration as establishing a distributed energy network instead of the traditional centralized energy system. Data science attempts to elucidate a system through data analysis and takes a bottom-up approach. This explanation is data science in the narrow sense. Computational science, on the other hand, attempts to depict the behavior of a system using mathematical models and takes a top-down approach. Both bottom-up and top-down approaches are needed to understand a system sufficiently. In this book, we use the term "data science" broadly, including computational science. This book covers a wide range of topics. Topics addressed in the book include fluctuation and correlation of renewable energy, unit commitment model to capture intermittency of renewable energy and to operate interconnected power grid, linear programming model of smart grid, causality inference for finance in renewable energy and carbon pricing, network model and coupled oscillator model of evolving micro-grids, hydrogen production using nuclear power. This book also explains the desired innovation to reduce the integration cost significantly using innovative technologies, e.g., energy storage with hydrogen production and vehicle-to-grid technology. Illustrated by analyzing carefully selected examples of renewable integration using different types of grid flexibility, this book is indispensable to readers who make policy recommendations to establish the distributed energy network integrated with large-scale renewable energy by disentangling the nexus of energy, environment, and economic growth.

Kyoto, Japan
September 2023

Yuichi Ikeda

Acknowledgments

I would like to express my deep and sincere gratitude to the following individuals equally. This book would not have been possible without the collaborations, cooperations, advices, and other forms of support of the following individuals: Nobuo Tanaka (Chair of the Steering Committee of Innovation for Cool Earth Forum, Executive Director Emeritus of International Energy Agency, and CEO, Tanaka Global, Inc.), Prof. Peter Taylor (University of Leeds), Dr. Uwe Remme (International Energy Agency), David Elzinga (Asian Development Bank), Dr. Steve Heinen (Vector Ltd.), Prof. Kazuhiko Ogimoto (University of Tokyo), Kazuto Kataoka (University of Tokyo), Prof. Takashi Ikegami (Tokyo University of Agriculture and Technology), Akira Yabumoto (J-Power), Keisuke Inoue (Tokyo Electric Power Company), Hidetoshi Nishimura (Economic Research Institute for ASEAN and East Asia, ERIA), Shigeru Kimura (Economic Research Institute for ASEAN and East Asia, ERIA), Prof. Alexander Keeley (Kyushu University), Hitomi Tanaka (DMG Mori Co., Ltd), Emi Yoshiakwa (Ripple Labs Inc.), Lauren Weymouth (Ripple Labs Inc.), Ken Weber (Ripple Labs Inc.), Zelda Marquardt (Kyoto University), Yu Kimura (Kyoto University, Interoperability Labs Ltd.), Sena Omura (Interoperability Labs Ltd.), and Takeru Shimojima (Interoperability Labs Ltd.).

Contents

Part I
Current World

Chapter 1
Introduction: Today's Our System

1.1 Man-Made Systems

1.1.1 Sustainability

On the Earth, there is an inflow and outflow of energy, such as the intake of sunlight (ultraviolet, visible, and infrared light) and the emission of thermal radiation into space. On the other hand, very little matter enters or leaves the Earth. Although it is not a perfect isolated system because of the energy flow in and out, the Earth can be regarded as an isolated system as a rough approximation. Thermodynamics tells us that, in the equilibrium state of an isolated system, a homogeneous state with no structure that maximizes entropy is realized.

In the past ten years of research on the global economy from a microscopic viewpoint by network analysis with many students and collaborators, we have accumulated examples that show that the global economy is not a homogeneous state but has a characteristic pattern structure [1]. For example, international trade is actively conducted among nearby firms in similar industries. Capital ownership ties among firms are also marked by investment among firms in similar industries. International trade in intermediate goods and investment for their production facility form the world production system. The same pattern is observed in migration, with migrants sending international remittances at the same level as the investment amount among international firms. This characteristic pattern structure is evidence that implies the global economy is not in an ideal equilibrium state.

Let us consider a more realistic system than the isolated one discussed above. The Earth has a geographically uneven distribution of various resources, such as raw materials and various factors of production. For example, crude oil is unevenly distributed in the Middle East, North America, and South Asia. The population is unevenly distributed among countries in the Northern Hemisphere's mid-latitudes. By maximizing configuration entropy with the geographic maldistribution of these

Y. Ikeda, *Data Science of Renewable Energy Integration*,
Evolutionary Economics and Social Complexity Science 30,
https://doi.org/10.1007/978-981-99-8779-5_1

resources as given constraints, some structure reflecting the maldistribution of resources will emerge. However, the uneven distribution of resources alone cannot fully explain the observed characteristic pattern structure of the global economy.

Let us recall that the global economy is embedded in the global environment. The global economy takes in various resources from the environment, referred to as an external system. It emits greenhouse gases and waste to the environment as the external system based on specific values and social norms. This indicates that the global economy is not an isolated system but an open nonequilibrium system in which energy and materials flow in and out of the external system. Furthermore, the global economy can be treated as a steady state since economic growth is small in the short term on an annual scale. Therefore, the global economy can be considered a nonequilibrium steady state driven by the exchange of energy and materials with the external system. In such a state, entropy need not be maximized, which is consistent with having a characteristic structure.

The concept of general equilibrium in economics states that an economic equilibrium state is realized by maximizing the utility of all economic agents with the supply–demand equilibrium as a given constraint. This concept corresponds to entropy maximization in an isolated system and does not explicitly consider the system is nonequilibrium. If the global economy is not in equilibrium as realized by utility maximization, what is the basic principle that explains the global economy? This question provides a new research topic for the statistical physics of nonequilibrium stationary systems. The research for a new fundamental principle to replace general equilibrium is the research in econophysics [2, 3], which I have been working on with my collaborators for the past 17 years, and in network science, where I have been working with many students for the past ten years to clarify examples of pattern structures in the global economy.

Earlier, we pointed out that the global production system takes in various resources from the external system and vents greenhouse gases and waste to the external system based on specific values and social norms. Humans constitute this economic system. Humans and animals take in oxygen and grain and emit CO_2 and waste; plants take in CO_2 and waste emitted by humans and animals and generate oxygen and grain. Thus, we can see that the human–animal and plant systems are complementary, as shown in Fig. 1.1a. In the past, the Earth has evolved as the

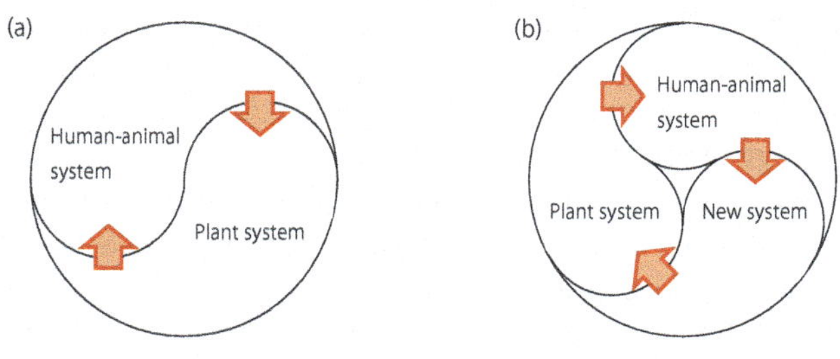

Fig. 1.1 Complementarity: (**a**) Two way and (**b**) Three way, "New system" takes in the global issues represented by global warming and generates what could be useful inputs to the global economic system

totality of these human–animal and plant systems, achieving a state of equilibrium under the given constraints. This is the meaning of sustainability. In recent years, however, global warming caused by excessive CO_2 emissions from the economic system and other global issues resulting from economic globalization have been pointed out as increasingly severe threats to sustainability.

It would be convenient if there were a "new system" to take in the global issues represented by global warming and generate what could be valuable inputs to the global economic system. A "new system" that is complementary to the economic system, like animal–plant systems, would make it possible to improve sustainability, as shown in Fig. 1.1b. There is a very useful new tool for presenting solutions to global issues. It is a distributed ledger based on blockchain technology. Some people may have a negative impression of blockchain technology, thinking of speculation on crypto assets with bursting prices, money laundering, and fraud related to such speculation. However, it is unwise not to take great advantage because of the drawbacks. By entirely using network science, topology, machine learning, quantum informatics, and computational science, it is possible to predict the price burst caused by anomalies such as money laundering fraud to some extent. By using the power of mathematics to suppress shortcomings and taking full advantage of distributed ledgers based on blockchain technology, we can create a "new system" that can solve global issues.

Renewable energy is an effective way to reduce CO_2 emissions. However, solar power and wind energy have the disadvantage of fluctuating power generation depending on the weather. An energy storage device is needed to compensate for unintended fluctuations in power generation. However, having an energy storage device to compensate for fluctuations in renewable energy output is cost-prohibitive. Fortunately, electric vehicles are becoming increasingly popular. Most vehicles spend little time in motion and much of their time parked. Therefore, we consider connecting the energy storage devices of these parked electric vehicles to the power grid and using them as a countermeasure against output fluctuations of renewable energies. Suppose a market for buying and selling the right to use electricity generated by renewable energy or the right to use the energy storage devices of electric vehicles is popularized by a distributed ledger based on blockchain technology. In that case, it will be possible to integrate more renewable energy and electric vehicles into the power grid without significantly increasing costs. This would curb CO_2 emissions and provide a solution to the global problem of global warming. At this time, new and nearly forgotten traditional values will arise, such as acting not only for one's own benefit but also for the benefit of all, that is, large-scale use of renewable energy and lending one's unused capital, which is storage devices in parked electric vehicles, to be used for other people's purposes. This "new system" incorporates the global warming problem and generates new values complementary to the global economic system.

As a prototype of such a "new system," we are currently developing an energy trading system using blockchain technology and conducting a demonstration experiment at a dormitory with the cooperation of many students at our graduate school. Furthermore, in addition to measures against global warming using the energy trading system, we are also developing a life monitoring system that regularly measures electricity consumption, temperature, humidity, and PM2.5 concentration in the living environment using the smart contract function of blockchain technology and is developing measures to prevent depopulation in rural areas by ensuring safe living conditions. We also consider developing a system to manage personal identification to medical care based on personal records for health care using blockchain technology to improve wellbeing for an aging society.

1.1.2 Beyond Reductionism

To further improve sustainability, we need a major change in our thinking. To achieve this goal, first, we consider the simplicity. The concept of simplicity has its origin in the reductionism. Reductionism is the idea that the object under consideration is composed of many lower elements and that if the elements are simple enough, we can understand the nature of the object under consideration from the properties of its elements. Simplicity means that the object under consideration can be divided into sufficiently small pieces and the divided elements can be understood.

From elementary particles to the universe, it constitutes a reductionism hierarchy. The orders of length and energy of each hierarchy are "particle and nuclei" (10^{-15} m, >GeV), "atom and molecule" (10^{-9} m, eV), "macromolecule" (10^{-6} m, meV), "society and economy" (m, 10^{-3}eV), "planet" (10^{7} m, 10^{-2} eV), "solar system" (10^{12} m, eV-MeV), and "universe" (10^{28} m, eV – 10^{20} eV). The reductionism is based on a linear theory. The laws of the higher hierarchy with a larger scale are expressed differently from the laws of the lower hierarchy with a smaller scale. When the laws of the higher hierarchy can be explained by deduction based on the laws of the lower hierarchy, we consider that we have understood the phenomenon. Each level of theory is linear, and it is possible to predict another phenomenon by superposition of solutions that explain one phenomenon at the same level. Historically, the theories at each hierarchical level have constituted a separate body of academic discipline.

However, we note that the trend of academic research has grown since the middle of the twentieth century to overcome academic barriers beyond the boundaries of a specific academic discipline in expanding the research fields of various sciences. This trend began with a theory for discussing the formation of living organisms using "the negative entropy", i.e., the emission of entropy outside a system [4], game theory to explain the rationale and deductive aspects of human decision-making [5], and cybernetics to attempt to establish a general theory of control in nature, biological systems, and society [6].

In 1970, nonlinear phenomena called deterministic chaos, characterized by sensitive dependence on initial conditions, strange attractors, and self-similarity, were found in various hierarchical layers of the universe. The existence of chaos is considered to be the most substantial evidence of the breakdown of reductionism.

At the end of the twentieth century, the study of nonlinear and nonequilibrium systems evolved into complex systems and complex adaptive systems [7, 8]. Today, a quiet revolution to overcome academic barriers beyond a specific academic discipline's boundaries is in progress by adding a new strong perspective of complex networks [9, 10]. This was the first era the complexity was explicitly recognized.

Looking at the history of the development of complexity from a different angle, it is also the history of the development of systems science. Systems science is the study of natural, artificial, and social systems. Natural systems are mainly the subject of natural science. It includes systems in physical science. Artificial systems are the systems mainly targeted by engineering. These include power systems, information systems, and transportation systems. The artificial system is often called the man-made system. Social systems are systems that are mainly targeted by social sciences. They include economic systems, such as labor markets, financial markets, and political systems. Of the social system, international trade and foreign direct investment, in particular, constitute the global production system. In global issues, various social and natural sciences factors are intertwined in a complicated manner. However, we need a general methodology for integrating social and natural sciences. The difficulty in tackling global issues is distilled by the need for a general methodology for integrating social sciences and natural sciences.

In the modern history of humanity, those who establish the man-made system have dominated the world. For example, the following people are well known as such: Thomas Edison (the founder of General Electric Company), Alexander Graham Bell (the founder of the company later known as AT&T), Henry Ford (the founder of Ford Motor Company), Jeff Bezos (the founder of Amazon), Larry Page (one of the founders of Google), and Mark Zuckerberg (the founder of Facebook, now renamed Meta Platforms). Here, we exclude religious and political thinkers and practitioners who contributed to the construction of social systems. Engineers who do not aim to build systems are rather recognized as craftsmen.

This book discusses the electric power system as a man-made system classified as a complex system. It examines the nexus of energy, environment, and economic growth, which is a characteristic of this man-made system.

1.2 Centrally Managed Power System

1.2.1 Energy Balance Table

The energy flow from the primary energy supply, through the conversion process, to final consumption is shown in a tabular form and is called an energy balance table. It usually shows the energy flow over a year. Globally, the International Energy Agency (IEA) of the Organization for Economic Cooperation and Development (OECD) produces an annual table by country for OECD member countries and many non-member countries. Different countries use different units of energy

Table 1.1 Energy Conversion Factors

From \To	TJ	Gcal	Mtoe	MBtu	GWh
TJ	1	238.8	2.388×10^{-5}	947.8	0.2778
Gcal	4.2868×10^{-3}	1	10^{-7}	3.968	1.163×10^{-3}
Mtoe	4.1868×10^{4}	107	1	3.968×10^{7}	11630
MBtu	1.0551×10^{-3}	0.252	2.52×10^{-8}	1	2.931×10^{-4}
GWh	3.6	860	8.6×10^{-5}	3412	1

depending on their statistical data. The conversion factors of the units are summarized in Table 1.1.

The energy balance table shows primary energy production, imports, exports, inventory changes, secondary energy conversion by the energy industry, and consumption by each final energy consumption sector by energy type [11]. For energy industries such as petroleum refining, power generation, and heat supply, energy used as raw materials is shown minus, and energy produced as products is shown plus. For power generation by nuclear power, hydroelectric power, primary energy production is calculated based on the average thermal efficiency of thermal power generation. A basic understanding of different energies [12, 13] facilitates a detailed understanding of energy balance tables. The energy balance tables for Japan, Germany, and the USA in 2020 are shown in Tables 1.2, 1.3, and 1.4. In those tables, TPES and TFC are abbreviations of "Total Primary Energy Supply" and "Total Final Consumption", respectively.

Each column corresponds to the production of each energy, and each row shows the energy flow. Production comes from coal, gas, and oil mines. Nuclear is the estimated value, which is the primary heat equivalent of the electricity produced by a nuclear power plant with an average thermal efficiency of 33%. The flow is negative for stock building and positive for stock use. Two rows under TPES are transformation: "Electricity, CHP and heat plants" and "Oil refineries, transformation". Transformation means a change from the primary energy to the secondary energy. Transformation is negative for input and positive for output. In Tables 1.2, 1.3, and 1.4, there are only two items for transformation, but in reality, detailed data items are maintained. Five rows under TFC are end use: "Industry", "Transport", "Residential", "Commercial and public services", and "Other final consumption". End use means energy consumption by sector for each energy.

Using the energy balance tables, important figures are readily calculated. For instance, oil dependence of Japan in 2020 is calculated as $(5076 + 1118)/16{,}109 \times 100 = 38.45\%$. Further, the import dependence of Japan in 2020 is calculated as $14{,}941/(14{,}941 + 1816) \times 100 = 89.16\%$. The oil dependence of Germany in 2020 is calculated as $(3026 + 323)/11{,}654 \times 100 = 28.74\%$. Further, the import dependence of Germany in 2020 is calculated as $8981/(8981 + 4046) \times 100 = 68.94\%$. Oil dependence of the USA in 2020 is calculated as $(36{,}261 - 6860)/85{,}324 \times 100 = 34.46\%$. Further, the import dependence of the USA in 2020 is calculated as $19{,}852/(19{,}852 + 90{,}437) \times 100 = 18.00\%$.

Table 1.2 Japan's Energy Balance Table in 2020 (PJ)

Flow \Product	Coal, peat, and oil shale	Crude, NGL, and feedstocks	Oil products	Natural gas	Nuclear	Renewables and waste	Electricity	Heat	Total
Production	18	18	0	81	423	1277	0	.	1816
Imports	4351	4904	1841	3761	0	84	0	0	14,941
Exports	−94	0	−403	0	0	0	0	0	−497
TPES	4274	5076	1118	3858	423	1360	0	0	16,109
Electricity, CHP, and heat plants	−2705	−11	−351	−2738	−423	−1087	3633	15	−3667
Oil refineries, transformation	0	−5150	5270	0	0	0	0	0	119
TFC	784	0	5548	1123	0	265	3267	23	11,009
Industry	748	0	689	424	0	140	1145	0	3146
Transport	0	0	2534	1	0	18	62	0	2615
Residential	0	0	491	390	0	9	952	1	1844
Commercial and public services	5	0	452	298	0	75	1095	22	1947
Other final consumption	30	0	1382	10	0	22	12	0	1456

Table 1.3 Germany's Energy Balance Table in 2020 (PJ)

Flow \Product	Coal, peat, and oil shale	Crude, NGL, and feedstocks	Oil products	Natural gas	Nuclear	Renewables and waste	Electricity	Heat	Total
Production	979	130	0	169	702	2065	0	0	4046
Imports	883	3516	1481	2783	0	145	172	0	8981
Exports	−59	0	−929	0	0	−127	−241	0	−1356
TPES	1865	3626	323	3123	702	2083	−69	0	11,654
Electricity, CHP, and heat plants	−1361	0	−48	−804	−702	−1339	2038	427	−1789
Oil refineries, transformation	0	−4048	3989	0	0	0	0	0	−59
TFC	240	0	3614	2275	0	722	1727	376	8955
Industry	210	0	145	812	0	173	760	168	2269
Transport	0	0	1950	20	0	140	40	0	2150
Residential	13	0	544	910	0	288	457	160	2374
Commercial and public services	0	0	115	416	0	86	451	48	1116
Other final consumption	17	0	860	116	0	34	19	0	1045

Table 1.4 The USA's Energy Balance Table in 2020 (PJ)

Flow \Product	Coal, peat, and oil shale	Crude, NGL, and feedstocks	Oil products	Natural gas	Nuclear	Renewables and waste	Electricity	Heat	Total
Production	10,788	30,208	0	32,984	8980	7477	0	0	90,437
Imports	111	13,947	3021	2471	0	82	221	0	19,852
Exports	−1720	−7555	−8569	−5203	0	−146	−51	0	−23,244
TPES	9280	36,261	−6860	30,105	8980	7387	170	0	85,324
Electricity, CHP, and heat plants	−8341	0	−358	−12,105	−8980	−3881	15,260	457	−17,947
Oil refineries, transformation	0	−31,595	30,692	0	0	0	0	0	−903
TFC	556	381	27,963	14,925	0	3503	13,600250	61,178	
Industry	542	0	761	5697	0	1336	2557	197	11,090
Transport	0	0	20,495	1040	0	1408	37	0	22,980
Residential	0	0	587	4603	0	529	5331	..	11,049
Commercial and public services	14	0	478	3122	0	166	4574	52	8407
Other final consumption	0	381	5643	464	0	64	1101	..	7653

Table 1.5 Carbon Emission Factors

Fuel	Carbon Emission Factor (tC/TJ)
Crude Oil	20.0
Natural Gas	17.2
Gasoline	18.9

Other important figures are CO_2 emission derived from energy use (CO_2). This quantity is calculated as "Energy consumption (TJ) $\times CO_2$ emission factors (tCO_2/TJ)". This is identical to "Energy consumption (TJ) $\times$ C emission factors (tC/TJ) $\times$ 44/12". Carbon emission factors are shown in Table 1.5 for fossil fuels. CO_2 emission from oil consumption of Japan in 2020 is $5076 \times 20.0 = 101{,}520tC$. CO_2 emission from natural gas consumption of Japan in 2020 is $3858 \times 17.2 = 66{,}358tC$. CO_2 emission from oil consumption of Germany in 2020 is $3026 \times 20.0 = 60{,}520tC$. CO_2 emission from natural gas consumption of Germany in 2020 is $3123 \times 17.2 = 53{,}716tC$. CO_2 emission from oil consumption of the USA in 2020 is $36{,}261 \times 20.0 = 725{,}220tC$. CO_2 emission from natural gas consumption of the USA in 2020 is $30{,}105 \times 17.2 = 517{,}806tC$.

1.3 Elementary Power System Engineering

1.3.1 Mechanism of Frequency Control

Since the demand for electric power changes from time to time, the output from the generators must be adjusted to these changes. To do so, the amount of change in demand must be detected in some way. In a power system based on AC transmission, changes in power demand can be detected from the rotational speed, or frequency of rotation, of the generator. Consider a simple power system consisting of one generator and one load.

$$M\frac{df}{dt} = P_m - P_e, \tag{1.1}$$

where P_m is, for example, the rotational power of a steam turbine in a thermal power generator, P_e is the generator's electrical output, M is the inertia constant, and f is the AC frequency. An imbalance between the supply and demand of electricity causes a difference between P_m and P_e, which appears as a change in f.

Table 1.6 summarizes the demand–supply relationship and frequency change. Based on the relationship between the supply–demand relationship and the frequency change, the feedback control to keep the supply–demand balance of the power system is called frequency control. Frequency control is performed by two types of control loops: local control loops and global control loops.

Table 1.6 Supply–Demand Balance and Frequency

Supply–Demand Balance	Rotor Acceleration	Frequency Change
$P_m > P_e$	Positive (Acceleration)	Increase $f > f_0$
$P_m = P_e$	0 (No acceleration/deceleration)	Constant $f = f_0$
$P_m < P_e$	Negative (Deceleration)	Decrease $f < f_0$

In the local control loop, the supply–demand balance $P_m = P_e$ can be maintained by measuring the rotor frequency f (equal to the AC frequency) at individual generators and adjusting the amount of steam entering the thermal generator using a regulator. However, the deviation from the reference frequency F_0 (frequency deviation) cannot be returned to 0. To eliminate the frequency deviation, a global control loop is required.

A load dispatching center can issue commands to change the rotating power to multiple generators and adjust the rotating power by speed changers at each generator to keep the AC frequency f of the system equal to the reference frequency f_0.

The deviation of the AC frequency f to the reference frequency f_0 is written as

$$\Delta f = f - f_0. \tag{1.2}$$

When the governor detects a frequency deviation, the rotational force is changed accordingly.

$$\Delta P_m = -\frac{1}{r}\Delta f, \tag{1.3}$$

where r is a parameter that describes the characteristics of the governor and is called the speed regulation ratio. If the AC frequency falls below the reference value, Δf will be negative, meaning insufficient rotational power. The local control loop using the governor and the global control loop using the speed changer are shown in Fig. 1.2.

We consider a power system with two generators rotating synchronously and supplying power at load P_L.

$$M_1\frac{df}{dt} = P_{m1} - P_{e1} \tag{1.4}$$

$$M_2\frac{df}{dt} = P_{m2} - P_{e2} \tag{1.5}$$

$$P_{e1} + P_{e2} = P_L. \tag{1.6}$$

Here, for generators 1 and 2, the generator inertia constants are M_1 and M_2, respectively, the outputs are P_{e1} and P_{e2}, respectively, and the rotational forces are P_{m1} and P_{m2}, respectively.

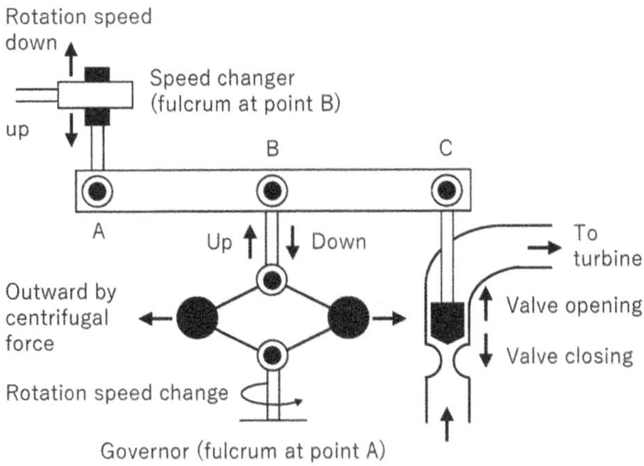

Fig. 1.2 Control Loops: The Governor provides local control, and the speed changer provides global control

Suppose the load changes stepwise from P_{L0} to ΔP_L.

$$P_L \to P_{L0} + \Delta P_L. \tag{1.7}$$

The AC frequency of the supplied power is accordingly increased from f_0 by Δf, and the rotational power is increased from P_{m10} to P_{m10}. The rotational force changes from P_{m10} by ΔP_{m1} and from P_{m20} by ΔP_{m2}.

$$f \to f_0 + \Delta f \tag{1.8}$$

$$P_{m1} \to P_{m10} + \Delta P_{m1} \tag{1.9}$$

$$P_{m2} \to P_{m20} + \Delta P_{m2}. \tag{1.10}$$

From equations (1.4), (1.5), (1.6), the following equations describe the entire power system:

$$(M_1 + M_2)\frac{df}{dt} = P_{m1} + P_{m2} - P_L. \tag{1.11}$$

Also, substituting equations (1.7), (1.9), and (1.10) into the above equation, we obtain the following equation for the frequency deviation:

$$(M_1 + M_2)\frac{d\Delta f}{dt} = -\left(\frac{1}{r_1} + \frac{1}{r_2}\right)\Delta f - \Delta P_L. \tag{1.12}$$

This equation can be solved immediately and yields the following solution:

$$\Delta f(t) = e^{-At} + B \tag{1.13}$$

$$A = \frac{\frac{1}{r_1} + \frac{1}{r_2}}{M_1 + M_2} \tag{1.14}$$

$$B = \frac{\Delta P_L}{M_1 + M_2}. \tag{1.15}$$

This result shows that the frequency deviation Δf does not become 0 after an infinite time and that the frequency cannot be returned to the reference frequency only by adjusting the rotational power by the governor. From the viewpoint of power quality, it is necessary to keep the frequency at the standard reference value. For this purpose, global frequency control using a speed changer called Load Frequency Control (LFC) is used.

1.3.2 Power System Stability

Power system stability is one of the most critical issues for integrating fluctuating renewable energy into a power grid. Consider a mass point in potential energy shown in Fig. 1.3, and explain its dynamical stability. Figure 1.3a shows that a mass point placed at point S is stationary. Even if its position is slightly changed, it will return to point S. On the other hand, a quality point placed at point U can be made to remain stationary if sufficient care is taken. However, if its position is changed even

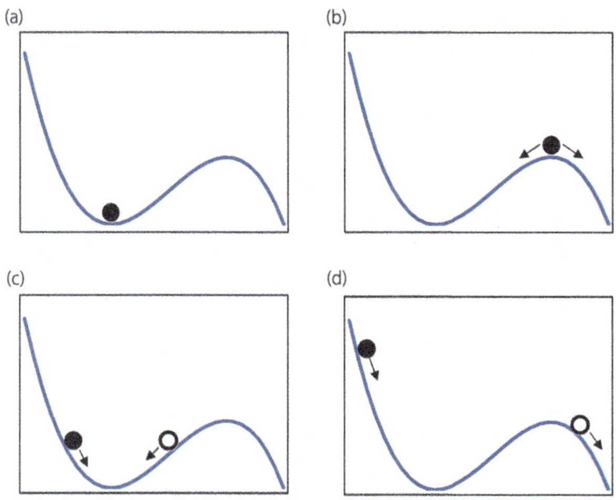

Fig. 1.3 System stability: A mass point placed in potential energy is a simplification of a single machine connected to an infinite bus. (**a**) Stable equilibrium. (**b**) Unstable equilibrium. (**c**) Transiently stable. (**d**) Transiently unstable

slightly, as in Fig. 1.3b, it will fall off the potential energy pile. Point S is stable equilibrium, and point U is unstable equilibrium.

Consider the motion of masses placed at different positions as in Fig. 1.3c. The mass placed at point A oscillates near point S, dissipates kinetic energy due to friction, and eventually comes to rest at point S. The mass placed at point A' oscillates near point S, dissipates kinetic energy due to friction, and eventually rests at point S. The same is true for the mass point placed at point A'. Such a system is transiently stable. However, Fig. 1.3d shows that the mass point placed at point B passes through point S, overcomes point U, and tumbles down to the proper valley. Such a system is transiently unstable.

In the following, we will discuss the stability of the power system based on this discussion of dynamical stability. Power systems made up of many synchronous machines and transmission lines exhibit synchronization. The stability and control of this synchronization of the machines have been studied in power system engineering [14].

The dynamical property of a single synchronous machine is often studied in the single machine connected to the infinite bus shown in Fig. 1.4a, where V_g, $\angle\delta$, I, x_l, r_l, V_∞, and $\angle 0$ are the generator voltage, generator phase, current, transmission line reactance, transmission line resistance, infinite bus voltage, and infinite bus phase, respectively. Here, the infinite bus is characterized by constant voltage and constant frequency. The active power P_e and reactive power Q_e are calculated by

$$P_e + i\,Q_e = V_g \exp(i\delta) I^*, \tag{1.16}$$

where $*$ indicates the complex conjugate, and the current I is given by

$$I = \frac{V_g \exp(i\delta) - V_\infty}{r_l + ix_l}. \tag{1.17}$$

By substituting Eq. (1.17) into Eq. (1.16), the active power P_e is obtained as the real part of the equation

$$P_e = \frac{V_g V_\infty}{x_l} \sin\delta = P^{\max} \sin\delta, \tag{1.18}$$

by considering relation $r_l \ll x_l$. P_e is plotted as a function of the generator phase δ in Fig. 1.4b. It should be noted that P_e is the electric power transmitted through the transmission line and works as a synchronizing force in power systems.

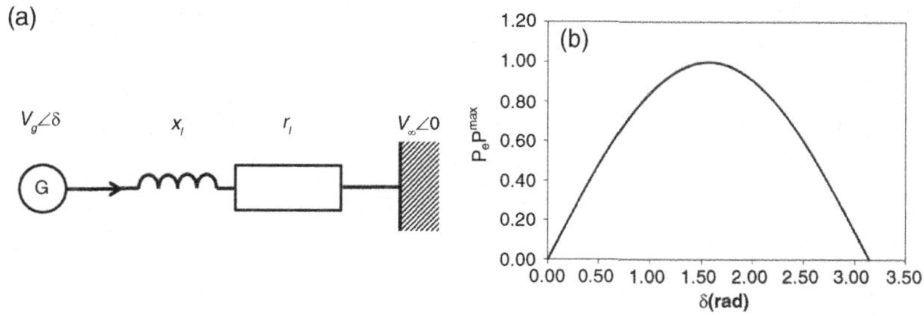

Fig. 1.4 **Power system stability** (**a**) single machine connected to infinite bus and (**b**) synchronizing force are shown

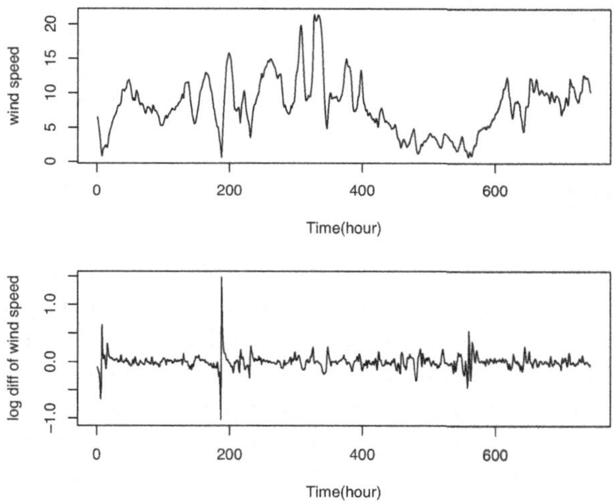

Fig. 1.5 **NASA The POWER Project Data:** Wind speed (m/s) and its change rate at 50 m from the surface in London: Jan. 2020

1.4 Intermittent Renewable Energy

Hourly changes of wind speed at 50 m from the surface in London for January 2020 and July 2020 are shown in the following figures: Figs. 1.5 and 1.6, respectively.

The top panel shows wind speed s_t in m/s, and the bottom panel shows its change rate defined as the log-difference of the wind speed: $r_t = \log s_t - \log s_{t-1}$. As seen in the upper panel, the wind continues to blow throughout the day. This contrasts sharply with the sunshine, which is always zero at night. However, wind speed constantly fluctuates and behaves erratically. This trend is consistent throughout the year. The change rate in wind speed shown in the lower panel shows long periods with the same level of magnitude but occasionally substantial changes. The change rate is greater in winter than in summer in London.

For a time series of each month in 2020, see Appendix A.

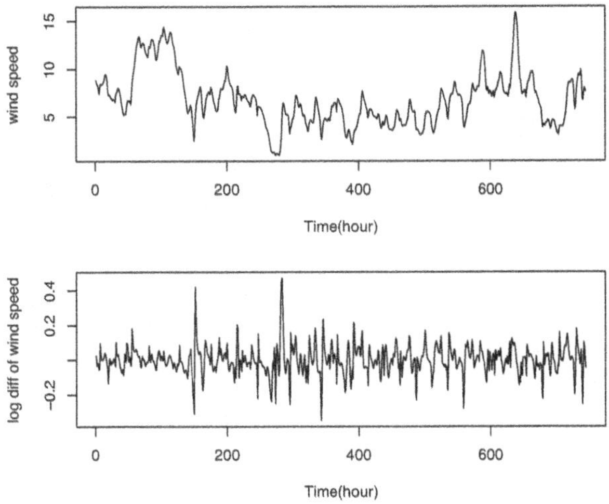

Fig. 1.6 NASA The POWER Project Data: Wind speed (m/s) and its change rate at 50 m from the surface in London: Jul. 2020

1.5 Summary

In this chapter, I have attempted to give an overview of the characteristics of the socio-technical systems we use daily. In particular, the electric power system is a typical example of a system based on central control, and advanced control technologies have been developed to ensure stable operation. Today, however, we are required to integrate renewable energies such as solar and wind power into the power grid to reduce greenhouse gas emissions. Renewable energies are known to have output fluctuations, and to understand this in practice, a time series of wind speeds in London is presented.

References

1. Yuichi Ikeda, Hiroshi Iyetomi, Takayuki Mizuno, "Big Data Analysis on Global Community Formation and Isolation -Sustainability and Flow of Commodities, Money, and Humans", Springer Nature, 2021, ISBN: 9789811549434.
2. Hideaki Aoyama, Yoshi Fujiwara, Yuichi Ikeda, Hiroshi Iyetomi, Wataru Souma, "Econophysics and Companies –Statistical Life and Death in Complex Business Networks–", Cambridge University Press, 2010, ISBN: 9780521191494.
3. Hideaki Aoyama, Yoshi Fujiwara, Yuichi Ikeda, Hiroshi Iyetomi, Wataru Souma, Hiroshi Yoshikawa, "Macro-Econophysics –New Studies on Economic Networks and Synchronization–", Cambridge University Press, 2017, ISBN: 9781107198951.
4. Erwin Schrëdinger, "What Is Life? The Physical Aspect of the Living Cell", Cambridge University Press, 1944, ISBN: 0-521-42708-8
5. John von Neumann, Oskar Morgenstern, "Theory of Games and Economic Behavior", Princeton University Press, 1944, ISBN: 978-0691130613
6. Norbert Wiener, "Cybernetics Or Control and Communication in the Animal and the Machine", MIT Press, 1961, ISBN: 026273009X, 9780262730099

7. John H. Holland, "Adaptation in Natural and Artificial Systems: An Introductory Analysis with Applications to Biology, Control, and Artificial Intelligence", MIT Press, 1992, ISBN: 9780262581110, 0262581116
8. Murray Gell-Mann, "The Quark and the Jaguar: Adventures in the Simple and the Complex", Little, Brown & Company, 1994, ISBN-13: 978-0316907415
9. D. J. Watts , S. H. Strogatz, "Collective dynamics of 'small-world' networks", Nature. 1998 Jun 4;393(6684):440-2. https://doi.org/10.1038/30918.
10. Albert-László Barabási, "Linked: How Everything Is Connected to Everything Else and What It Means for Business, Science, and Everyday Life", Basic Books, 2014, ISBN-13: 978-0465085736
11. IEA world energy balances highlights (2022 edition) https://www.iea.org/data-and-statistics/data-product/world-energy-balances
12. David J. c. Mackay, "Sustainable Energy – Without the Hot Air", UIT Cambridge Ltd; Illustrated edition, May 2009, ISBN-13: 978-0954452933
13. Robert L. Jaffe and Washington Taylor, "The Physics of Energy", Cambridge University Press, January 2018, ISBN-13: 978-1107016651
14. P. Kunder, "Power System Stability and Control" (McGraw-Hill, New York, 1993), p. 221.

Chapter 2
Statistical Analysis

This chapter briefly explains the basics of statistics, regression analysis, and time series analysis. There are excellent textbooks published on these basics [1, 2].

2.1 Elementary Statistics

This section reviews elementary statistics as a first step in the data science needed to study various man-made systems.

2.1.1 Statistical Quantities

Consider data x with n components.

$$x = \{x_1, x_2, \cdots, x_n\}. \tag{2.1}$$

Data x can be divided into the following m bins:

$$X_0 = \min[x], \tag{2.2}$$

$$X_m = \max[x], \tag{2.3}$$

$$\Delta X = \frac{X_m - X_0}{m}, \tag{2.4}$$

Y. Ikeda, *Data Science of Renewable Energy Integration*,
Evolutionary Economics and Social Complexity Science 30,
https://doi.org/10.1007/978-981-99-8779-5_2

Fig. 2.1 Frequency distribution: Each bin contains milk up to the red line

where X_0 is the lower limit of the smallest bin and X_m is the upper limit of the largest bin. ΔX is a bin size. The smallest bin is $i = 1$, and the largest bin is $i = m$. The frequency F_i is the number of data samples in each bin $X_i - X_{i-1}$. The frequency distribution

$$\{(X_1, F_1), (X_2, F_2), \cdots, (X_m, F_m)\} \tag{2.5}$$

is shown in Fig. 2.1.

The mean is the value obtained by dividing the sum of each element by the number of elements belonging to a population,

$$\mu = \frac{1}{N}\sum_{i=1}^{N} x_i. \tag{2.6}$$

The deviation is the difference between the value of each element belonging to the mean of the population,

$$\delta_i = x_i - \mu. \tag{2.7}$$

The variance is the mean square of the deviations,

$$\sigma^2 = \frac{1}{N}\sum_{i=1}^{N} \delta_i^2, \tag{2.8}$$

where σ is called the standard deviation.

We apply the transformation that makes the mean zero and the standard deviation one.

$$z_i = \frac{x_i - \mu}{\sigma}(i = 1, \cdots, n). \tag{2.9}$$

Skewness is a measure of the symmetry of the distribution and is 0 for a normal distribution. Compared to the normal distribution, it takes a positive value when the distribution is shifted to the left and the tail extends to the right, and a negative value

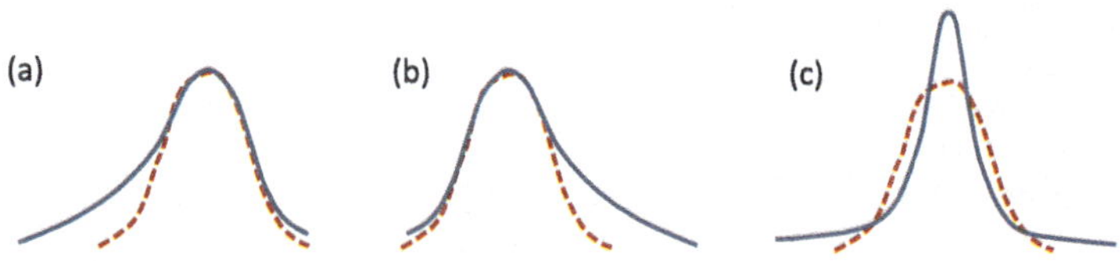

Fig. 2.2 Skewness and kurtosis: The red dotted curve represents the normal distribution being compared. (**a**) Negative skewness. (**b**) Positive skewness. (**c**) Positive kurtosis

when the distribution is shifted to the right and the tail extends to the left, as shown in Fig. 2.2a and b:

$$skewness = \mathrm{E}[\mathrm{z}^3] = \frac{1}{\mathrm{n}}\sum_{\mathrm{i}=1}^{\mathrm{n}} \mathrm{z}_\mathrm{i}^3. \tag{2.10}$$

The kurtosis is a measure of the sharpness of a distribution. Compared to a normal distribution, kurtosis is positive for distributions with sharp peak and long tail, as shown in Fig. 2.2c, and negative for distributions with more rounded peak and short tail:

$$kurtosis = \frac{\mathrm{E}[\mathrm{z}^4]}{\mathrm{E}[\mathrm{z}^2]^2} - 3. \tag{2.11}$$

Next, we discuss the inequality among the elements of the population. We obtain $y = \{y_1 < \cdots < y_k < \cdots < y_n\}$ by $y = Sort[x]$ in ascending order. A two-dimensional plot with

$$\frac{\mathrm{rank}[\mathrm{y_k}]}{n} \tag{2.12}$$

on the horizontal axis and

$$\frac{\sum_{i=1}^{k} y_i}{\sum_{i=1}^{n} y_i} \tag{2.13}$$

on the vertical axis yields a Lorenz curve. The Gini coefficient is defined as the area between the Lorenz curve and a diagonal line multiplied by two. The Gini coefficient is an indicator of inequality, and its range is from 0 to 1. The larger the Gini coefficient means, the more inequality (Fig. 2.3).

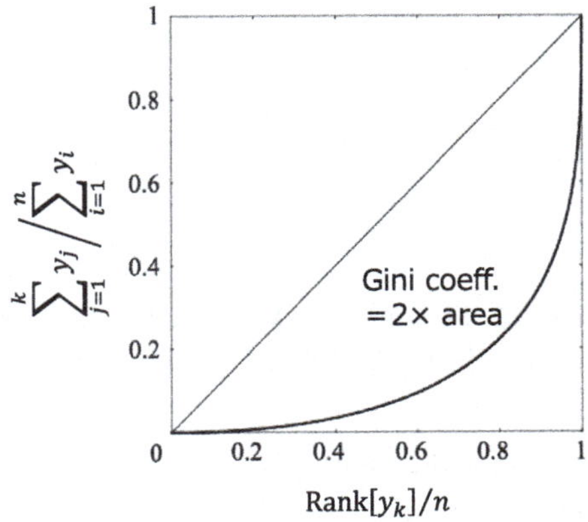

Fig. 2.3 Lorenz curve and Gini coefficient: Gini coefficient is an area between the Lorenz curve and a diagonal line multiplied by two

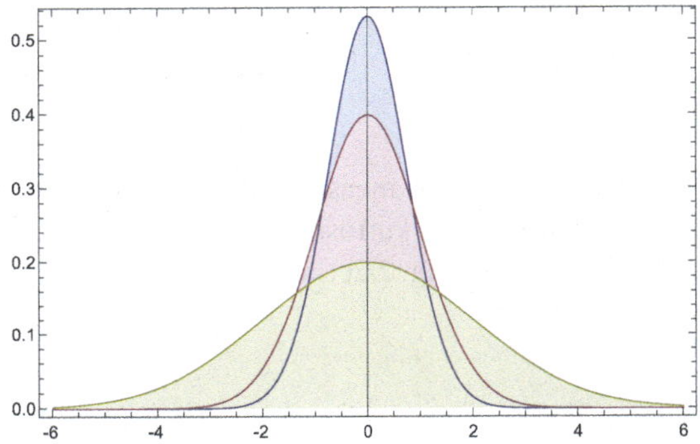

Fig. 2.4 Probability density function of the normal distribution: $\mu = 0, \sigma = \{0.75, 1.0, 2.0\}$

2.1.2 *Probability Distribution*

In many statistical analyses, the reference probability density is the normal distribution or Gaussian distribution:

$$p(x) = N(\mu, \sigma^2) = \frac{1}{\sqrt{2\pi}\sigma} \exp\left[-\frac{(x-\mu)^2}{2\sigma^2} \right], \tag{2.14}$$

where μ and σ are a mean and a standard deviation, respectively. We plot the probability density function of the normal distribution $N(\mu, \sigma^2)$ for $\mu = 0$, $\sigma = 0.75, 1.0, 2.0$ in Fig. 2.4.

The binomial distribution f_x is a discrete distribution of the number of success x in n-times trials, when the probability of success is equal to p:

$$f_x = \binom{n}{x} p^x (1-p)^{n-x} \tag{2.15}$$

$$\binom{n}{x} = \frac{n!}{(n-x)!x!} \tag{2.16}$$

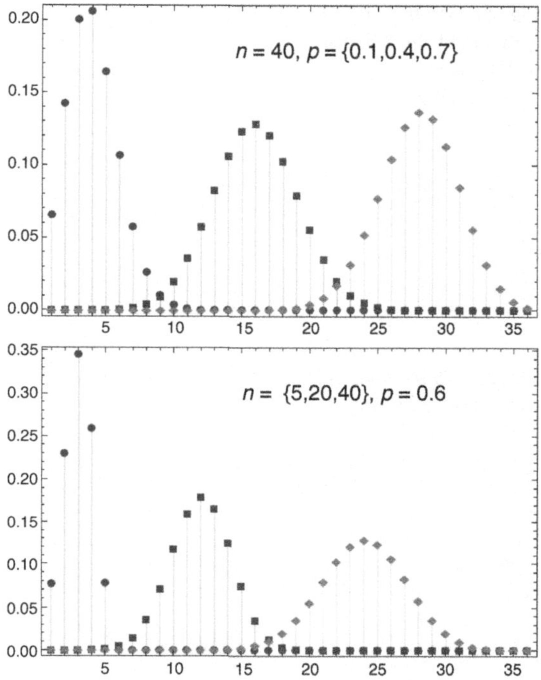

Fig. 2.5 The binomial distribution: In the upper panel, n is fixed and plotted for different p. In the lower panel, p is fixed and plotted for different n

$$\mu = np \tag{2.17}$$

$$\sigma^2 = np(1 - p). \tag{2.18}$$

Here, both x and n are integers. Figure 2.5 shows the parameter dependences of the binomial distributions. The distribution shifts toward the right as p or n increases.

When p is small, the binomial distribution can be approximated by the Poisson distribution, a continuous probability density function.

$$p(x) = \frac{\lambda^x}{x!} e^{-\lambda}, \tag{2.19}$$

where $\mu = \lambda$ and $\sigma^2 = \lambda$. Note that the mean is equal to the variance. Figure 2.6 shows the parameter dependences of the Poisson distributions for $\lambda = 5.0, 10.0, 20.0$.

Using the probability density function $p(x)$, the probability P can be written as $P = p(x)dx$. The height of the elongated rectangular area in Fig. 2.7 is the probability density $p(x)$, and the width is dx, so the area equals the probability P.

The cumulative probability $P_<(x)$ is given by

$$P_<(x) = \int_{-\infty}^{x} p(x')dx', \tag{2.20}$$

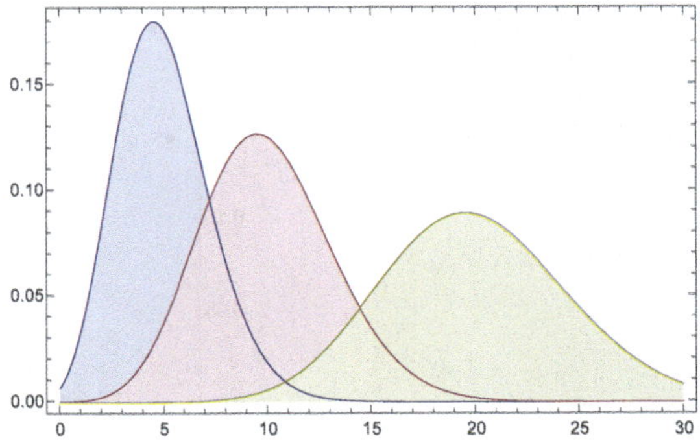

Fig. 2.6 Probability density function of the Poisson distribution: $\lambda = \{5.0, 10.0, 20.0\}$

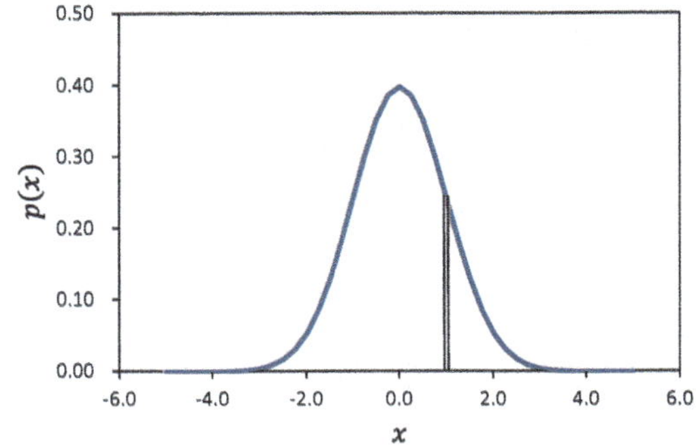

Fig. 2.7 Probability density $p(x)$ **and probability** $P = p(x)dx$: The width of the rectangle in the vicinity of $x = 1$ is dx, the height is $p(x)$, and the area of the rectangle is equal to the probability P

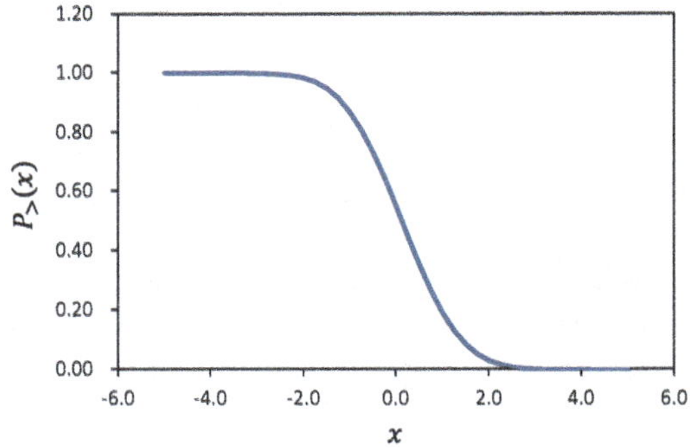

Fig. 2.8 Complementary cumulative probability $P_>(x)$: See Eq. (2.21) for definition

and the complementary cumulative probability $P_>(x)$ is given by

$$P_>(x) = \int_x^\infty p(x')dx'. \tag{2.21}$$

The complementary cumulative probability $P_>(x)$ calculated from the probability density $p(x)$ shown in Fig. 2.7 using Eq. (2.21) is shown in Fig. 2.8.

Therefore, we obtain the probability density $p(x)$ by calculating the derivative of $P_>(x)$:

$$p(x) = -\frac{P_>(x)}{x}. \tag{2.22}$$

The probability of event x and event y occurring simultaneously is described by a joint probability density p_{xy}:

$$p_{xy} = p(x, y). \tag{2.23}$$

In this case, the probability of event x occurring, regardless of event y, is given by the marginal probability p_x:

$$p_x = \int_{-\infty}^{\infty} p_{xy} dy. \tag{2.24}$$

Similarly, the probability of event y occurring, regardless of event x, is given by the marginal probability p_y:

$$p_y = \int_{-\infty}^{\infty} p_{xy} dx. \tag{2.25}$$

2.1.3 *Correlation Coefficients*

The correlation coefficient is an indicator that relates the change in the first random variable with the change in the second random variable.

Figure 2.9 shows a scatter plot for the relationship between the US dollar (USD) and Aussie dollar (AUD) change rates. When a straight line is applied to the relationship between these variables, the square of the correlation coefficient shows

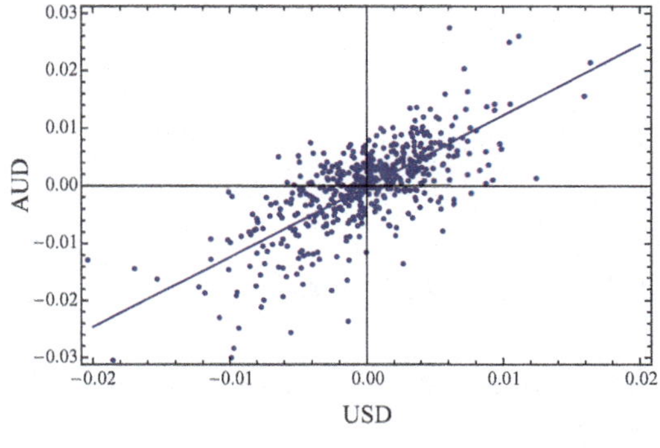

Fig. 2.9 Scatter plot of USD and AUD: The data are distributed near the regression line, showing a moderate positive correlation

how well the line represents the relationship. Thus, we can tell that the correlation coefficient indicates linear correlation.

Even if there is a correlation between two variables, we should not assume that one variable caused the other. Sometimes, there is no causal link when we obtain a definite correlation coefficient.

When there is a relationship between USD and AUD change rates, we can think of the following three cases:

1. USD changes cause AUD changes.
2. AUD changes cause USD changes.
3. Changes in a third variable X cause USD and AUD changes.

The first and second cases show a causal link between the two variable changes. The correlation coefficient, however, cannot be used to specify the cause of the causal link. In the third case, a finite correlation coefficient might be obtained even if there is no causal link between the two variables, called spurious correlation.

2.2 Statistical Estimation

2.2.1 *Parent Population and Parameter*

The parent population is a set consisting of all individuals. Parameter θ is a generic name for statistical quantity in a parent population. For instance, population mean μ is the mean value in a parent population, and population variance σ^2 is the variance in a parent population.

On the other hand, the sample is a set consisting of sampled individuals. Sampled parameter Θ is a generic name of statistical quantity in a sample set. For instance, sample average $\bar{X}$ is the mean value in a sample set, and sample variance S^2 is the variance in a sample set.

$$\bar{X} = \frac{1}{n}\sum_{i=1}^{n} x_i \tag{2.26}$$

$$S^2 = \frac{1}{n}\sum_{i=1}^{n}(x_i - \bar{X})^2. \tag{2.27}$$

2.2.2 *Point Estimation*

Statistical estimation is an estimation of parameters for a parent population using a sample set. Point estimation is an estimation of a parameter as a scalar value. An unbiased estimator is an estimation for a parameter without bias. Sampled parameter Θ is called unbiased estimator of parameter θ, if $\theta = \mathrm{E}[\Theta]$.

For instance, sampled parameter $\bar{X}$ is called unbiased estimator of mean μ, if $\mu = \mathrm{E}[\bar{\mathrm{X}}]$:

$$\begin{aligned} \mathrm{E}[\bar{\mathrm{X}}] &= \mathrm{E}\left[\frac{1}{\mathrm{n}}\sum_{\mathrm{i}=1}^{\mathrm{n}} \mathrm{x_i}\right] \\ &= \frac{1}{n}\sum_{i=1}^{n} \mathrm{E}[\mathrm{x_i}] \\ &= \frac{1}{n}\sum_{i=1}^{n} \mu \\ &= \mu. \end{aligned} \tag{2.28}$$

Sampled parameter S^2 is called unbiased estimator of variance σ^2, if $\sigma^2 = \frac{n}{(n-1}\mathrm{E}[\mathrm{S}^2]$:

$$\begin{aligned} S^2 &= \frac{1}{n}\sum_{i=1}^{n}(x_i - \bar{X})^2 \\ &= \frac{1}{n}\sum_{i=1}^{n}\{(x_i - \mu) - (\bar{X} - \mu)\}^2 \\ &= \frac{1}{n}\sum_{i=1}^{n}(x_i - \mu)^2 - (\bar{X} - \mu)^2, \end{aligned} \tag{2.29}$$

$$\begin{aligned} \mathrm{E}[\mathrm{S}^2] &= \frac{1}{n}\sum_{i=1}^{n} \mathrm{E}[(\mathrm{x_i} - \mu)^2] - \mathrm{E}[(\bar{\mathrm{X}} - \mu)^2] \\ &= \frac{1}{n}\sum_{i=1}^{n} \mathrm{E}[(\mathrm{x_i} - \mu)^2] - \mathrm{E}\left[\left(\frac{1}{\mathrm{n}}\sum_{\mathrm{i}=1}^{\mathrm{n}} \mathrm{x_i} - \mu\right)^2\right] \\ &= \frac{1}{n}\sum_{i=1}^{n} \mathrm{E}[(\mathrm{x_i} - \mu)^2] - \frac{1}{\mathrm{n}^2}\sum_{\mathrm{i}=1}^{\mathrm{n}} \mathrm{E}[(\mathrm{x_i} - \mu)^2] \\ &= \frac{n-1}{n}\sigma^2, \end{aligned} \tag{2.30}$$

where the independent relation $\mathrm{E}[(\mathrm{x_i}-\mu)(\mathrm{x_j}-\mu)] = \mathrm{E}[\mathrm{x_i}-\mu]\mathrm{E}[\mathrm{x_j}-\mu] = 0$ is used in the second term of the L.H.S.

2.2.3 Maximum Likelihood Estimation

Maximum likelihood estimation is a method to estimate parameters for a parent population using a sample set when a probability distribution for the parent population is known. The estimated parameter is called a maximum likelihood estimator. The likelihood means probability and is often written using L.

We show some examples of the maximum likelihood estimation as follows. The first example is the Poisson distribution. Suppose n samples were randomly sampled from a parent population with the Poisson distribution, $\{x_1, x_2, \cdots, x_n\}$. The likelihood L is written as

$$\begin{aligned} L &= p(x_1, x_2, \cdots, x_n) = p(x_1) \cdot p(x_2) \cdots p(x_n) \\ &= e^{-\mu} \frac{\mu^{x_1}}{x_1!} \cdot e^{-\mu} \frac{\mu^{x_2}}{x_2!} \cdots e^{-\mu} \frac{\mu^{x_n}}{x_n!} \\ &= e^{-n\mu} \frac{\mu^{\sum_{i=1}^{n} x_i}}{\prod_{i=1}^{n} x_i!}. \end{aligned} \tag{2.31}$$

The likelihood L is maximized when the condition,

$$\frac{dL}{d\mu} = \frac{L}{\mu}\left(-n\mu + \sum_{i=1}^{n} x_i\right) = 0, \tag{2.32}$$

is satisfied. Therefore, we have $\mu = \sum_{i=1}^{n} x_i/n$. The concept of the maximum likelihood estimation for the Poisson distribution is shown in the upper panel of Fig. 2.10.

The second example is the normal distribution. Suppose n samples were randomly sampled from a parent population with the normal distribution, $\{x_1, x_2, \cdots, x_n\}$. The likelihood L is written as

$$\begin{aligned} L &= p(x_1, x_2, \cdots, x_n) = p(x_1) \cdot p(x_2) \cdots p(x_n) \\ &= \frac{1}{\sqrt{2\pi}\sigma} \exp\left[-\frac{(x_1-\mu)^2}{2\sigma^2}\right] \cdot \frac{1}{\sqrt{2\pi}\sigma} \exp\left[-\frac{(x_2-\mu)^2}{2\sigma^2}\right] \cdot \\ &\quad \cdots \frac{1}{\sqrt{2\pi}\sigma} \exp\left[-\frac{(x_n-\mu)^2}{2\sigma^2}\right] \\ &= \left(\frac{1}{\sqrt{2\pi}\sigma}\right)^n \exp\left[-\frac{\sum_{i=1}^{n}(x_i-\mu)^2}{2\sigma^2}\right]. \end{aligned} \tag{2.33}$$

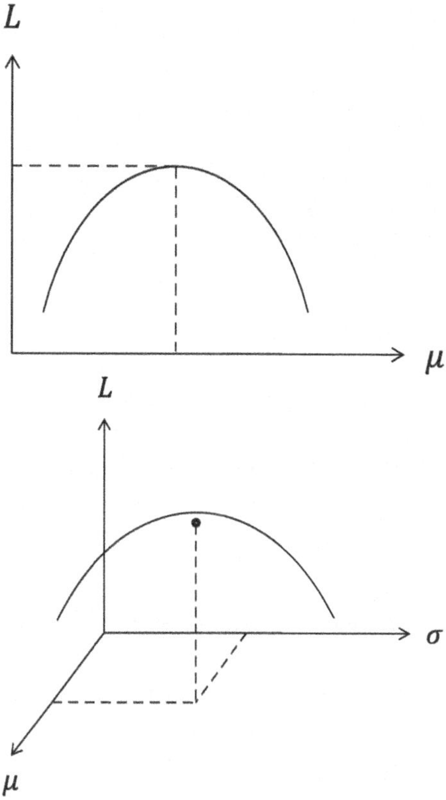

Fig. 2.10 Maximum likelihood estimation: The upper panel is the Poisson distribution, and the lower is the normal distribution

The likelihood L is maximized when the condition,

$$\frac{dL}{d\mu} = -\frac{n\mu - \sum_{i=1}^{n} x_i}{\sigma^2} L = 0, \tag{2.34}$$

is satisfied. Therefore, we have $\mu = \sum_{i=1}^{n} x_i / n$. The concept of the maximum likelihood estimation for the normal distribution is shown in the lower panel of Fig. 2.10.

2.2.4 Interval Estimation

Next, we explain the interval estimation. The interval estimation estimates a range of a parameter θ with a given confidence level γ.

Estimate Θ_1 and Θ_2 for the sampled values in order that the probability where the unknown parameter θ is in the range $\Theta_1 < \theta < \Theta_2$ is equal to the given confidence level γ. Here $[\Theta_1, \Theta_2]$ is the confidence interval. The confidence level γ is defined as the probability in the confidence interval and is often quite close to 1, such as $\gamma = 95\%$, or 98%.

An example of the interval estimation is explained. n samples were chosen from a parent population with $N(\mu, \sigma^2)$. The sample average $\bar{X}$ is distributed according to $N(\mu, \sigma^2/n)$, where the mean and the variance of $\bar{X}$ are calculated as $n\mu/n$ and $(\sqrt{n}\sigma/n)^2$, respectively. When the population variance σ^2 is known, we estimate a confidence interval with a given confidence level γ. The sample average $\bar{X}$ is standardized by

$$Z = \frac{\bar{X} - \mu}{\sigma/\sqrt{n}}, \tag{2.35}$$

and the variable Z is distributed with the standardized normal distribution $N(0, 1)$, where the confidence intervals are $[-Z_1, Z_1] = [-1.282, 1.282]$ and $[-2.576, 2.576]$ for the confidence levels $\gamma = 98\%$ and 99.9%, respectively. For a given confidence interval, we have

$$-Z_1 < Z < Z_1. \tag{2.36}$$

Thus we have the confidence interval of the parent population with $N(\mu, \sigma^2)$,

$$-\bar{X} - \frac{\sigma}{\sqrt{n}} Z_1 < \mu < \bar{X} + \frac{\sigma}{\sqrt{n}} Z_1. \tag{2.37}$$

2.2.5 *Hypothesis and Statistical Test*

A statistical hypothesis is a hypothesis about a distribution in a parent population. Statistical test means true–false decision of the hypotheses. Significance level, or risk rate, α is the criterion for the decision and is often chosen as 1%, or 5%.

We explain an example of the statistical test below. Assume that a distribution of the length of radish harvested in village A is known, as shown in Fig. 2.11. When

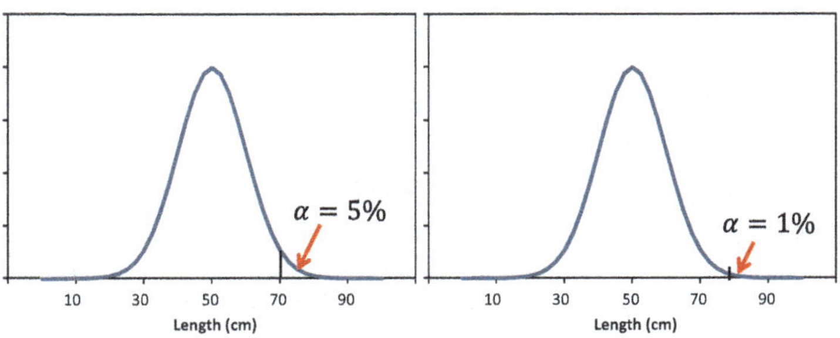

Fig. 2.11 Statistical Test: The left panel is with significance level $\alpha = 5\%$ and the right $\alpha = 1\%$

a radish is 71 cm, we will test a hypothesis "This radish is harvested in village A" with significance level $\alpha = 5\%$. The 71 cm radish is located in the tail region with $\alpha = 5\%$. Therefore, the hypothesis H_0 is rejected. Then, we will test the hypothesis with significance level $\alpha = 1\%$. The 71 cm radish is not located in the tail region with $\alpha = 1\%$. Therefore, the hypothesis H_0 is adapted.

Adoption of the hypothesis does not mean that the hypothesis is proven but means that it is not specific hypothesis is incorrect. That is double negation. Therefore, null hypothesis H_0 should be designed to be rejected in the statistical test. If the null hypothesis is rejected, alternative hypothesis H_1 is given to be adopted. On the other hand, if the null hypothesis is adapted, it comes to nothing. We have two kinds of errors. One is the type I error or false positive. This means rejection of the null hypothesis H_0, despite the correctness of H_0. The other is the type II error or false negative. This means the adoption of the null hypothesis H_0, despite the correctness of H_1.

The t-statistic was introduced in 1908 by William Sealy Gosset, a chemist working for the Guinness brewery in Ireland [3]. "Student" was his pen name. Using sample average $\bar{X}$ and sample variance s^2, we define variable t:

$$\bar{X} = \frac{x_1 + \cdots + x_n}{n} \tag{2.38}$$

$$s^2 = \frac{1}{n-1} \sum_{i=1}^{n} (X_i - \bar{X})^2 \tag{2.39}$$

$$t = \frac{\bar{X} - \mu}{s/\sqrt{n}}. \tag{2.40}$$

The variable t follows a probability density $p(t)$, called Student's t-distribution:

$$p(t) = \frac{\Gamma((m+1)/2)}{\sqrt{m\pi}\,\Gamma(m/2)} (1 + t^2/m)^{-(m+1)/2}, \tag{2.41}$$

where $m = n‘1$ is degrees of freedom and $\Gamma(\cdot)$ is the gamma distribution. The t-distribution with a large degree of freedom m can be approximated by the standard normal distribution $N(0, 1)$.

The hypothesis "parameter β is equal to 0" is tested with a given significance level using the following t,

$$t = \frac{\mathrm{E}[\beta] - 0}{\mathrm{SE}[\beta]}. \tag{2.42}$$

Here $\mathrm{E}[\beta]$ is estimation and $\mathrm{SE}[\beta]$ is standard error. If $t = 1.8$, the hypothesis is adopted at significance level 1%. We cannot say anything. If $t = 2.9$, the hypothesis is rejected. The estimated parameter is statistically significant at a significant level 1%. The probability of taking a value greater than t in a t-distribution $Pr(> |t|)$ is

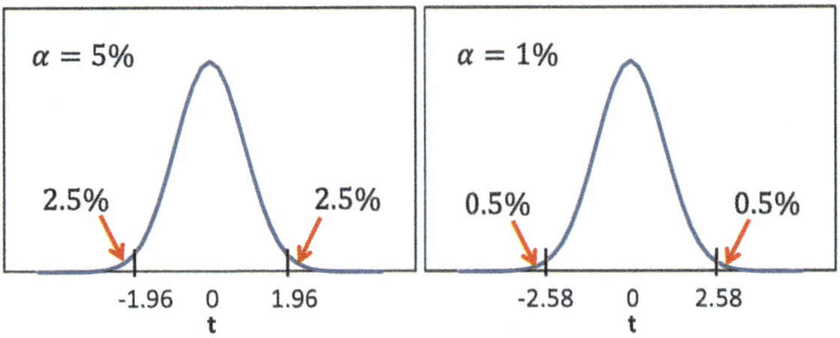

Fig. 2.12 Two-sided test with t-distribution: The significance level α is allocated half to each side

called the p-value. The hypothesis is rejected at a significant level α, if $\alpha >$ p-value. Therefore, the p-value has to be small.

Consider the following simple example. You rolled a dice 400 times and obtained even numbers 224 times. Is this a good dice? $\mu = np = 400 \times 0.5 = 200$ $\sigma^2 = np(1-p) = 400 \times 0.5 \times 0.5 = 10^2$ $t = (x-\mu)/\sigma = (224-200)/10 = 2.4$. Assume that a distribution of the even number is t-distribution, as shown in Fig. 2.12. We test the null hypothesis H_0 "Population mean is $t = 2.4$".

In the one-sided or one-tailed test, alternative hypothesis H_1 "Population mean $t \geq 2.4$" or H_1 "Population mean $t \leq 2.4$". In the two-sided or two-tailed test, the alternative hypothesis H_1 "Population mean $t \neq 2.4$". Figure 2.12 depicts the significance level α of two-sided test; the left panel is for $\alpha = 5\%$ and the right panel is for $\alpha = 1\%$.

$t = 2.4$ is located in the tail region with $\alpha = 5\%$. Therefore, the hypothesis H_0 is rejected. Then, we will test the hypothesis with significance level $\alpha = 1\%$. $t = 2.4$ is not located in the tail region with $\alpha = 1\%$. Therefore, the hypothesis H_0 is adapted.

2.3 Regression Analysis

2.3.1 *Single Regression Analysis*

An explained variable, or an objective variable $Y_i (i = 1, \cdots, N)$, is modeled using a linear relation

$$Y_i = \alpha + \beta X_i + \epsilon_i, \tag{2.43}$$

where $X_i (i = 1, \cdots, N)$ is an explanatory variable or an independent variable, and $\epsilon_i (i = 1, \cdots, N)$ is residual or error. In this model, the squared sum of error, or the sum of squared error

$$SS_E = \sum_{i=1}^{N} \epsilon_i^2 = \sum_{i=1}^{N} (Y_i - \alpha - \beta X_i)^2, \tag{2.44}$$

is minimized to estimate regression coefficients, or regression parameters α and β. This is called the single regression analysis. The conditions to minimize the sum of square SS_E are

$$\frac{\partial SS_E}{\partial \alpha} = -2 \sum_{i=1}^{N} (Y_i - \alpha - \beta X_i) = 0, \tag{2.45}$$

$$\frac{\partial SS_E}{\partial \beta} = -2 \sum_{i=1}^{N} (Y_i - \alpha - \beta X_i) X_i = 0. \tag{2.46}$$

These conditions are rewritten as the normal equation,

$$\alpha N + \beta \sum_{i=1}^{N} X_i = \sum_{i=1}^{N} Y_i, \tag{2.47}$$

$$\alpha \sum_{i=1}^{N} X_i + \beta \sum_{i=1}^{N} X_i^2 = \sum_{i=1}^{N} X_i Y_i. \tag{2.48}$$

The regression parameters α and β are obtained by solving Eqs. (2.47) and (2.48),

$$\alpha = \frac{1}{N} \sum_{i=1}^{N} Y_i - \beta \frac{1}{N} \sum_{i=1}^{N} X_i = \bar{Y} - \beta \bar{X}, \tag{2.49}$$

$$\beta = \frac{\sum_{i=1}^{N} (X_i - \bar{X})(Y_i - \bar{Y})}{\sum_{i=1}^{N} (X_i - \bar{X})^2} = \frac{S_{XY}}{SS_X}. \tag{2.50}$$

This method of parameter estimation is often called the least square method.

2.3.2 *Analysis of Variance*

The analysis of variance (ANOVA) is a methodology to decompose the sum of square SS_Y into the sum of the square of each factor and to compare the magnitude of each square of sum using the ANOVA table shown in Table 2.1 (Fig. 2.13).

The residual ϵ_i is given by rewriting Eq. (2.43) using the predicted value $\hat{Y}_i$,

$$\epsilon_i = Y_i - \hat{Y}_i = Y_i - (\alpha + \beta X_i). \tag{2.51}$$

Table 2.1 ANOVA table

Factor	Squared Sum	Degree of Freedom	Mean Square
Regression	SS_R	$\phi_1 = 1$	$MS_R = SS_R$
Error	SS_E	$\phi_2 = N - 2$	$MS_E = \frac{SS_E}{N-2}$
–	SS_Y	$\Phi = N - 1$	–

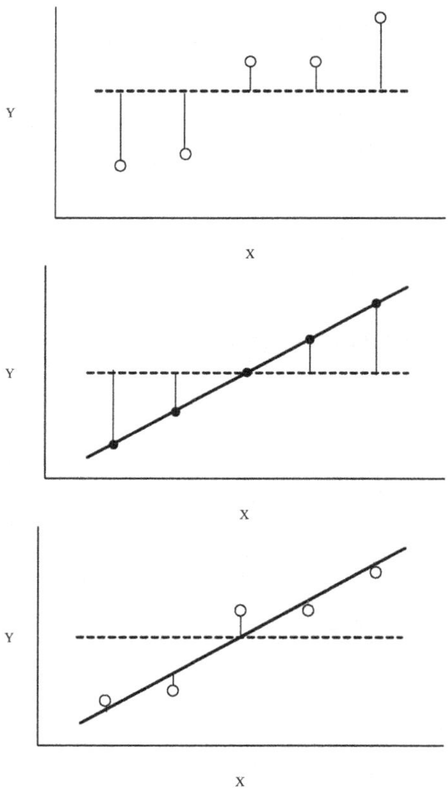

Fig. 2.13 ANOVA: Deviation (the upper panel) equals the sum of regression (the middle panel) and residual (the lower panel). See Eq. (2.56)

Residuals $\epsilon_i (i = 1, \cdots, N)$ are not independent of each other because of the relation

$$\sum_{i=1}^{N} \epsilon_i = \sum_{i=1}^{N} (Y_i - \hat{Y}_i) = 0. \tag{2.52}$$

Then we rewrite residual ϵ_i as a sum of deviation and regression as

$$Y_i - \hat{Y}_i = (Y_i - \bar{Y}) - (\hat{Y}_i - \bar{Y}). \tag{2.53}$$

By taking squared sum of both sides of Eq. (2.53), we have the relation

$$\sum_{i=1}^{N} (Y_i - \hat{Y}_i)^2 = \sum_{i=1}^{N} (Y_i - \bar{Y})^2 + \sum_{i=1}^{N} (\hat{Y}_i - \bar{Y})^2 - 2\sum_{i=1}^{N} 2(Y_i - \bar{Y})(\hat{Y}_i - \bar{Y}). \tag{2.54}$$

Using the relation

$$\sum_{i=1}^{N} (Y_i - \bar{Y})(\hat{Y}_i - \bar{Y}) = \sum_{i=1}^{N} (\hat{Y}_i - \bar{Y})^2 \tag{2.55}$$

with Eq. (2.54), we obtain the decomposition for the sum of squares,

$$\sum_{i=1}^{N} (Y_i - \bar{Y})^2 = \sum_{i=1}^{N} (\hat{Y}_i - \bar{Y})^2 + \sum_{i=1}^{N} (Y_i - \hat{Y}_i)^2, \tag{2.56}$$

where the left-hand side of the equation, the first term of the right-hand side of the equation, and the second term of the right-hand side of the equation are called the whole sum of squares SS_Y, the squared sum of regression SS_R, and the squared sum of error SS_E, respectively. SS_R corresponds to the component explained by the regression, and SS_E corresponds to the component that is not explained by the regression. We note that SS_R is rewritten as

$$SS_R = \sum_{i=1}^{N} (\hat{Y}_i - \bar{Y})^2 = \beta \sum_{i=1}^{N} (X_i - \bar{X})(Y_i - \bar{Y}). \tag{2.57}$$

We have the relation of the sum of squares SS_Y. Therefore, the whole degree of freedom Φ is equal to $N - 1$. On the other hand, the squared sum of regression SS_R is determined by a single parameter α. Therefore, we have $\phi_1 = 1$. Consequently, we have $\phi_2 = \Phi - \phi_1 = N - 2$. The mean square of regression MS_R and mean square of error MS_E shown in the most right column of Table 2.1 are defined as the squared sum of regression SS_R divided by the degree of freedom ϕ_1 and the squared sum of error SS_E divided by the degree of freedom ϕ_2.

2.3.3 Interval Estimation of Parameters

From Eq. (2.57), we have

$$\beta = \frac{\sum_{i=1}^{N}(X_i - \bar{X})(Y_i - \bar{Y})}{SS_X} = \sum_{i=1}^{N} \frac{X_i - \bar{X}}{SS_X} Y_i - \sum_{i=1}^{N} \frac{X_i - \bar{X}}{SS_X} \bar{Y} = \sum_{i=1}^{N} \frac{X_i - \bar{X}}{SS_X} Y_i. \tag{2.58}$$

As $Y_i (i = 1, \cdots, N)$ are independent variables and $\frac{X_i - \bar{X}}{SS_X}$ are regarded as constant numbers, the variance of β is calculated as follows:

$$V[\beta] = \sum_{i=1}^{N} \left(\frac{X_i - \bar{X}}{SS_X}\right)^2 V[Y_i] = \sum_{i=1}^{N} \left(\frac{X_i - \bar{X}}{SS_X}\right)^2 \sigma^2 = \frac{\sigma^2}{\sum_{i=1}^{N}(X_i - \bar{X})^2}. \tag{2.59}$$

Similarly, the variance of α is calculated from Eq. (2.49),

$$V[\alpha] = \left(\frac{\partial \alpha}{\partial \beta}\right)^2 V[\beta] = \frac{\bar{X}^2}{\sum_{i=1}^{N}(X_i - \bar{X})^2} \sigma^2. \tag{2.60}$$

2.3.4 Multiple Regression Analysis

An explained variable or an objective variable $y_i (i = 1, \cdots, N)$ is modeled using a linear relation,

$$y_i = \theta_0 + \theta_1 x_{i1} + \theta_2 x_{i2} + \cdots + \theta_p x_{ip} + \epsilon_i, \tag{2.61}$$

where $x_{ip} (i = 1, \cdots, N)$ are explanatory variables or independent variables. The number of model parameters is not equal to two but equal to $p + 1$. This model is called multiple regression analysis.

We introduce a vector of explained variable

$$\boldsymbol{y} = \begin{pmatrix} y_1 \\ y_2 \\ \vdots \\ y_N \end{pmatrix}, \tag{2.62}$$

a vector of residual

$$\boldsymbol{\epsilon} = \begin{pmatrix} \epsilon_1 \\ \epsilon_2 \\ \vdots \\ \epsilon_N \end{pmatrix}, \tag{2.63}$$

a vector of model parameters

$$\boldsymbol{\theta} = \begin{pmatrix} \theta_1 \\ \theta_2 \\ \vdots \\ \theta_p \end{pmatrix}, \tag{2.64}$$

and a matrix of explanatory variables

$$X = \begin{pmatrix} 1 & x_{11} & \cdots & x_{1p} \\ 1 & x_{21} & \cdots & x_{2p} \\ \vdots & \vdots & \vdots & \vdots \\ 1 & x_{N1} & \cdots & x_{Np}. \end{pmatrix}. \tag{2.65}$$

Using these vectors and matrix, Eq. (2.61) is briefly rewritten as

$$\boldsymbol{y} = X\boldsymbol{\theta} + \boldsymbol{\epsilon}. \tag{2.66}$$

We write parameter $\boldsymbol{\theta} = \hat{\boldsymbol{\theta}}$ when $\boldsymbol{\epsilon} = \boldsymbol{0}$. In order to obtain a formal solution for parameters $\hat{\boldsymbol{\theta}}$, we multiply X^T from the left-hand side to Eq. (2.66),

$$X^T \boldsymbol{y} = X^T X \hat{\boldsymbol{\theta}}, \tag{2.67}$$

where a matrix with suffix ^{T} indicates transposed matrix. Then a formal solution for parameters $\hat{\boldsymbol{\theta}}$ is obtained by multiplying $(X^T X)^{-1}$ from the left-hand side to Eq. (2.67),

$$\hat{\boldsymbol{\theta}} = (X^T X)^{-1} X^T \boldsymbol{y}, \tag{2.68}$$

where a matrix with suffix $^{-1}$ indicates inverse matrix.

The squared sum of errors of parameter $\boldsymbol{\theta}$ is given by

$$SS_E(\boldsymbol{\theta}) = \boldsymbol{\epsilon}^2 = (\boldsymbol{y} - X\boldsymbol{\theta})^T (\boldsymbol{y} - X\boldsymbol{\theta}) \tag{2.69}$$

and rewritten as

$$\begin{aligned}SS_E(\boldsymbol{\theta}) &= (\boldsymbol{y} - X\boldsymbol{\theta})^T(\boldsymbol{y} - X\boldsymbol{\theta}) \\ &= \{\boldsymbol{y} - X\hat{\boldsymbol{\theta}} + X(\hat{\boldsymbol{\theta}} - \boldsymbol{\theta})\}^T\{\boldsymbol{y} - X\hat{\boldsymbol{\theta}} + X(\hat{\boldsymbol{\theta}} - \boldsymbol{\theta})\} \\ &= (\boldsymbol{y} - X\hat{\boldsymbol{\theta}})^T(\boldsymbol{y} - X\hat{\boldsymbol{\theta}}) + (\boldsymbol{y} - X\hat{\boldsymbol{\theta}})^T X(\hat{\boldsymbol{\theta}} - \boldsymbol{\theta}) \\ &+ \{X(\hat{\boldsymbol{\theta}} - \boldsymbol{\theta})\}^T(\boldsymbol{y} - X\hat{\boldsymbol{\theta}}) + \{X(\hat{\boldsymbol{\theta}} - \boldsymbol{\theta})\}^T X(\hat{\boldsymbol{\theta}} - \boldsymbol{\theta}).\end{aligned} \tag{2.70}$$

By substituting the relations that the second and third terms of the r.h.s. of Eq. (2.70) are equal to zero,

$$(\boldsymbol{y} - X\hat{\boldsymbol{\theta}})^T X(\hat{\boldsymbol{\theta}} - \boldsymbol{\theta}) = \{X^T(\boldsymbol{y} - X\hat{\boldsymbol{\theta}})\}^T(\hat{\boldsymbol{\theta}} - \boldsymbol{\theta}) = 0, \tag{2.71}$$

and

$$\{X(\hat{\boldsymbol{\theta}} - \boldsymbol{\theta})\}^T(\boldsymbol{y} - X\hat{\boldsymbol{\theta}}) = (\hat{\boldsymbol{\theta}} - \boldsymbol{\theta})^T X^T(\boldsymbol{y} - X\hat{\boldsymbol{\theta}}) = 0. \tag{2.72}$$

We have the relation

$$SS_E(\boldsymbol{\theta}) = (\boldsymbol{y} - X\hat{\boldsymbol{\theta}})^T(\boldsymbol{y} - X\hat{\boldsymbol{\theta}}) + \{X(\hat{\boldsymbol{\theta}} - \boldsymbol{\theta})\}^T X(\hat{\boldsymbol{\theta}} - \boldsymbol{\theta}). \tag{2.73}$$

Then we obtain the squared sum of errors of the estimated parameter $\hat{\boldsymbol{\theta}}$

$$SS_E(\hat{\boldsymbol{\theta}}) = (\boldsymbol{y} - X\hat{\boldsymbol{\theta}})^T(\boldsymbol{y} - X\hat{\boldsymbol{\theta}}), \tag{2.74}$$

by applying

$$\{X(\hat{\boldsymbol{\theta}} - \boldsymbol{\theta})\}^T X(\hat{\boldsymbol{\theta}} - \boldsymbol{\theta}) \geq 0, \tag{2.75}$$

for Eq. (2.73). Therefore, we have inequality:

$$SS_E(\boldsymbol{\theta}) \geq SS_E(\hat{\boldsymbol{\theta}}). \tag{2.76}$$

We can note that the estimated parameters $\hat{\boldsymbol{\theta}}$ minimize the squared sum of errors $SS_E(\boldsymbol{\theta})$. This means that the estimated parameters $\hat{\boldsymbol{\theta}}$ are equal to the least square solution.

2.3.5 Test of Parameter Estimation

First we describe $\hat{Y}_i - \bar{Y}$ or $Y_i - \hat{Y}_i$ as x_i. The variables $x_1, x_2, \cdots, x_N$ are independent of each other and obey to normal distribution $N(0, 1)$. Then MS_R or

MS_E is described by $z = x_1^2 + x_2^2 + \cdots + x_N^2$, and it is known that variable z obeys to χ^2 distribution,

$$T_n(z) = \begin{cases} \frac{1}{2^{n/2}\Gamma(n/2)} z^{\frac{n-2}{n}} e^{-\frac{z}{2}} & (z > 0) \\ 0 & (z \leq 0) \end{cases}, \tag{2.77}$$

where n and $\Gamma(x)$ are the degree of freedom and the gamma function, respectively. $\Gamma(x)$ is defined by

$$\Gamma(x) = \int_0^\infty y^{x-1} e^{-y} dy. \tag{2.78}$$

In the case that x is an integer or half-integer, it is written respectively as

$$\Gamma(n+1) = n!, \tag{2.79}$$

$$\Gamma\left(n + \frac{1}{2}\right) = \frac{(2n)!}{2^{2n} n!}\sqrt{\pi}. \tag{2.80}$$

Next, we describe F-value F, which is defined by a ratio,

$$F = \frac{MS_R}{MS_E}, \tag{2.81}$$

and discuss the distribution to which F-value F obeys.

When z_1 and z_2 are independent of each other and obey to χ^2 distributions with degree of freedom ϕ_1 and ϕ_2, respectively, variable $r = \frac{\phi_2 z_1}{\phi_1 z_2}$, defined as the ratio of $\frac{z_1}{\phi_1}$ and $\frac{z_2}{\phi_2}$, obeys the distribution called F-distribution,

$$f_{\phi_1,\phi_2}(r) = \begin{cases} \frac{\phi_1^{\frac{\phi_1}{2}} n^{\frac{\phi_2}{2}} r^{\frac{\phi_1-2}{2}}}{B(\frac{\phi_1}{2}, \frac{\phi_2}{2})(\phi_1 r + \phi_2)^{\frac{\phi_1+\phi_2}{2}}} & (r > 0) \\ 0 & (r \leq 0) \end{cases}, \tag{2.82}$$

where $B(s, t)$ is the beta function defined using $\Gamma(\cdot)$,

$$B(s, t) = \frac{\Gamma(s)\Gamma(t)}{\Gamma(s+t)}. \tag{2.83}$$

Then, we test the null hypothesis H_0 that all parameters are equal to zero. If $F > t$, H_0 is rejected at a given significant level, e.g., $\alpha = 0.5\%$. Thus, we say that all parameters are not equal to zero. On the other hand, if $F < t$, H_0 is accepted at a given significant level, e.g., $\alpha = 0.5\%$. Thus, we do not say that all parameters are not equal to zero. F-value F is the most important statistical value to show the validity of the regression analysis (Fig. 2.14).

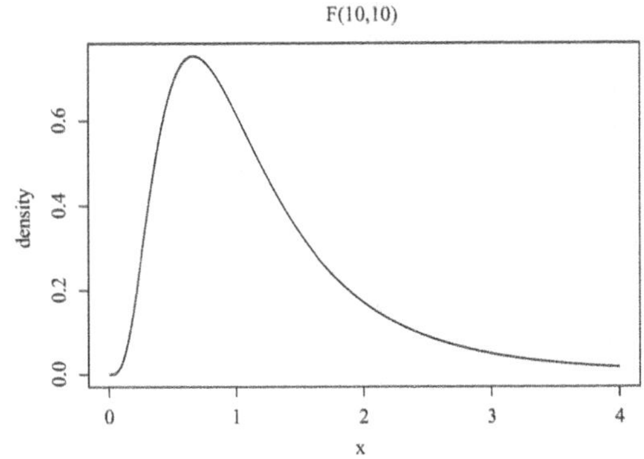

Fig. 2.14 F-distribution: $f_{\phi_1,\phi_2}(x)$ with $\phi_1 = 10$ and $\phi_2 = 10$. See Eq. (2.82)

Finally, we can write the coefficient of determination R^2 from Eq. (2.56) as follows:

$$R^2 = \frac{SS_R}{SS_Y} = 1 - \frac{SS_E}{SS_Y}. \tag{2.84}$$

Here, the smaller error SS_E means larger R^2.

2.4 Time Series Analysis

2.4.1 Origins of Temporal Variations

Time series of economic indicator $x(t)$, such as GDP or index of industrial production, shows temporal variations. Some variation shows a cyclical change, which is often referred to as a business cycle. Business cycles are categorized into four types: (1) the Kitchin inventory cycle (3–5 years), (2) the Juglar fixed investment cycle (7–11 years), (3) the Kuznets infrastructural investment cycle (15–25 years), and (4) the Kondratiev wave or long technological cycle (45–60 years).

The following variables are used to characterize the temporal variation of a time series. The index, change rate, and growth rate of $x(t)$ are defined by

$$x'(t) = \frac{x(t)}{x(t_0)}, \tag{2.85}$$

$$s(t) = \frac{x(t) - x({}_t0)}{x(t_0)}, \tag{2.86}$$

$$r(t) = \frac{x(t) - x(t-1)}{x(t-1)}, \tag{2.87}$$

respectively. Here t_0 is a base year.

Suppose the case where the time series $x(t)$ is monthly. Consider decomposing a time series into multiple components. Most monthly time series include seasonal variations. Therefore, it is necessary to decompose the time series into its components. The original time series $x(t)$ is decomposed into trend variation $T(t)$, cyclical variation (cycle due to inventory adjustment) $C(t)$, seasonal variation $S(t)$, and random fluctuation $I(t)$. Two different models of decomposition are used; one is the additive model $x(t) = T(t) + C(t) + S(t) + I(t)$ and the other is the multiplicative model $x(t) = T(t) \cdot C(t) \cdot S(t) \cdot I(t)$. Note that the log transform of the multiplicative model is regarded as the log-additive model.

Moving average is a smoothing method by replacing data with an average value. The central, backward, and the forward-moving averages are defined by

$$\bar{x}(t) = \frac{x(t-k) + \cdots + x(t) + \cdots + x(t+k)}{2k+1}, \tag{2.88}$$

$$\bar{x}(t) = \frac{x(t-2k) + \cdots + x(t)}{2k+1}, \tag{2.89}$$

$$\bar{x}(t) = \frac{x(t) + \cdots + x(t+2k)}{2k+1}, \tag{2.90}$$

respectively.

A simple time series $x(t)$ decomposition is possible using the additive model and the central moving average. Here, we assume that the ranges of cycles of variations are known as follows. The cycle of T is longer than 39 months. The cycle of C is between 39 months and 13 months. The cycle of S is between 12 months and 3 months. The cycle of I is shorter than 3 months. If we calculate moving average of $x(t)$ with period $p = 2k + 1$, the variation with cycle shorter than p will be removed. Thus the moving averages with periods 39, 13, and 3 obtain time series $T(t)$, $a(t) \equiv T(t) + C(t)$, and $b(t) \equiv T(t) + C(t) + S(t)$, respectively. Then we obtain cyclical variation $C(t)$ by subtracting $T(t)$ from $a(t)$, and seasonal variation $S(t)$ by subtracting $a(t)$ from $b(t)$, and random fluctuation $I(t)$ by subtracting $b(t)$ from $x(t)$. However, we note that the moving average time series decomposition is explained only for educational purposes, but actual time series decomposition or seasonal adjustment is based on the X-12-ARIMA model.

2.4.2 *Stationary Process and Auto-correlation*

The log-growth rate $r(t)$ is defined by

$$\begin{aligned} r(t) &= \log x(t) - \log x(t-1) \\ &= \log\left(1 + \frac{x(t) - x(t-1)}{x(t-1)}\right) \\ &\approx \frac{x(t) - x(t-1)}{x(t-1)}. \end{aligned} \tag{2.91}$$

Fig. 2.15 Exchange rate: The original time series for (**a**) JPY/USD, (**b**) JPY/EUR, (**c**) JPY/AUD are plotted

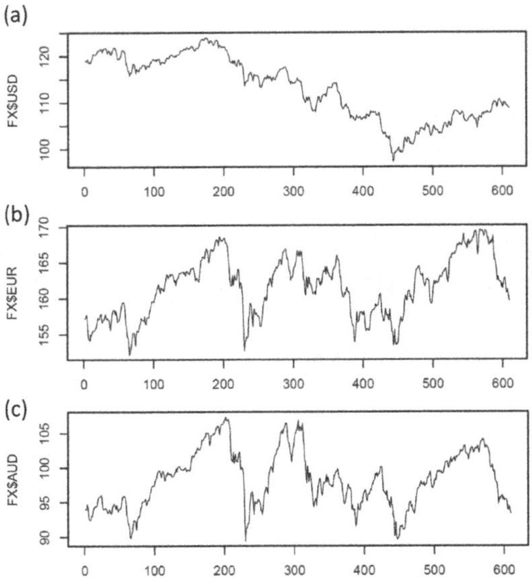

Fig. 2.16 The log-growth rate of exchange rate: The time series of the log-growth rate for (**a**) JPY/USD, (**b**) JPY/EUR, (**c**) JPY/AUD are plotted

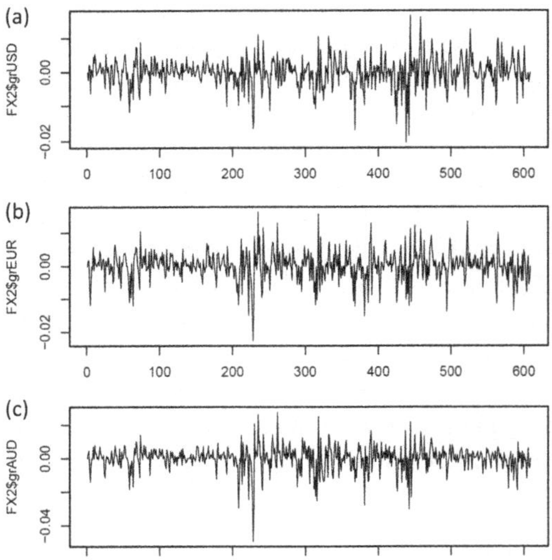

The log-growth rate $r(t)$ is a good approximation to the growth rate defined by Eq. (2.87). In order to calculate the growth rate, we take a time difference, which allows us to remove the linear trend. Similarly, if we calculate the second-order finite difference, we can remove the quadratic trend (Figs. 2.15 and 2.16).

Next, we consider the conditions of the stationary process. We divide the time series of interest into multiple time series of interval T. We calculate mean value

μ_x, variance σ_x^2, and covariance $\mathrm{Cov}[x(t), x(t+\tau)]$ for each time series of interval T, as follows:

$$\mu_x = \frac{1}{T}\sum_{t=1}^{T} x(t) = \mathrm{const} \tag{2.92}$$

$$\sigma_x^2 = \frac{1}{T-1}\sum_{t=1}^{T}(x(t) - \mu_x)^2 = \mathrm{const} \tag{2.93}$$

$$\mathrm{Cov}[x(t), x(t+\tau)] = \frac{1}{T-1}\sum_{t=1}^{T}(x(t) - \mu_x)(x(t+\tau) - \mu_x) \equiv \gamma(\tau) \tag{2.94}$$

$$\gamma(0) = \sigma_x^2. \tag{2.95}$$

If mean value μ_x is the same independent of the time series, and variance σ_x^2 is the same independent of the time series, covariance $\mathrm{Cov}[x(t), x(t+\tau)]$ is the same independent of the time series and depends only on lag τ, time series $x(t)$ is called stationary process.

Here, we defined the auto-correlation function $\rho(\tau)$ using $\gamma(\tau)$:

$$\rho(\tau) = \frac{\gamma(\tau)}{\gamma(0)} = \frac{\mathrm{Cov}[x(t), x(t+\tau)]}{\sigma_x^2}. \tag{2.96}$$

Specifically, if the auto-correlation function $\rho(\tau)$ satisfies the following relation:

$$\sum_{\tau=-\infty}^{\infty} \rho(\tau) = \mathrm{const}, \tag{2.97}$$

time series is called the short-term memory process. Contrary, if $\rho(\tau)$ satisfies the following relation:

$$\sum_{\tau=-\infty}^{\infty} \rho(\tau) = \infty, \tag{2.98}$$

time series is called the long-term memory process.

The unit root test is another known method to check whether a process is stationary. Assume that the time series x_t evolves according to the following equation:

$$x_t = a x_{t-1} + u_t. \tag{2.99}$$

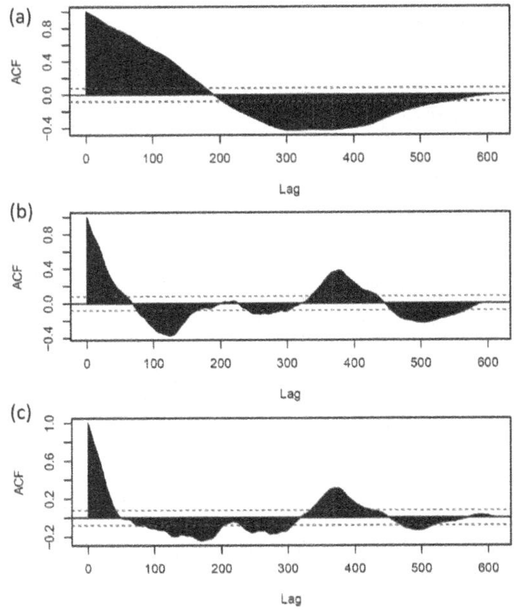

Fig. 2.17 Auto-correlation function of exchange rate: (**a**) JPY/USD, (**b**) JPY/EUR, (**c**) JPY/AUD

Here a is a model parameter, and u_t is the residual. In this case, it is known that the time series exhibits the following properties. When parameter $a < 1$, time series x_t is stationary. When $a = 1$, time series x_t is non-stationary. $a = 1$ is called a unit root. Our null hypothesis H_0 is "a unit root is present", which means that the time series is non-stationary. Using t value $t = (\mathrm{E}[a] - 1)/(\mathrm{SE}[a])$, p-value $p = Pr(> |t|)$, and significance level (risk rate) α, we perform the t test. If $p < \alpha$, H_0 is rejected. This means that "a unit root is not present". Therefore, time series is a stationary process. On the other hand, if $p > \alpha$, H_0 is adopted. This means that "a unit root is present". Therefore, time series is a non-stationary process.

Auto-correlation functions of exchange rate time series for (a) JPY/USD, (b) JPY/EUR, (c) JPY/AUD are shown in Fig. 2.17. Auto-correlation functions do not converge to a specific value; therefore, time series is the long-term memory process. The unit root test obtains p-values 0.5827, 0.1933, and 0.317 for (a) JPY/USD, (b) JPY/EUR, (c) JPY/AUD, respectively. We obtained a large p-value, so H_0 was adopted. Thus, the time series is a non-stationary process. The non-stationary process is modeled by a random walk (standard Brownian motion with H = 1/2).

Auto-correlation functions of the log-growth rate of exchange rate time series for (a) JPY/USD, (b) JPY/EUR, (c) JPY/AUD are shown in Fig. 2.18. Auto-correlation functions converge to a specific value; therefore, time series is the short-term memory process. The unit root test obtains a p-value less than 2.2×10^{-16} for all cases (a) JPY/USD, (b) JPY/EUR, (c) JPY/AUD. We obtained a small p-value, so H_0 was rejected. Thus, the time series is a stationary process.

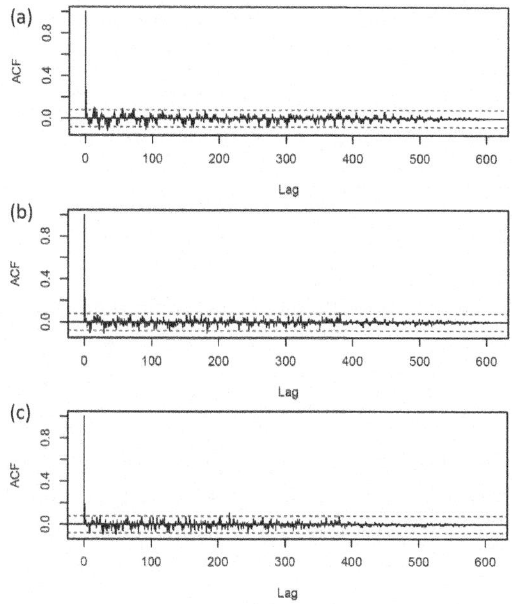

Fig. 2.18 Auto-correlation function of the log-growth rate of exchange rate: (**a**) JPY/USD, (**b**) JPY/EUR, (**c**) JPY/AUD

2.4.3 AR(p) Model

The pth order auto-regression model of time series x_t is given by

$$x_t = \sum_{m=1}^{p} a_m x_{t-m} + u_t, \tag{2.100}$$

where a_m is the model parameter and u_t is the residual. If u_t satisfies the following equations:

$$\mathrm{E}[u_t] = 0 \tag{2.101}$$

$$\mathrm{E}[u_t u_s] = \begin{cases} \sigma^2 \ (t = s) \\ 0 \ (t \neq s) \end{cases}. \tag{2.102}$$

u_t is called the white noise. a_m satisfies the characteristic equation:

$$1 = \sum_{m=1}^{p} a_m z^m. \tag{2.103}$$

$|a_m| < 1$ for all m is the condition of stationary process.

We write the auto-covariance function for the AR model as follows:

$$\begin{aligned}\gamma(s) = \gamma_s &= \mathrm{E}[x_{t+s}x_t]\ (s \geq 0)\\&= \mathrm{E}[(u_{t+s} + au_{t+s-1} + \cdots + a^k u_{t+s-k} + \cdots)\\&\quad \times (u_t + au_{t-1} + \cdots + a^k u_{t-k} + \cdots)]\\&= a^s[1 + a^2 + \cdots]\sigma^2.\end{aligned} \tag{2.104}$$

Using $\gamma(s)$, we define the auto-correlation function for the AR model as follows:

$$\rho_s := \rho(s) = \frac{\gamma(s)}{\gamma(0)} = a^s\ (s = 0, 1, 2, \cdots). \tag{2.105}$$

For the AR(1) model, the auto-correlation function $\rho(s)$ decreases at a geometric rate when $|a| < 1$ for the stationary process. Therefore, time series with the stationary process exhibits the properties of a short-term process.

2.4.4 Model Selection Using Box–Jenkins Method

Auto-correlation function p_k, defined as follows, is the indicator to show the strength of correlation at different lag k:

$$p_k = \begin{pmatrix} 1 & \rho_1 & \rho_2 & \cdots & \rho_{k-1} \\ \rho_1 & 1 & \rho_1 & \cdots & \rho_{k-2} \\ \vdots & \vdots & \ddots & \ddots & \vdots \\ \rho_{k-1} & \rho_{k-2} & \rho_{k-3} & \cdots & 1 \end{pmatrix}. \tag{2.106}$$

We remove the effect from lag k by replacing the kth column by $(\rho_1, \rho_2, \cdots, \rho_k)$:

$$p_k^* = \begin{pmatrix} 1 & \rho_1 & \rho_2 & \cdots & \rho_1 \\ \rho_1 & 1 & \rho_1 & \cdots & \rho_2 \\ \vdots & \vdots & \ddots & \ddots & \vdots \\ \rho_{k-1} & \rho_{k-2} & \rho_{k-3} & \cdots & \rho_k \end{pmatrix}. \tag{2.107}$$

Using p_k and p_k^*, we define the partial auto-correlation function:

$$\Phi_k = \frac{|p_k^*|}{|p_k|}\ (k = 1, 2, \cdots). \tag{2.108}$$

Φ_k is the indicator to show the strength of correlation after removing the effect from lag k.

If Φ_k is non-zero, the past point (lag k) affects the present point. This implies that if Φ_k is 0, then the kth parameter a_k of the AR model is 0. In AR(p) model,

$$\Phi_k = 0 \text{ for } k > p. \tag{2.109}$$

Box–Jenkins method is the method to determine order p and parameter a_i of the AR (p) model using the partial auto-correlation function ϕ_s. For AR(1) model $x_t = a_1 x_{t-1} + u_t$, the auto-correlation function is $\rho_s = a_1^s$. The partial auto-correlation functions for different k are obtained as follows:

$$\Phi_1 = \frac{\rho_1}{1} = a_1 \tag{2.110}$$

$$\Phi_2 = \frac{\begin{vmatrix} 1 & \rho_1 \\ \rho_1 & \rho_2 \end{vmatrix}}{\begin{vmatrix} 1 & \rho_1 \\ \rho_1 & 1 \end{vmatrix}} = \frac{\rho_2 - \rho_1^2}{1 - \rho_1^2} = \frac{a_1^2 - a_1^2}{1 - a_1^2} = 0 \tag{2.111}$$

$$\Phi_3 = \Phi_4 = \cdots = 0. \tag{2.112}$$

We confirmed $\Phi_k = 0$ for $k > 1$ in AR(1) model.

2.4.5 *MA(q) Model*

AR(1) model is rewritten using only the residual terms u_{t-k}:

$$x_t = u_t + a_1 x_{t-1} \tag{2.113}$$

$$= u_t + a_1 (u_{t-1} + a_1 x_{t-2}) \tag{2.114}$$

$$= u_t + a_1 u_{t-1} + \cdots + a_1^k u_{t-k} + \cdots. \tag{2.115}$$

This is obtained by repeatedly substituting the time-shifted AR(1) model for x_{t-1} on the right-hand side of the AR(1) model. By generalizing this expression, we define the following model:

$$x_t = u_t + \sum_{k=1}^{q} b_k u_{t-k}, \tag{2.116}$$

which is called the qth order moving average model, MA(q) model.

2.4.6 ARIMA(p,d,q) Model

Auto-regressive moving average model ARMA(p,q) is defined by combining the AR (p) model and the MA (q) model, as follows:

$$x_t = u_t + \sum_{m=1}^{p} a_m x_{t-m} + \sum_{k=1}^{q} b_k u_{t-k}. \tag{2.117}$$

Applying ARMA(p,q) to the dth difference time series, we obtain the auto-regressive integrated moving average model ARIMA(p,d,q). This is the general form of the primary time series model. The Box–Jenkins method cannot be used to determine the order of this model. In the following, we explain how to determine the order of the model.

2.4.7 Model Selection Using Information Criterion

For a model of time series x_t,

$$x_t = \epsilon_t + \sum_{m=1}^{p} \alpha_m x_{t-m} + \beta, \tag{2.118}$$

we define the likelihood $L(\theta)$ as follows:

$$L(\theta) = \Pi_t p(x_t, \theta), \tag{2.119}$$

where θ is the model parameter:

$$\theta = \{\alpha_m, \beta\} \ (m = 1, 2, \cdots, p). \tag{2.120}$$

Specifically, the likelihood is written down as

$$L(\theta) = \left(\frac{1}{\sqrt{2\pi}\sigma}\right)^T \exp\left[-\frac{\sum_{t=1}^{T}(x_t - \sum_{m=1}^{p} \alpha_m x_{t-m} + \beta)^2}{2\sigma^2}\right]. \tag{2.121}$$

Furthermore, we define the log-likelihood $l(\theta)$ by taking the logarithm of $L(\theta)$:

$$l(\theta) := \log L(\theta) = T \log\left(\frac{1}{\sqrt{2\pi}\sigma}\right) - \frac{\sum_{t=1}^{T}(x_t - \sum_{m=1}^{p} \alpha_m x_{t-m} + \beta)^2}{2\sigma^2}. \tag{2.122}$$

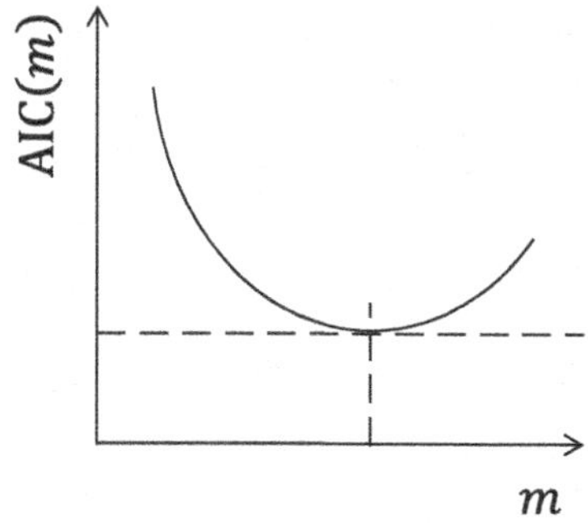

Fig. 2.19 Akaike information criterion: The log-likelihood dominates for small m and the number of model parameters m dominates for large m

Model order p is selected using the Akaike information criterion:

$$\mathrm{AIC}(p) = -2 \log L(\theta) + 2p = -2l(\theta) + 2p, \tag{2.123}$$

where p is the number of model parameters. For a small number of p, AIC(p) is dominated by the log-likelihood:

$$\mathrm{AIC}(p) \approx -2l. \tag{2.124}$$

On the other hand, for a large number of p, AIC(p) is dominated by the number of model parameters p:

$$\mathrm{AIC}(p) \approx 2p. \tag{2.125}$$

Both the model order and model parameters are determined to minimize the AIC. In general, the larger the model order, the larger the number of parameters, and thus the larger the likelihood. Minimizing the AIC means trying to obtain as large a likelihood as possible with as few parameters as possible. This is called the principle of parsimony (Fig. 2.19).

For a given number of p, the model parameters are estimated using the maximum likelihood estimation:

$$\frac{\partial l}{\partial \theta_m} = 0 \ (m = 1, 2, \cdots, p). \tag{2.126}$$

Specifically, the likelihood equations are written down as

$$\frac{\partial l}{\partial \sigma} = 0, \tag{2.127}$$

$$\frac{\partial l}{\partial \alpha_i} := S(\alpha_i) = 0, \ (i = 1, 2, \cdots, p) \tag{2.128}$$

$$\frac{\partial l}{\partial \beta} := S(\beta) = 0. \tag{2.129}$$

We use the Newton–Raphson algorithm to solve the likelihood equations in an iterative way:

$$\alpha_i^{(k+1)} = \alpha_i^{(k)} - \frac{S(\alpha_i^{(k)})}{H(\alpha_i^{(k)})}, \quad (i = 1, 2, \cdots, p) \tag{2.130}$$

$$\beta^{(k+1)} = \beta^{(k)} - \frac{S(\beta^{(k)})}{H(\beta^{(k)})}, \tag{2.131}$$

where function $H(\cdot)$ is defined as follows:

$$H(\alpha_i) := -\frac{\partial S(\alpha_i)}{\partial \alpha_i}, \quad (i = 1, 2, \cdots, p) \tag{2.132}$$

$$H(\beta) := -\frac{\partial S(\beta)}{\partial \beta}. \tag{2.133}$$

2.5 Summary

This chapter explained the basics of statistics needed to deal with the nature of renewable energy with fluctuating output. After reviewing basic statistical concepts and statistical tests, regression analysis and time series analysis were explained. In regression analysis, the data reproducibility for the model as a whole and the statistical significance of each explanatory variable are important. In time series analysis, the fundamental analysis was the subject of stationary processes, and the importance of parameter estimation by the maximum likelihood method and model selection by the amount of information was explained.

References

1. David A. Freedman, "Statistical Models: Theory and Practice", Cambridge University Press; 2nd edition, June, 2009, ISBN 978-0521743853
2. Charles Wheelan, "Naked Statistics: Stripping the Dread from the Data", W W Norton & Co Inc; Reprint edition, January, 2014, ISBN 978-0393347777
3. J. F. Box, "Guinness, Gosset, Fisher, and Small Samples", Statistical Science. 2 (1), pp.45–52, (1987). https://doi.org/10.1214/ss/1177013437.

Chapter 3
Fluctuation and Correlation of Renewable Energy

3.1 Principal Component Analysis

3.1.1 Basic Concept of PCA

We are interested in multi-dimensional data for various practical purposes. However, it is often seen that the multi-dimensional data are hard to understand intuitively due to correlations between some of the dimensions of the data. The purpose of the principal component analysis (PCA) is information contraction by converting a set of the original correlated data into a small set of uncorrelated variables. The uncorrelated variables are called the principal components. Suppose the data you are interested in are multi-dimensional, and the dimension of the data is higher than the data's number. In that case, multi-regression analysis cannot be applied.

A multi-dimensional data $\boldsymbol{x}_i (i = 1, \cdots, n)$ is given by a column vector,

$$\boldsymbol{x}_i = \begin{pmatrix} x_{i1} \\ x_{i2} \\ \vdots \\ x_{ip} \end{pmatrix}, \tag{3.1}$$

or a row vector,

$$\boldsymbol{x}_i^T = (x_{i1}, x_{i2}, \cdots, x_{ip}), \tag{3.2}$$

Y. Ikeda, *Data Science of Renewable Energy Integration*,
Evolutionary Economics and Social Complexity Science 30,
https://doi.org/10.1007/978-981-99-8779-5_3

where p is the dimension of data. Similarly, a p-dimensional weight vector $\boldsymbol{w}$ is given by a column vector,

$$\boldsymbol{w} = \begin{pmatrix} w_1 \\ w_2 \\ \vdots \\ w_p \end{pmatrix}, \tag{3.3}$$

or a row vector,

$$\boldsymbol{w}^T = (w_1, w_2, \cdots, w_p). \tag{3.4}$$

Here, we define the principal component by

$$f_i = \sum_{j=1}^{p} w_j x_{ij} = \boldsymbol{w}^T \boldsymbol{x}_i = \boldsymbol{x}_i^T \boldsymbol{w}. \tag{3.5}$$

In the PCA, we determine the weight $\boldsymbol{w}$ to maximize the variance of the prinipal component $V[f]$ with the constraint $|\boldsymbol{w}|^2 = \boldsymbol{w}^T\boldsymbol{w} = 1$.

$$E[f] = \bar{f} = \frac{1}{n}\sum_{i=1}^{n} f_i = \frac{1}{n}\sum_{i=1}^{n}\sum_{j=1}^{p} w_j x_{ij} = \sum_{j=1}^{p} w_j \bar{x}_j = \boldsymbol{w}^T \bar{\boldsymbol{x}} \tag{3.6}$$

$$\begin{aligned} V[f] &= \frac{1}{n-1}\sum_{i=1}^{n}(f_i - \bar{f})^2 = \frac{1}{n-1}\sum_{i=1}^{n}\{\sum_{j=1}^{p} w_j(x_{ij} - \bar{x}_j)\}^2 \\ &= \frac{1}{n-1}\sum_{i=1}^{n}\{\boldsymbol{w}^T(\boldsymbol{x}_i - \bar{\boldsymbol{x}})\}^2 = \frac{1}{n-1}\sum_{i=1}^{n}|\boldsymbol{w}^T(\boldsymbol{x}_i - \bar{\boldsymbol{x}})|^2. \end{aligned} \tag{3.7}$$

We obtain the following relation using the Pythagorean theorem (see Fig. 3.1),

$$\begin{aligned} |\boldsymbol{x}_i - \bar{\boldsymbol{x}}|^2 &= |\boldsymbol{x}_i - \hat{\boldsymbol{x}}_i|^2 + |\hat{\boldsymbol{x}}_i - \bar{\boldsymbol{x}}|^2 \\ &= |\boldsymbol{x}_i - \hat{\boldsymbol{x}}_i|^2 + |\boldsymbol{w}^T(\boldsymbol{x}_i - \bar{\boldsymbol{x}})|^2. \end{aligned} \tag{3.8}$$

The variance of the principal component $V[f]$ is rewritten using the above relation,

$$V[f] = \frac{1}{n-1}\sum_{i=1}^{n} |\boldsymbol{x}_i - \bar{\boldsymbol{x}}|^2 - |\boldsymbol{x}_i - \hat{\boldsymbol{x}}_i|^2. \tag{3.9}$$

Maximizing variance $V[f]$ means minimizing the sum of foot of perpendicular $|\boldsymbol{x}_i - \hat{\boldsymbol{x}}_i|^2$ because each term in the r.h.s. of Eq. (3.9) is positive. Minimizing the sum

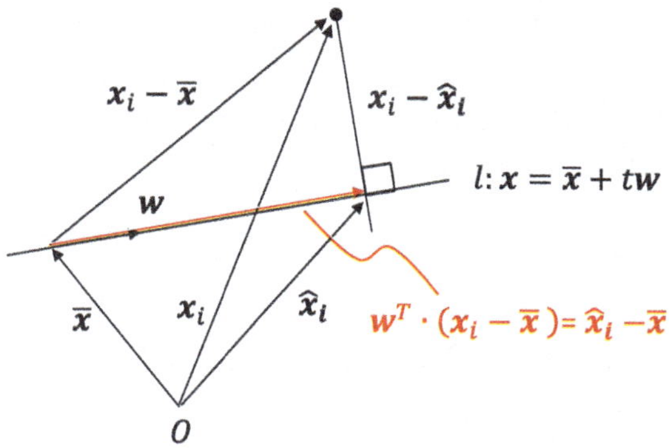

Fig. 3.1 The variance of the principal component $V[f]$ is rewritten using the relation obtained by the Pythagorean theorem

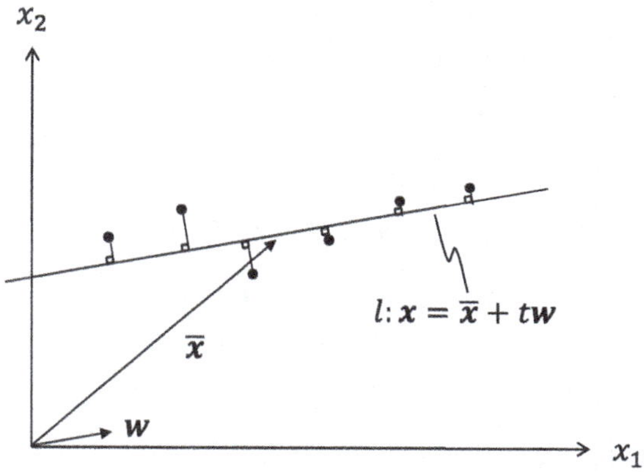

Fig. 3.2 The principal component analysis minimizes the sum of foot of perpendicular

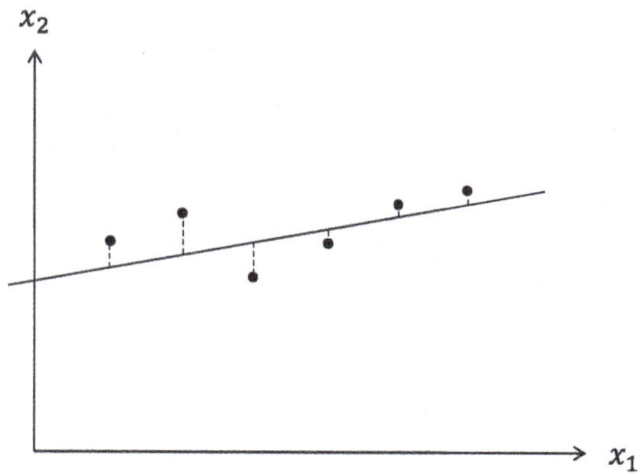

Fig. 3.3 The least square method minimizes the sum of the square root of residuals

of the foot of perpendicular $|\boldsymbol{x}_i - \hat{\boldsymbol{x}}_i|^2$ is depicted in Fig. 3.2 for the case of $n = 6$ and $p = 2$. It is noted that maximizing the variance of the principal component $V[f]$ is different from minimizing the sum of the square root of residuals in the least square method (see Fig. 3.3).

3.1.2 Maximization of $V[f]$

For two-dimensional data, we write the expectation value and variance of the principal component f

$$E[f] = \bar{f} = w_1\bar{x}_1 + w_2\bar{x}_2, \tag{3.10}$$

$$\begin{aligned} V[f] &= \frac{1}{n-1}\sum_{i=1}^{n}(f_i - \bar{f})^2 = \frac{1}{n-1}\sum_{i=1}^{n}\{(w_1x_{i1} + w_2x_{i2}) - (w_1\bar{x}_1 + w_2\bar{x}_2)\}^2 \\ &= \frac{1}{n-1}\sum_{i=1}^{n}\{(w_1(x_{i1} - \bar{x}_1) + w_2(x_{i2} - \bar{x}_2)\}^2 \\ &= \frac{1}{n-1}\sum_{i=1}^{n}\{(w_1^2(x_{i1} - \bar{x}_1)^2 + 2w_1w_2(x_{i1} - \bar{x}_1)(x_{i2} - \bar{x}_2) + w_2^2(x_{i2} - \bar{x}_2)^2\} \\ &= w_1^2S_{11} + 2w_1w_2S_{12} + w_2^2S_{22} \\ &= (w_1, w_2)\begin{pmatrix} S_{11} & S_{12} \\ S_{12} & S_{22} \end{pmatrix}\begin{pmatrix} w_1 \\ w_2 \end{pmatrix} = \boldsymbol{w}^T S\boldsymbol{w}, \end{aligned} \tag{3.11}$$

where $S_{11} = \frac{1}{n-1}\sum_{i=1}^{n}(x_{i1} - \bar{x}_1)^2$, $S_{12} = \frac{1}{n-1}\sum_{i=1}^{n}(x_{i1} - \bar{x}_1)(x_{i2} - \bar{x}_2)$, $and\, S_{22} = \frac{1}{n-1}\sum_{i=1}^{n}(x_{i2} - \bar{x}_2)^2$.

We explain the method of Lagrange multiplier. We find x_0 and y_0 to maximize $f(x, y) = 1 - x^2 - y^2$ with constraint $g(x, y) = x + y - 1 = 0$ (see Fig. 3.4). By introducing Lagrange multiplier λ, we define function $Q(x, y)$ by

$$Q(x, y, \lambda) = f(x, y) - \lambda g(x, y). \tag{3.12}$$

Point $(x, y) = (x_0, y_0)$ at which function $f(x, y)$ has maximum value satisfies equations,

$$\frac{Q}{x} = -2x - \lambda = 0, \tag{3.13}$$

$$\frac{Q}{y} = -2y - \lambda = 0, and \tag{3.14}$$

$$\frac{Q}{\lambda} = x + y - 1 = 0. \tag{3.15}$$

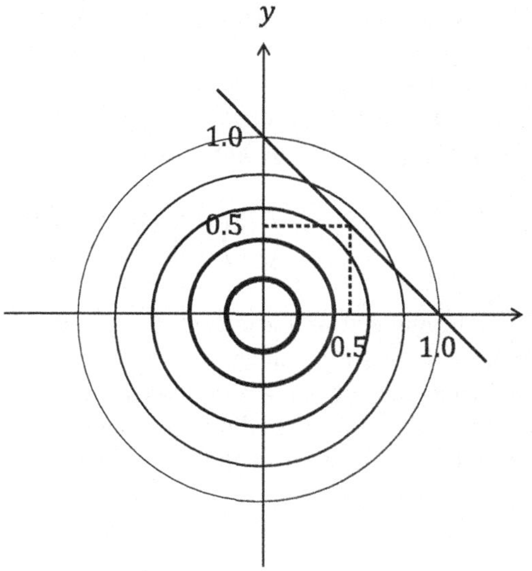

Fig. 3.4 The method of Lagrange multiplier maximizes $f(x, y) = 1 - x^2 - y^2$ with constraint $g(x, y) = x + y - 1 = 0$

By solving the above equations, we obtain solution $(x_0, y_0) = (\frac{1}{2}, \frac{1}{2})$. Multiplier $\lambda = -1$ is equal to ratio of $\frac{f}{x}$ to $\frac{g}{x}$, or to ratio of $\frac{f}{y}$ to $\frac{g}{y}$.

We maximize $V[f]$ with constraint $w_1^2 + w_2^2 = 1$ using the method of Lagrange multiplier. Function $Q(x, y)$ and conditions satisfied by solutions are

$$Q(w_1, w_2, \lambda) = V[f] - \lambda(w_1^2 + w_2^2 - 1). \tag{3.16}$$

$$\frac{Q}{w_1} = 2w_1 S_{11} + 2w_2 S_{12} - 2\lambda w_1 = 0, \tag{3.17}$$

$$\frac{Q}{w_2} = 2w_2 S_{22} + 2w_1 S_{12} - 2\lambda w_2 = 0, and \tag{3.18}$$

$$\frac{Q}{\lambda} = w_1^2 + w_2^2 - 1 = 0. \tag{3.19}$$

Equations (3.17) and (3.18) are rewritten as a matrix form by

$$\begin{pmatrix} S_{11} & S_{12} \\ S_{12} & S_{22} \end{pmatrix} \begin{pmatrix} w_1 \\ w_2 \end{pmatrix} = \lambda \begin{pmatrix} w_1 \\ w_2 \end{pmatrix}, or \tag{3.20}$$

$$S\boldsymbol{w} = \lambda \boldsymbol{w}, or \tag{3.21}$$

$$(S - \lambda I)\boldsymbol{w} = 0, \tag{3.22}$$

which is referred as the eigenvalue problem.

We explain the eigenvalue problem. We can analyze the eigenvalue of the correlation matrix. Multiplying a vector by a matrix generally produces a vector with a different magnitude and direction. A vector with only a different magnitude without any change in direction is called an eigenvector. A multiplying factor that can only change the magnitude is called the eigenvalue of the matrix.

$$\begin{pmatrix} x_1' \\ x_2 \end{pmatrix} = \begin{pmatrix} \cos\theta & -\sin\theta \\ \sin\theta & \cos\theta \end{pmatrix} \begin{pmatrix} x_1 \\ x_2 \end{pmatrix}. \tag{3.23}$$

PCA is formulated as the eigenvalue problem. It focuses on the covariance or correlation between variables in terms of N-dimensional data and represents the overall characteristics of several indicators. In other words, PCA is another typical multivariate analysis method for information contraction.

From Eq. (3.22), we have the characteristic equation

$$|S - \lambda I| = \begin{vmatrix} S_{11} - \lambda & S_{12} \\ S_{12} & S_{22} - \lambda \end{vmatrix} = 0, or \tag{3.24}$$

$$\lambda^2 - (S_{11} + S_{22})\lambda + S_{11}S_{22} - S_{12}^2 = 0. \tag{3.25}$$

We note that two roots λ_1 and λ_2 are real number because

$$D = (S_{11} + S_{22})^2 - 4(S_{11}S_{22} - S_{12}^2) = (S_{11} - S_{22})^2 + 4S_{12}^2 \geq 0. \tag{3.26}$$

Thus, we have two principal components. The first component is

$$f_1 = \boldsymbol{w}_1^T \boldsymbol{x} \tag{3.27}$$

and the second component is

$$f_2 = \boldsymbol{w}_2^T \boldsymbol{x}. \tag{3.28}$$

Variance of the principal components is calculated as follows:

$$V[f_1] = \boldsymbol{w}_1^T S \boldsymbol{w}_1 = \lambda_1 \boldsymbol{w}_1^T \boldsymbol{w}_1 = \lambda_1, \tag{3.29}$$

$$V[f_2] = \boldsymbol{w}_2^T S \boldsymbol{w}_2 = \lambda_2 \boldsymbol{w}_2^T \boldsymbol{w}_2 = \lambda_2 \tag{3.30}$$

and are equal to the eigenvalues (see Fig. 3.5).

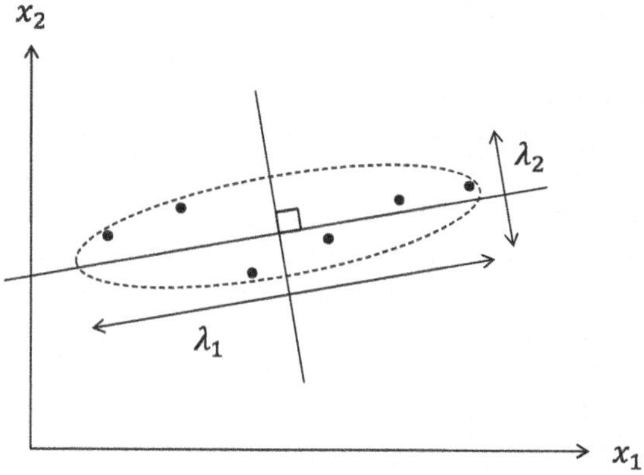

Fig. 3.5 Eigenvalue: Variances of the principal components are equal to the eigenvalues

3.1.3 Important Indices

The following are the important indices of PCA:

Contribution ratio

$$r_j = \frac{\lambda_j}{tr(S)} \tag{3.31}$$

Cumulative contribution ratio

$$C_j = \sum_{j'=1}^{j} r'_j = \frac{\lambda_1 + \lambda_2 + \cdots + \lambda_j}{tr(S)} \tag{3.32}$$

The jth principal component score of the ith element

$$f_{ij} = w_{j1}x_{i1} + w_{j2}x_{i2} + \cdots + w_{jp}x_{ip} = \sum_{k=1}^{p} w_{jk}x_{ik} \tag{3.33}$$

3.2 Solar Photovoltaic Power

3.2.1 Cross-correlation of PV Output Fluctuation

Restructuring of the electric utility industry and large-scale grid integration of PV systems were intensively discussed after the East Japan Earthquake of 2011. The former includes separating electrical power generation from power distribution and transmission, establishing the retail power market, and revitalizing the wholesale power markets. Although the institutional design of the power markets is still an open question in Japan, the market has to be designed to have an optimal operation schedule, which is obtained using a unit commitment calculation through competitions between generation companies.

The large-scale grid integration of PV systems brings another kind of problem, namely, the PV output fluctuation, into the power system operation. The planned installation capacity of PV systems will be 100 GW in 2030 in the whole of Japan [1], and about one-third of the capacity will be in the Tokyo area. A significant fraction of the PV system will be installed on the rooftop of the consumer's residential houses and office buildings, which are widely distributed in Tokyo. Therefore, the forecast of PV output with high spatial resolution is a crucial problem to be considered, and the cross-correlations of the PV outputs will be essential quantities to estimate the forecast error of PV output.

Concerning the above discussion, the concept of "local production for local consumption of renewable energy" has been proposed in Japan. Because electric power is in great demand in the Tokyo area, the area price could be high enough to be close to the feed-in tariff price for PV power. For this reason, the concept of "local production for local consumption of renewable energy" of PV power is considered to be economically feasible [2]. This concept is also advantageous because of the mitigation of transmission loss. However, it is to be noted that this concept needs careful consideration for PV and wind power because of the inherent nature of output fluctuation, even though it is suitable for geothermal and biomass energies [3].

In this chapter, we analyzed the cross-correlation of PV output fluctuation for the actual PV output time series data [4] in both the Tokyo area and the whole of Japan using the principal component analysis with the random matrix theory. Based on the obtained cross-correlation coefficients, the forecast error for PV output was estimated for some extreme cases. Then, the operation schedule of thermal plants was calculated to integrate PV output using our unit commitment model [5, 6] with the estimated forecast error. The system-balancing cost of the PV system was also estimated with or without demand response. Finally, the validity of the concept of "local production for local consumption of renewable energy" and alternative policy implications were also discussed.

3.2.2 System-Wide Output Fluctuation

The forecast of system-wide PV output is decomposed as

$$pv_t^{(f)} \equiv X(t) = \sum_{i=1}^{N} x_i(t) = \sum_{i=1}^{N} c_i y_i(t), \tag{3.34}$$

where $y_i(t) = x_i(t)/c_i$ and c_i are the forecast of PV output per installed capacity (load factor) and the installed capacity in the ith site, respectively. Our unit commitment model [5, 6] requires the PV output forecast time series and the forecast error to estimate the optimal operation schedule considering the PV output fluctuation. Suppose both the accuracy and spatial resolution of the PV forecasting are high. In that case, the forecasted time series is similar to a moving average of actual PV output for each PV site. Consequently, the cross-correlation of residual time series, equal to subtracting the actual output from forecast output at each time point, is expected to be a white noise. Thus, the forecast error of system-wide PV output σ_X is

$$\sigma_p^2 = \sum_{i=1}^{N} \left(\frac{\partial X}{\partial y_i}\right)^2 \sigma_i^2 = \sum_{i=1}^{N} c_i^2 \sigma_i^2, \tag{3.35}$$

where σ_i is the forecast error of PV output per installed capacity in the ith site. On the other hand, if the spatial resolution of the forecast is low and, for example, we have just a few forecasted sites in the Tokyo area, the residual time series includes the cross-correlation between the various PV sites located in different places. In this case, we have a larger forecasting error due to the cross-correlations. The forecast error of system-wide PV output σ_X is written as

$$\begin{aligned} \sigma_p^2 &= \sum_{i=1}^{N} \left(\frac{\partial X}{\partial y_i}\right)^2 \sigma_i^2 + 2 \sum_{i=2}^{N} \sum_{j<i} \left(\frac{\partial X}{\partial y_i}\right)\left(\frac{\partial X}{\partial y_j}\right) \sigma_{ij} \\ &= \sum_{i=1}^{N} c_i^2 \sigma_i^2 + 2 \sum_{i=2}^{N} \sum_{j<i} c_i c_j \sigma_{ij} \end{aligned} \tag{3.36}$$

$$\sigma_{ij} = \sigma_i \sigma_j \rho_{ij} \tag{3.37}$$

by including covariance among different sites σ_{ij}. Here, ρ_{ij} is the cross-correlation coefficient among different sites. Generally, it is expected that the number of forecasted sites is smaller than that of the installed sites N. For instance, we cannot forecast PV output for each rooftop PV of all the residential houses and office buildings with high accuracy in the Tokyo area due to both technological and economic reasons. Therefore, it is required to consider the cross-correlation σ_{ij} to estimate the forecast error of system-wide PV output σ_X.

3.2.3 *Random Matrix Theory*

We analyzed the de-trended PV output $z_i(t)$ obtained by filtering the actual PV output time series per installed capacity $y_i(t)$ using the Fourier series expansion.

In general, correlation coefficients are expected to be associated with random noise for a fluctuating time series such as PV output. The correlation coefficient between points i and j is calculated by

$$C_{ij} = \frac{\langle (z_i(t) - \langle z_i \rangle)(z_j(t) - \langle z_j \rangle) \rangle}{\sqrt{(\langle z_i^2 \rangle - \langle z_i \rangle^2)(\langle z_j^2 \rangle - \langle z_j \rangle^2)}}, \tag{3.38}$$

where $z_i(t)$ is the de-trended PV output at the site $i(= 1, \cdots, N)$ and time $t(= 1, \cdots, L)$ and $\langle \cdot \rangle$ indicates the time average for the time series.

Now we consider the eigenvalue problem

$$C|\alpha\rangle = \lambda_\alpha |\alpha\rangle \tag{3.39}$$

for the correlation matrix C. λ_α and $|\alpha\rangle$ are the eigenvalues and the corresponding eigenvector, respectively. We assume that the eigenvalues are arranged in decreasing order ($\alpha = 0, \cdots, N-1$). Once the eigenvalues are calculated using Eqs. (3.38) and (3.39), the distribution of eigenvalue $\rho(\lambda)^E$ is obtained.

According to the random matrix theory [7–10], distribution of the eigenvalue for the matrix $\frac{1}{T}HH^T$ where all elements of the matrix H are given as a random number $N(0, \sigma^2)$ is given by

$$\rho(\lambda)^T = \frac{Q}{2\pi} \frac{\sqrt{(\lambda_{\max} - \lambda)(\lambda - \lambda_{\min})}}{\lambda}, \tag{3.40}$$

where

$$Q = \frac{L}{N}, \tag{3.41}$$

$$\lambda = [\lambda_{\min}, \lambda_{\max}], \tag{3.42}$$

$$\lambda_{\min} = (1 - \frac{1}{\sqrt{Q}})^2, \text{ and} \tag{3.43}$$

$$\lambda_{\max} = (1 + \frac{1}{\sqrt{Q}})^2. \tag{3.44}$$

Equation (3.40) is exact at the limit $N, L \to \infty$. For a randomly fluctuating time series such as PV output, the distribution $\rho(\lambda)^E$ obtained by data analysis is expected to agree to the distribution $\rho(\lambda)^T$ calculated using Eqs. (3.40) to (3.44) for $\lambda \leq \lambda_{\max}$. Therefore, only the small number of eigenvalues for $\lambda > \lambda_{\max}$ has the information of genuine correlation.

In order to extract the genuine correlation, we rewrite the correlation matrix C using eigenvalue λ_α and the corresponding eigenvector $|\alpha\rangle$ [11]. First, we define the complex conjugate vector of the eigenvector $|\alpha\rangle$ by

$$\langle\alpha| = |\alpha^*\rangle^t. \tag{3.45}$$

For the real symmetric matrix, such as the correlation matrix C, all elements of the eigenvector $|\alpha\rangle$ are real. Thus, the complex conjugate denotes the transpose t.

Then, the correlation matrix C is rewritten as

$$C = \sum_{\alpha=0}^{N-1} \lambda_\alpha |\alpha\rangle\langle\alpha| \tag{3.46}$$

by multiplying Eq. (3.39) with the transposed vector $\langle\alpha|$ from the left-hand side and taking summation over α. Here, the property of the projection operator $|\alpha\rangle\langle\alpha|$

$$\sum_{\alpha=0}^{N-1} |\alpha\rangle\langle\alpha| = 1 \tag{3.47}$$

was used. As a result, the correlation matrix C of Eq. (3.46) is divided in the following components:

$$C = C^t + C^r = \sum_{\alpha=0}^{N_t} \lambda_\alpha |\alpha\rangle\langle\alpha| + \sum_{\alpha=N_t+1}^{N-1} \lambda_\alpha |\alpha\rangle\langle\alpha|. \tag{3.48}$$

The first term C^t corresponds to the genuine correlation component ($\lambda > \lambda_{\max}$). The second term C^r corresponds to the random component ($\lambda \leq \lambda_{\max}$). The term $\lambda_0|0\rangle\langle 0|$ is interpreted as the change as a whole system, such as the weather change.

We introduce the vector $|z(t)\rangle$, which consists of the time series of PV output $z_i(t) (i = 1, \cdots, N)$. Then the vector $|z(t)\rangle$ is expanded on the basis of the eigenvectors $|\alpha\rangle$ [11] :

$$|z(t)\rangle = \sum_{\alpha=0}^{N-1} a_\alpha(t)|\alpha\rangle. \tag{3.49}$$

The expansion coefficient $a_\alpha(t)$ is obtained using the orthogonality of the eigenvectors:

$$a_\alpha(t) = \langle\alpha|z(t)\rangle. \tag{3.50}$$

The time series corresponding to the genuine correlation C^t is extracted by truncating the summation up to N_t in Eq. (3.49):

$$|z(t)\rangle = \sum_{\alpha=0}^{N_t} a_\alpha(t)|\alpha\rangle. \tag{3.51}$$

3.2.4 Data Analysis

The genuine components of cross-correlation of the de-trended PV output per installed capacity were studied using the random matrix theory. The analyzed data are the output time series acquired every hour for each prefecture [4]. Before analyzing the data, two preprocessing were made. First, the data during nighttime were removed. Then, the trend was removed from the time series by filtering out the components with a period longer than six hours using the Fourier series expansion. Therefore, only the short-term fluctuation is the component that remained in the time series. The auto-correlation function and fluctuation distribution for Tokyo in May are shown in Fig. 3.6. The memory in the auto-correlation function gets lost within a few hours. This means that the trend component is well removed. The kurtosis of the fluctuation distribution is 5.0849, which is significantly larger than the value expected for the normal distribution, i.e., 3.0. This means the actual fluctuation distribution has a longer tail than the normal distribution. The two different types of functional forms of fluctuation distribution are shown in Fig. 3.7. If the fluctuation is distributed according to the normal distribution, the probability density function is

$$p(x) = \frac{1}{\sqrt{2\pi\sigma^2}} \exp\Big[-\frac{(x-\mu)^2}{2\sigma^2}\Big], \tag{3.52}$$

and the cumulative distribution function is written using the error function erf[·] as

$$\phi(x) = \frac{1}{2}\Big(1 + \mathrm{erf}\Big[\frac{x-\mu}{\sqrt{2\sigma^2}}\Big]\Big), \tag{3.53}$$

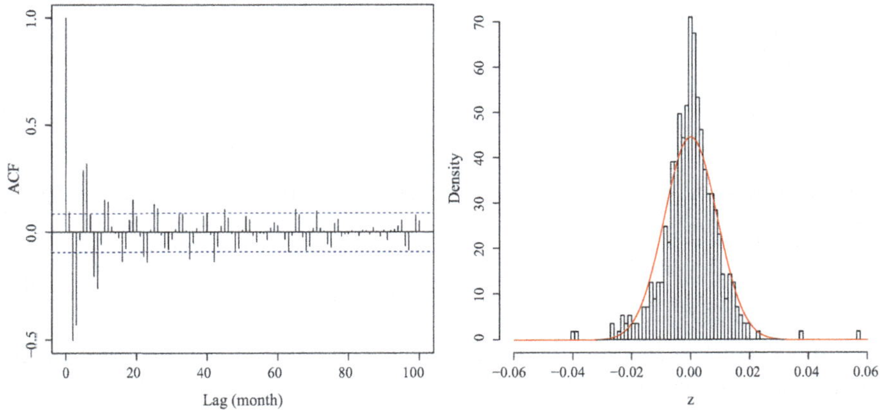

Fig. 3.6 Auto-correlation function and fluctuation distribution for Tokyo in May are shown in the left and right panels, respectively. The red curve in the right panel is the normal distribution. The kurtosis of the actual distribution is positive

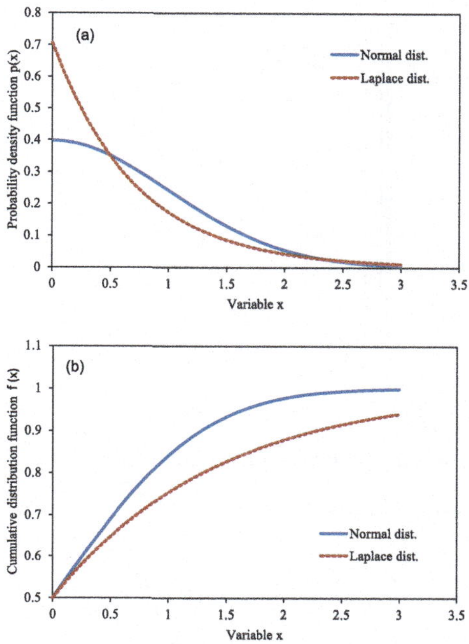

Fig. 3.7 Functional Form of Fluctuation Distribution: The positive kurtosis is reproduced for the Laplace distribution

where μ and σ are the mean and standard deviation, respectively. However, if the probability density function $p(x)$ is a Laplace distribution

$$p(x) = \frac{1}{2b} \exp\left[-\frac{|x-\mu|}{b}\right], \tag{3.54}$$

then the cumulative distribution function $\phi(x)$ is

$$\phi(x) = \frac{1}{2}\left(1 + \mathrm{sgn}(x-\mu)\left(1 - \exp\left[-\frac{|x-\mu|}{b}\right]\right)\right). \tag{3.55}$$

Here, a standard deviation is given by $\sigma = \sqrt{2}b$ and $\mathrm{sgn}(x-\mu) = +(x \geq \mu), -(x < \mu)$. The functional forms for these distributions are depicted for $\mu = 0$ and $\sigma = 1$ in Fig. 3.7. It is to be noted here that the Laplace distribution shows a distribution tail longer than the normal distribution.

Eigenvalue distribution for the Tokyo area and Japan in May is shown in Fig. 3.8. For the Tokyo area, we calculate $\lambda_{\max} = 1.35$ using Eq. (3.44) with $N = 9$ and $L = 420$. The upper panel of Fig. 3.8 depicts that only the largest eigenvalue is larger than $\lambda_{\max}$. On the other hand, for the whole of Japan, we calculate $\lambda_{\max} = 1.88$ with $N = 47$ and $L = 420$. The lower panel of Fig. 3.8 depicts that the five largest eigenvalues are larger than $\lambda_{\max}$.

We show the distribution of genuine correlation coefficients calculated for the detrended PV output time series in both the Tokyo area and the whole of Japan. The cross-correlation coefficients for the Tokyo area are shown in Fig. 3.9. Panels (a) and (c) are genuine correlation C^t, and panels (b) and (d) are the random components

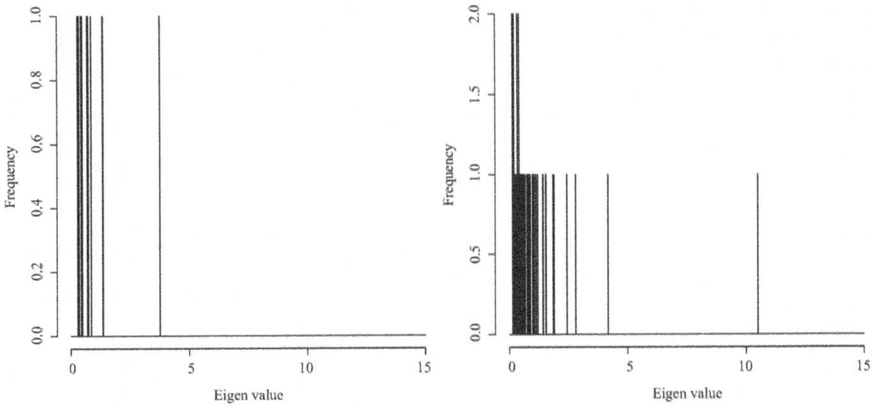

Fig. 3.8 Eigenvalue distribution for the Tokyo area (left) and the entire Japan (right) in May

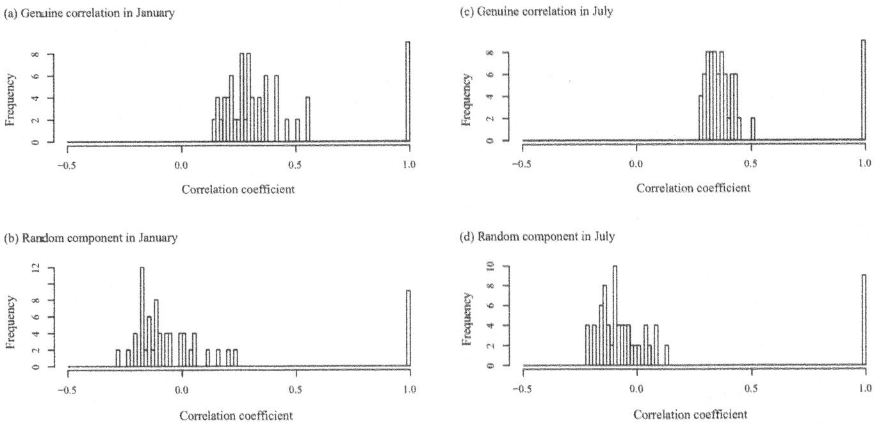

Fig. 3.9 Cross-correlation coefficients for the Tokyo area in January and July: Panels (**a**) and (**c**) are correlations, and panels (**b**) and (**d**) are random noise

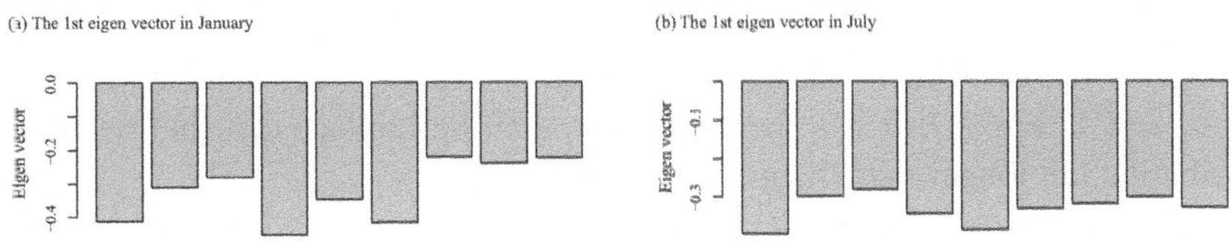

Fig. 3.10 The 1st eigenvector for the Tokyo area: panel (**a**) is January, and panel (**b**) is July

C^r. The genuine correlation C^t was calculated using only the largest eigenvalue and the corresponding eigenvector. Figure 3.9 depicts that the genuine correlation C^t has a positive correlation, and on the other hand, the random components C^r distribute around 0.0. The 1st eigenvector for the Tokyo area is shown in Fig. 3.10. The nine components correspond to eight prefectures, and Tokyo was included in the Tokyo

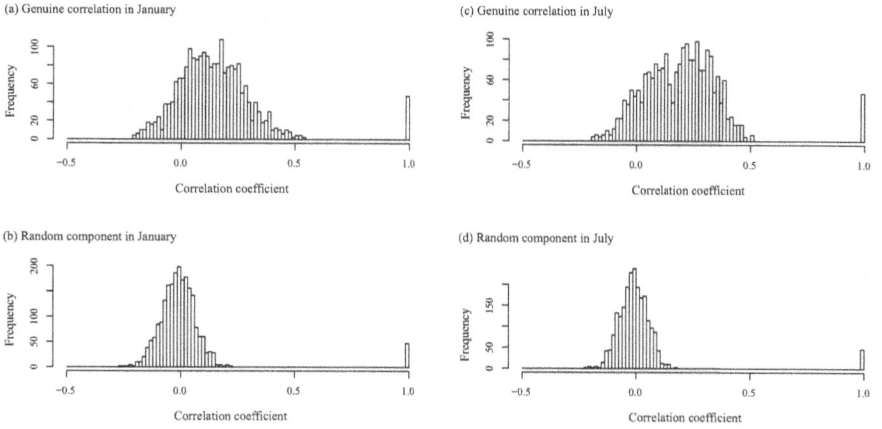

Fig. 3.11 Cross-correlation coefficients for the whole of Japan in January and July: Panels (**a**) and (**c**) are correlations, and panels (**b**) and (**d**) are random noise

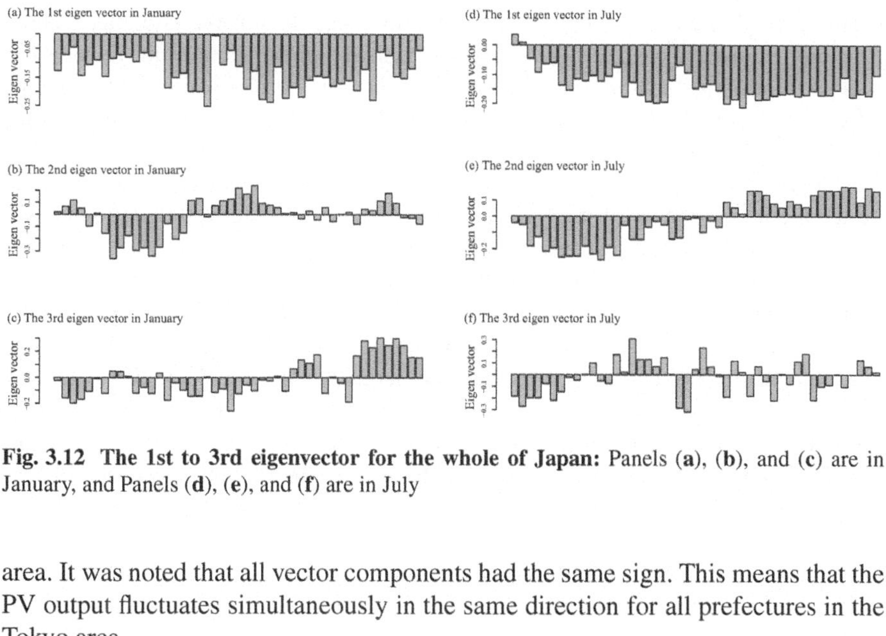

Fig. 3.12 The 1st to 3rd eigenvector for the whole of Japan: Panels (**a**), (**b**), and (**c**) are in January, and Panels (**d**), (**e**), and (**f**) are in July

area. It was noted that all vector components had the same sign. This means that the PV output fluctuates simultaneously in the same direction for all prefectures in the Tokyo area.

The cross-correlation coefficients for Japan are shown in Fig. 3.11. Panels (a) and (c) are genuine correlation C^t, and panels (b) and (d) are the random components C^r. The genuine correlation C^t was calculated using only the five largest eigenvalues and corresponding eigenvectors. Figure 3.11 depicts that the genuine correlation C^t has a positive correlation, and on the other hand, the random components C^r distribute around 0.0. The cross-correlation of PV output fluctuation in the Tokyo area was larger than the cross-correlation in Japan throughout the year. The 1st to 3rd eigenvectors for Japan are shown in Fig. 3.12. Forty-seven components correspond to all prefectures from Hokkaido to Okinawa in Japan. The 1st eigenvector has all components with the same sign. This means that the PV output fluctuates simultaneously in the same direction for all prefectures in Japan. The characteristic of the 1st eigenvector in Japan is similar to the Tokyo area. The 2nd eigenvector shows the weather change between eastern and western Japan. The 3rd eigenvector is more complicated. These characteristics of the 2nd to 5th eigenvectors correspond to the smaller correlation coefficients in Japan. However, it is noted that the coefficient of variation of PV output does not decrease proportionally to $N^{-1/2}$ as the number of PV sites N increased due to the observed cross-correlation among the sites. Thus, the so-called smoothing effect is expected to be smaller than the ideal case without cross-correlation.

3.2.5 Estimation of Forecast Error

We estimated the lower limit of the system-wide forecast error using the cross-correlation coefficients of the output fluctuation described in the previous section. Recently, numerical weather forecasting has gained higher accuracy due to meteorological information acquired by weather radars and meteorological satellites and the advancement of high-performance computers. Ultimately, as the forecast accuracy becomes higher, the forecast of the PV output time series at each site converges on the moving average trend of the site. Thus, we expect the short-term fluctuation to be the main component of the forecast error because short-term fluctuation cannot be forecasted. Therefore, we assume here that the lower limit of the forecast error is identical to the short-term fluctuation.

If the number of forecast sites is small, e.g., just one site in each prefecture, the system-wide forecast error involves the cross-correlation between the sites, and consequently, the system-wide error becomes large. On the other hand, if the number of forecast sites is large, the system-wide forecast error does not involve the cross-correlation between the sites, and consequently, the system-wide error becomes small. Shortly, installed PV systems will be widely distributed in various places. The actual system-wide forecast error is expected between the above two extreme cases.

We estimated the lower limit of the system-wide forecast errors and the coefficients of variation with/without considering the cross-correlations of the PV output fluctuation using Eqs. (3.36) and (3.37) with the genuine cross-correlation coefficient ρ_{ij} shown in Figs. 3.9 and 3.11. The installed capacity of PV systems in 2030 was estimated by dividing the 100 GW capacity in the whole of Japan [1] proportionally to the demand of each prefecture. The errors and variation coefficients estimations in the Tokyo area and Japan are shown in Tables 3.1 and 3.2, respectively. The 2nd to 5th columns of the tables represent error without correlation, coefficient of variation without correlation, error with correlation, and coefficient of variation with correlation, respectively. Both the system-wide forecast errors and the coefficients of variation are increased by considering the cross-

Table 3.1 Lower Limit of the System-Wide Forecast Errors in the Tokyo Area

Month	Error w/o cor	Var w/o cor	Error w cor	Var w cor
Jan	101.26	0.0168	181.36	0.0302
Feb	107.53	0.0167	206.78	0.0321
Mar	145.89	0.0208	296.55	0.0423
Apr	132.21	0.0195	246.95	0.0365
May	136.88	0.0194	262.23	0.0373
Jun	128.02	0.0226	221.59	0.0391
Jul	136.47	0.0222	256.34	0.0417
Aug	128.47	0.0187	236.19	0.0344
Sep	120.06	0.0202	239.98	0.0405
Oct	103.20	0.0191	183.09	0.0338
Nov	108.15	0.0222	236.34	0.0485
Dec	77.499	0.0144	154.74	0.0288

Table 3.2 Lower Limit of the System-Wide Forecast Errors in the Whole of Japan

Month	Error w/o cor	Var w/o cor	Error w cor	Var w cor
Jan	148.36	0.0093	352.85	0.0223
Feb	164.13	0.0090	408.97	0.0225
Mar	218.45	0.0103	655.99	0.0309
Apr	207.07	0.0094	532.03	0.0242
May	202.87	0.0089	579.35	0.0255
Jun	196.43	0.0108	449.46	0.0247
Jul	205.63	0.0104	567.38	0.0288
Aug	191.47	0.0087	532.09	0.0244
Sep	184.43	0.0099	557.29	0.0299
Oct	160.52	0.0092	409.99	0.0235
Nov	165.50	0.0113	558.46	0.0383
Dec	124.18	0.0083	313.99	0.0210

correlation of the fluctuation. The lower limit of the coefficients of variation in the Tokyo area is larger than the lower limit of the coefficients in Japan throughout the year.

3.3 Wind Power

In the previous section, we used solar irradiance fluctuations as an example to illustrate the possibility of grid-integrating more PV power by using correlations of irradiance over a wide area. In this section, we explain the random walk model of wind speed time series and the correlation of wind speed over a wide area. The broad region to be analyzed is Europe.

3.3.1 Random Walk Model of Wind Speed Time Series

When time series $\{x_t\}$ has a unit root, the time series is a non-stationary process. In the analysis of non-stationary processes, we explained in Sect. 2.4 how to model them using an AR model after converting them to stationary processes by taking time difference. However, modeling the original time series directly without taking time differences is also possible. Such a model is the random walk model. In the following, we examine the random walk model for wind speed time series through a simple auto-regression analysis. The following regression analysis is performed for the time series $\{x_t\}$:

$$x_{t+1} = a_0 x_t + a_1 + u_t. \tag{3.56}$$

In the random walk model, we have $a_0 = 1$. Let us check that this condition is satisfied in the actual time series. Here, the constant a_1 is called the drift term. In regression analysis, u_t is the residual, but in the random walk model, it is interpreted as noise. This noise is added to the system of interest from the outside, which drives the system. The analysis results for the time series of wind speed at 50 m above the surface in London in January 2020 are shown in Table 3.3. Since the estimated result for a_0 is 0.986541 and the standard error is 0.006027, the condition $a_0 = 1$ is satisfied. Furthermore, for the residual u_t, a QQ plot is shown in Fig. 3.13. This QQ plot shows that the distribution of u_t is not a normal distribution but a distribution with a long tail. A similar analysis was performed for wind speed data for July 2020. The results are shown in Table 3.4 and Fig. 3.14. These results show that the random walk model validates the July 2020 wind speed data. The long tail distribution for noise is represented as the Laplace distribution in the unit commitment model in later Sect. 6.1.

For the analysis of the random walk model of wind speed time series for each month in 2020, see Appendix B.

Table 3.3 Random Walk Model: Jan 2020

	Estimate	Std. Error	t value	Pr(>|t|)
a_1	0.113497	0.054065	2.099	0.0361*
a_0	0.986541	0.006027	163.693	<2e-16***

Residual standard error: 0.644 on 741 degrees of freedom
Multiple R-squared: 0.9731, Adjusted R-squared: 0.9731
F-statistic: 2.68e+04 on 1 and 741 DF, p-value: <2.2e-16

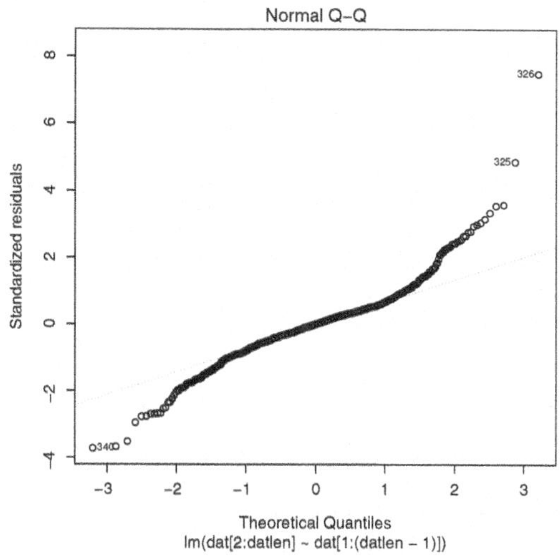

Fig. 3.13 QQ plot for residuals of fitting wind speed (m/s) at 50 m from the surface in London: Jan. 2020

Table 3.4 Random Walk Model: Jul 2020

	Estimate	Std. Error	t value	Pr(>|t|)
a_1	0.093125	0.044403	2.097	0.0363*
a_0	0.986066	0.006043	163.175	<2e-16***

Residual standard error: 0.4578 on 741 degrees of freedom

Multiple R-squared: 0.9729, Adjusted R-squared: 0.9729

F-statistic: 2.663e+04 on 1 and 741 DF, p-value: <2.2e-16

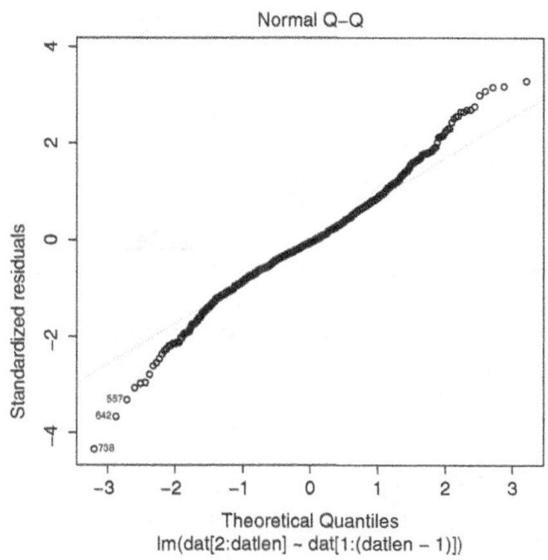

Fig. 3.14 QQ plot for residuals of fitting wind speed (m/s) at 50 m from the surface in London: Jul. 2020

3.3.2 *Correlation of Wind Speed in Europe*

Let us examine the seasonal variation of the correlation by analyzing NASA wind speed data for each month. In Fig. 3.15, the location of the capital cities of European

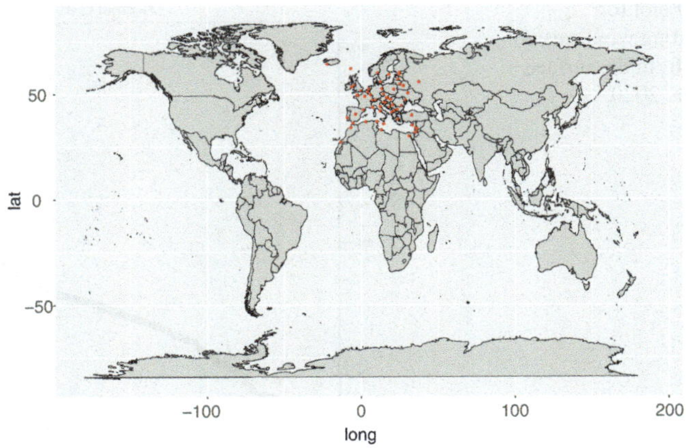

Fig. 3.15 Capital Cities in Europe

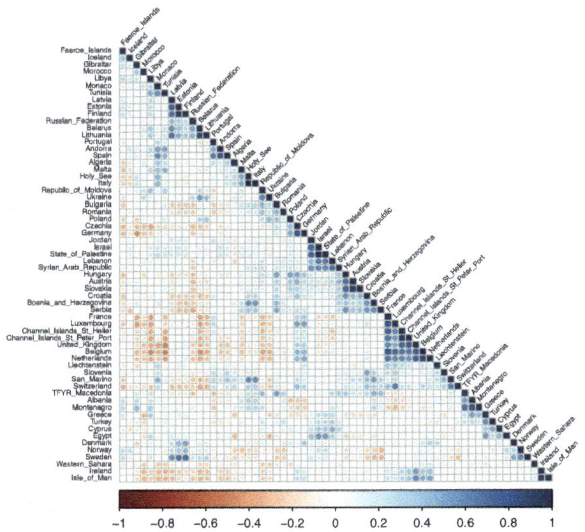

Fig. 3.16 NASA The POWER Project Data: Wind speed correlation among European countries at 50 m from the surface: Jan. 2020

countries is indicated by red dots. Figures 3.16 and 3.17 show the correlations of wind speeds at 50 m above the surface for each of these capitals for January and July, respectively. These figures show a positive correlation among relatively neighboring countries throughout the year and a negative correlation among geographically distant countries during winter. This suggests that more wind-generated power could be grid-integrated by utilizing the correlation of wind speeds across Europe.

For a correlation of each month in 2020, see Appendix C.

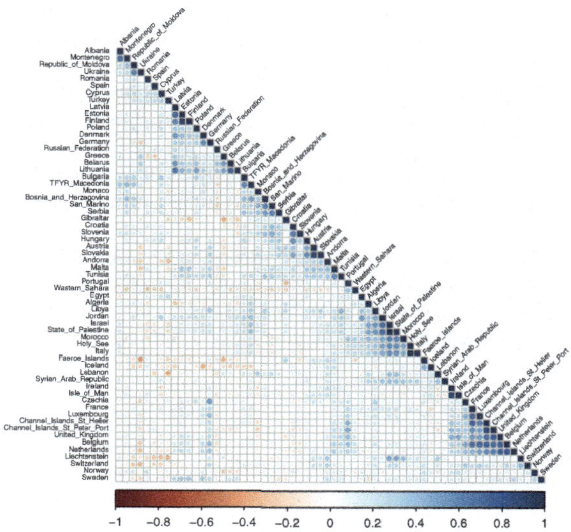

Fig. 3.17 NASA The POWER Project Data: Wind speed correlation among European countries at 50 m from the surface: Jul. 2020

3.4 Summary

In this chapter, after explaining principal component analysis, which performs information reduction based on correlations for multivariate data, we explained how to separate the signal and noise components of each mode of principal component analysis using random matrix theory. This method is discussed to extract statistically significant correlations and to smooth the output fluctuations of renewable energies installed over a wide area. Furthermore, the usefulness of the method described in this chapter is demonstrated by analyzing sunshine data in Japan and wind speed data in Europe.

References

1. Japan Photovoltaic Energy Association: "JPEA PV Outlook 2030", 2012. (in Japanese)
2. Hatta T., "Denryoku sisutemu kaikaku wo dou susumeruka (How do we step up the reform of the power system?)" (Nikkei Publishing Inc., Tokyo, 2012), ch. 5. (in Japanese)
3. Niitsuma H., Nakata T., "EIMY (energy in my yard)? A concept for practical usage of renewable energy from local sources",. European Geothermal Conf. 2003, Szeged, Hungary, 2003
4. Oozeki T., Fonseca J., Takashima T., Ogimoto K., "Dataset of photovoltaic system output for power system analysis", IEEJ, FTE11, 2011, pp. 1–5. (in Japanese)
5. Ikeda Y., Ikegami T., Kataoka K., Ogimoto K., "A unit commitment model with demand response for the integration of renewable energies", IEEE Power & Energy Society General Meeting 2012, 2012, vol. 1, no. 7, pp. 22–26. Available at http://www.arXiv:1112.4909v1 [cs.SY], accessed 21 December 2011

6. Ikeda Y., Ogimoto K., "A unit commitment model with demand response and electric storage device for the integration of variable renewable energies", IEEJ Trans. Power Energy, 2013 (in Japanese), 133, (7), pp. 598–605
7. Metha M.L., "Random matrices", Academic Press, Boston, 1991
8. Laloux L., Cizeau P., Bouchaud J.-P., Potters M., "Noise dressing of financial correlation matrices", Phys. Rev. Lett., 1999, 83, p. 1467
9. Pleroux V., Gopikrishnam P., Rosenow B., Amaral L.A.N., Stanley H.E, "Universal and nonuniversal properties of cross correlations in financial time series", Phys. Rev. Lett., 1999, 83, p. 1471
10. Pleroux V., Gopikrishnam P., Rosenow B., Amaral L.A.N., Guhr T., Stanley H.E., "Random matrix approach to cross correlations in financial data", Phys. Rev. E, 2002, 65, p. 066126
11. Iyetomi H., Nakayama Y., Yoshikawa H., et al., "What causes business cycles? Analysis of the Japanese industrial production data", J. Jpn. Int. Econ., 2011, 25, (3), pp. 246–272

Part II
Toward Decarbonized World

Chapter 4
Optimization and Dynamical System

4.1 Optimization Techniques

4.1.1 Linear Programming

Mathematical programming is a method for finding a solution that maximizes or minimizes an objective function while satisfying given constraints [1, 2]. In particular, when both the constraints and the objective function are linear, it is called linear programming [3]. Otherwise, it is called nonlinear programming. In this section, the most basic linear programming is explained.

First, let us consider a simple example of production planning. As shown in Table 4.1, we know the materials c and d, labor, and profit per kg of products A and B required to produce 1 kg of products. The constraints for materials c and d are less than 100 and 200 kg, respectively, and the constraint for labor is less than 150 h. Let products A and B output be $x(kg)$ and $y(kg)$, respectively. The linear programming problem to find the production volume $x(kg)$ and $y(kg)$ of products A and B to maximize the profit can be formulated as follows:

$$
\begin{aligned}
\text{maximize} \qquad & z = 8x + 6y \\
\text{subject to} \qquad & 3x + 5y \leq 100 \\
& 9x + 4y \leq 200 \\
& 10x + 50y \leq 150 \\
& x, y \geq 0.
\end{aligned}
\tag{4.1}
$$

Y. Ikeda, *Data Science of Renewable Energy Integration*,
Evolutionary Economics and Social Complexity Science 30,
https://doi.org/10.1007/978-981-99-8779-5_4

Table 4.1 Production planning of A and B

Product	*A*	*B*
Material *c* (kg)	3	5
Material *d* (kg)	9	4
Labor (hour)	10	50
Profit (USD)	8	6

The two-variable linear programming problem can be solved by drawing a diagram as shown in Fig. 4.1. The solution region is the lower overlapping region of the three constraints. The point in this region where the objective function is maximal gives the solution.

The formulation of the linear programming problem is given below:

$$\begin{aligned} &\text{maximize} \quad && z = \sum_{i=0}^{n-1} c_i x_i \\ &\text{subject to} \quad && \sum_{i=0}^{n-1} a_{ki} x_i \leq b_k \ (k = 0, 1, 2, \cdots, K-1) \\ & && x_i \geq 0 \ (i = 0, 1, 2, \cdots, n-1). \end{aligned} \tag{4.2}$$

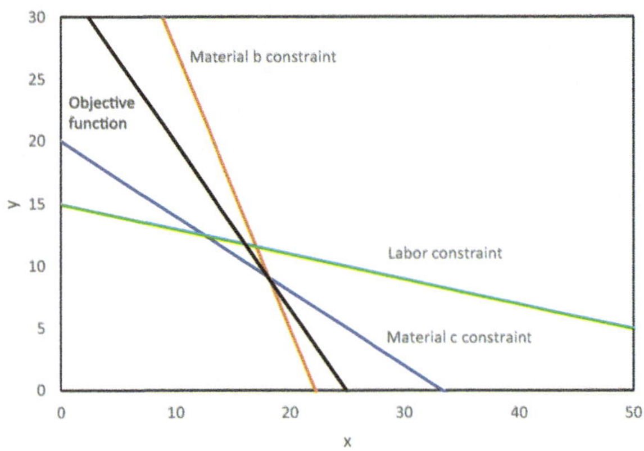

Fig. 4.1 Linear programming. In the region under all constraint equations, there exists an optimal solution

Add the slack variable $S_k (k = 0, 1, 2, \cdots, K-1)$ to the left-hand side of the inequality constraint to make it an equality constraint.

$$\begin{aligned}
&\text{maximize} && z = \sum_{i=0}^{n-1} c_i x_i \\
&\text{subject to} && \sum_{i=0}^{n-1} a_{ki} x_i + s_k = b_k \ (k = 0, 1, 2, \cdots, K-1) \\
& && x_i \geq 0 \ (i = 0, 1, 2, \cdots, n-1).
\end{aligned} \tag{4.3}$$

4.1.2 Simplex Method

Using the simplex method, let us solve a linear programming problem in Eq. (4.4).

$$\begin{aligned}
\text{maximize} \quad & z = 70x_1 + 120x_2 \\
\text{subject to} \quad & 9x_1 + 4x_2 \leq 1080 \\
& 4x_1 + 5x_2 \leq 600 \\
& 3x_1 + 10x_2 \leq 900 \\
& x, y \geq 0.
\end{aligned} \tag{4.4}$$

Now add the slack variable $s_k (k = 0, 1, 2, \cdots, K-1)$ to the left side of the inequality constraint to make it an equality constraint.

$$\begin{aligned}
\text{maximize} \quad & z = 70x_1 + 120x_2 \\
\text{subject to} \quad & 9x_1 + 4x_2 + s_1 = 1080 \\
& 4x_1 + 5x_2 + s_2 = 600 \\
& 3x_1 + 10x_2 + s_3 = 900 \\
& x, y \geq 0.
\end{aligned} \tag{4.5}$$

This problem can be represented in a simplex table as in Table 4.2. The value in the simplex table's bottom row is the objective function's coefficient multiplied by -1. If the bottom row has a negative value, select the column with the highest absolute value as the pivot column. Divide the column of constant terms by the value of the pivot column. Select the row with the lowest value as the pivot row.

Divide all the values of the pivot row by the values of the pivot row and the pivot column. Then, change the name of the basis variable of the pivot row from s_3 to the non-basis variable x_2 of the pivot column to obtain Table 4.3.

Table 4.2 Simplex Table 1

Basic variable	x_1	x_2	s_1	s_2	s_3	Constant	Upper limit
s_1	9	4	1	0	0	1080	1080/4 = 270
s_2	4	5	0	1	0	600	600/5 = 120
s_3	3	10	0	0	1	900	900/10 = 90
	−70	−120	0	0	0	0	

Table 4.3 Simplex Table 2

Basic variable	x_1	x_2	s_1	s_2	s_3	Constant
s_1	9	4	1	0	0	1080
s_2	4	5	0	1	0	600
x_2	3	10	0	0	1	900
	−70	−120	0	0	0	0

Next, in the pivot column, perform a sweep operation to set the values of the non-pivot rows to zero. The value of each row other than the pivot row is changed according to Eq. (4.6). The actual calculation is as follows (Table 4.4):

$$\text{New value in row } k = \text{current value in row } k - \text{value in pivot row} \times \text{value of pivot column in row } k. \tag{4.6}$$

This sweep operation results in the simplex table shown in Table 4.5. Since there are negative numbers in the bottom column, we return to the selection of the pivot column.

The column with the lowest value of −34 is the pivot column. Divide the column of constant terms by the value of the pivot column. The row with the lowest value (60) is the pivot row.

Divide all the values of the pivot row by the values of the pivot row and the pivot column. Then, by changing the name of the basis variable of the pivot row from s_2 to the non-basis variable x_1 of the pivot column, we obtain Table 4.6.

Next, a sweep operation is performed in the pivot column to set the non-pivot row values to zero according to Eq. (4.6). The actual calculation is as follows (Table 4.7):

With the sweep operation, the simplex table becomes as shown in Table 4.8.

Since there are no more negative numbers in the bottom row, we have reached the optimal solution. The optimal solution is given by the column of constant terms ($s_1 = 252, x_1 = 60$, and $x_2 = 72$) for the values of the base variables, and all the values of the non-base variables are zero ($s_2 = 0$ and $s_3 = 0$). The maximum value of the objective function is the bottom line value of the constant term ($z = 12{,}840$).

Table 4.4 Sweep Operation of Simplex Table 3

$9 - 0.3 \times 4 = 7.8$,	$4 - 1 \times 4 = 0$,	1,	0,	$0 - 0.1 \times 4 = -0.4$,	$1080 - 90 \times 4 = 720$
$4 - 0.3 \times 5 = 2.5$,	$5 - 1 \times 5 = 0$,	0,	1,	$0 - 0.1 \times 5 = -0.5$,	$600 - 90 \times 5 = 150$
–,	–,	–,	–,	–,	–
$-70 - 0.3 \times (-120) = -34$,	$-120 - (-120) = 0$,	0,	0,	$0 - 0.1 \times (-120) = 12$,	$0 - 90 \times (-120) = 10{,}800$

Table 4.5 Simplex Table 3

Basic variable	x_1	x_2	s_1	s_2	s_3	Constant	Upper limit
s_1	7.8	0	1	0	−0.4	720	720/7.8 = 92.3
s_2	2.5	0	0	1	−0.5	150	150/2.5 = 60
x_2	0.3	1	0	0	0.1	90	90/0.3 = 300
	−34	0	0	0	12	10,800	

Table 4.6 Simplex Table 4

Basic variable	x_1	x_2	s_1	s_2	s_3	Constant
s_1	7.8	0	1	0	−0.4	720
x_1	1	0	0	0.4	−0.2	60
x_2	0.3	1	0	0	0.1	90
	−34	0	0	0	12	10,800

4.1.3 Combinatorial Programming

When the optimal solution to a problem is discrete, it is called a combinatorial optimization problem. In particular, when variables can only take integer values, it is called integer programming, and when variables are limited to 0 or 1, it is called 0–1 integer programming. The unit commitment model described in Chapter 6 includes both real and integer variables and is called mixed integer programming.

Consider the following standard linear programming problem. The variable x_i representing economic activity is a real number. Assume that the cost of economic activity $c_i x_i$ is proportional to economic activity x_i, and minimize the total cost of this activity.

$$\begin{aligned} &\text{minimize} \quad z = \sum_{i=1}^{N} c_i x_i \\ &\text{subject to} \quad \sum_{i=1}^{N} a_{ki} x_i = b_k \ (k = 1, 2, \cdots, K) \\ &\qquad x_i \geq 0 \ (i = 1, 2, \cdots, N). \end{aligned} \tag{4.7}$$

However, there are two types of costs: variable costs and fixed costs. An example of a variable cost is employee wages, which are proportional to economic activity (hours worked). In the above problem, variable costs are taken into account, but fixed costs are not. An example of a fixed cost is the employee's insurance fee, which is required when the employment relationship arises. We assume that this insurance fee need not be paid when the working hours are 0, but a fixed amount of d_i must be paid when the working hours are more than 0, irrespective of the working hours.

Table 4.7 Sweep Operation of Simplex Table 5

$7.8 - 1 \times 7.8 = 0$,	$0 - 0 \times 7.8 = 0$,	$1 - 0 \times 7.8 = 1$,	$0 - 0.4 \times 7.8 = -3.12$,	$-0.4 + 0.2 \times 7.8 = 1.16$,	$720 - 60 \times 7.8 = 252$
–,	–,	–,	–,	–,	–
$0.3 - 1 \times 0.3 = 0$,	$1 - 0 \times 0.3 = 1$,	$0 - 0 \times 0.3 = 0$,	$0 - 0.4 \times 0.3 = -1.12$,	$0.1 + 0.2 \times 0.3 = 0.16$,	$90 - 60 \times 0.3 = 72$
$-34 - 1 \times (-34) = 0$,	$0 - 0 \times (-34) = 0$,	$0 - 0 \times (-34) = 0$,	$0 - 0.4 \times (-34) = 13.6$,	$12 + 0.2 \times (-34) = 5.2$,	$10{,}800 - 60 \times (-34) = 12{,}840$

Table 4.8 Simplex Table 5

Basic variable	x_1	x_2	s_1	s_2	s_3	Constant
s_1	0	0	1	−3.12	1.16	252
x_1	1	0	0	0.4	−0.2	60
x_2	0	1	0	−1.2	0.16	72
	0	0	0	13.6	5.2	12,840

With these fixed costs in mind, we formulate the problem as follows:

$$\begin{aligned} \text{minimize} \quad & z = \sum_{i=1}^{N} d_i y_i + c_i x_i \\ \text{subject to} \quad & \sum_{i=1}^{N} a_{ki} x_i = b_k \ (k = 1, 2, \cdots, K) \\ & x_i - M y_i \leq 0 \ (i = 1, 2, \cdots, N) \\ & x_i \geq 0 \ (i = 1, 2, \cdots, N) \\ & y_i \in \{0, 1\} \ (i = 1, 2, \cdots, N), \end{aligned} \tag{4.8}$$

where the variable y_i is a variable that takes on any value of 0 or 1. Also, let M be a sufficiently large positive constant. With such a setting of M, $y_i = 0$ when $x_i = 0$ and $y_i = 1$ when $x_i \geq 0$. This new formulation, which considers fixed costs, involves real and integer variables, indicating mixed integer programming.

Since integer programming does not have the concept of differentiation, it is impossible to use gradient information. Therefore, it is much more difficult to solve than optimization problems with continuous variables. Typical integer programming problems include the Knapsack problem, the assignment problem, and the traveling salesman problem. There is no systematization of general solution methods in integer programming, and an appropriate treatment is needed for these problems. The solution of integer programming is an area of active research. This section will briefly explain the branch-and-bound method as a solution method. For details, the readers may refer to specialized books on combinatorial optimization problems [4].

The integer programming problem P can be formulated in an n-dimensional space whose elements are integers $\mathbb{Z}$:

$$\begin{aligned} \text{maximize} \quad & f(x) \\ \text{subject to} \quad & g(x) \leq 0 \\ & x \leq \mathbb{Z}^n. \end{aligned} \tag{4.9}$$

In the branch-and-bound method, the allowed region satisfying the constraint $g(x) \leq 0$ is divided into multiple subregions by adding new constraints that limit the range of variables. By analyzing these subproblems with a limited range of variables, regions where optimal points cannot exist are identified and eliminated, thereby reducing the amount of computation in the search. Creating subproblems is "branching," and the process of eliminating regions where the optimal solution cannot exist is "bounding."

Branching is performed hierarchically using the dichotomy method. First, we divide the allowed region S satisfying $g(x) \leq 0$ into two subregions S_1 and S_2. Add a new constraint:

$$x_i \geq d_i + 1 \ (i = 1, 2, \cdots, n) \tag{4.10}$$

$$x_i \leq d_i \ (i = 1, 2, \cdots, n) \tag{4.11}$$

to each variable and denote the subregions S_1 and S_2 as those that satisfy Eqs. 4.10 and 4.11, respectively. Here d_i is a given integer. For the original problem P, let P_1 and P_2 be subproblems. If optimal solutions x_1^* and x_2^* are obtained for P_1 and P_2, then the larger of $f(x_1^*)$ and $f(x_2^*)$ is the optimal solution of problem P.

If an optimal solution cannot be obtained for P_1, S_1 is further divided into S_{11} and S_{12} to form subproblems P_{11} and P_{12} for P_1. Similarly for P_2, the subproblems P_{21} and P_{22} of P_2 are constructed. By repeatedly branching the region in this way, the subregion becomes smaller and smaller. A search tree can represent this branching process. In this branching process, the subregion that does not contain the optimal solution can be detected and excluded from the subsequent search, thereby reducing the amount of computation.

To find an approximate solution to the subproblem, consider a relaxation problem in which the variable condition is relaxed from an integer $x \leq \mathbb{Z}^n$ to a real number $x \leq \mathbb{R}^n$. The optimal solution of the relaxation problem $\overline{P}$ is larger than the optimal solution of the original problem P because of the relaxed condition. That is, if $\overline{x}^*$ and x^* are the optimal solutions of $\overline{P}$ and P, respectively, then $f(\overline{x}^*) \geq f(x^*)$.

Suppose the initial value x_0 is given for
$overline P$. If $f(x_0) \geq f(\overline{x}_1^*)$ for the optimal solution $\overline{x}_1^*$ of the relaxation problem $\overline{P}_1$, then $f(\overline{x}_1^*) \geq f(x_1^*)$ to $f(x_0) \geq f(x_1^*)$, and there is no need to search the subregion S_1 any further. This is because $f(x_0) \geq f(x)$ for any x belonging to S_1.

If $f(x_0) < f(\overline{x}_1^*)$, then we can separate the cases where $\overline{x}_1^*$ is an integer or not. If $\overline{x}_1^*$ is an integer, then $\overline{x}_1^*$ is a solution of P_1: $x_1^* = \overline{x}_1^*$. In that case, replace the initial value x_0 with x_1^* for the new initial value. If $\overline{x}_1^*$ is not an integer, split S_1 into S_{11} and S_{12}. To do so, a new additional condition on the components of $\overline{x}_1^*$ is

$$d_i < \overline{x}_{1i}^* < d_i + 1 \ (i = 1, 2, \cdots, n). \tag{4.12}$$

Since $\overline{x}_1^*$ is not an integer, it does not belong to S_1 or S_2.

This procedure is also performed for P_2. Furthermore, perform this procedure hierarchically and sequentially until all the optimal solutions of the valid subproblems are integer solutions. The optimal solution of the last remaining subproblem is the optimal solution of the original problem.

This branch-and-bound method strongly depends on how the initial value x_0 is chosen and how the branching is done. Therefore, combining these with another method of efficient selection has been studied. Metaheuristic optimization, such as the genetic algorithm described in the next section, is commonly used with the combination of the branch-and-bound method for combinatorial optimization problems [5].

4.1.4 Metaheuristic Optimization

The genetic algorithm (GA) is well known as a representative method of Metaheuristic Optimization [6, 7]. The algorithm of GA is described below. In GA, the state vector of a target characterized by a specific feature value is used as a variable. Consider the problem of searching for a target state vector whose feature values satisfy the design rule.

Input Data: Step (a)
Several input parameters have to be read in step (a). They are N_g, k_{max}, P_{∞}, P_{mut}, and c_{ls}. N_g is the number of genes. k_{max} is the number of generations to be calculated and is hence equal to the number of iterations from steps (d) to (j). P_{∞} and P_{mut} are probabilities of a cross-over and a mutation, respectively. P_{∞} and P_{mut} are normalized for a pair of genes and for each bit of gene, respectively. c_{ls} is a parameter for linear scaling in calculating the fitness value.

Initialization of Feature Parameter Database: step (b)
The physical quantity database is initialized in step (b). Parameters are used to calculate the feature values of the target based on the state vector.

Generation of Initial Gene: Step (c)
State vectors of objects are written in genes. Each gene represents an object with a specific state vector. Initial state vectors are generated using a random number, so each object has a random state. The state vector is written as a genotype. As a result, initial states are distributed over the entire state space. Such a distribution allows a global search to be made among all possible states.

Conversion of Genotype to Phenotype: Step (d)
A binary gene expression is called the genotype and is needed for a cross-over and a mutation. On the other hand, a decimal expression of the gene is called the phenotype. The genotype is converted to the phenotype in this step.

Calculation of Feature Value: Step (e)
For the obtained state vector, feature values are calculated using the state vector and the parameters obtained from the feature parameter database.

Fitness: Step (f)
Fitness indicates the degree of satisfaction with a given design rule.

Discrimination: Step (g)
The gene with the maximum fitness value is selected from all genes. If the selected value is larger than $f_{\max}$, then the state search is finished, and the process goes to step (k). Otherwise, the search continues to proceed to step (h).

Linear Scaling: Step (h)
An optimization using the genetic algorithm is very efficient if the fitness value is small. The efficiency, however, goes down for larger fitness values, so many searching steps are needed to search for the optimum fitness value. In order to prevent an inefficient search, fitness values are scaled using the linear scaling method. The scaled fitness value f' is obtained by scaling the raw fitness value f as below:

$$f' = a_{ls} f + b_{ls}, \tag{4.13}$$

where coefficients a_{ls} and b_{ls} are evaluated by two conditions: (i) the average value of the scaled fitness f'_{av} is equal to the average value of the raw fitness f_{av} and (ii) the maximum value of the scaled fitness f'_{max} is c_{ls} times of the average value of the raw fitness f_{av}.

$$a_{ls} = (c_{ls} - 1)\,\frac{f_{av}}{f_{max} - f_{av}} \tag{4.14}$$

$$b_{ls} = (1 - a_{ls})\, f_{av}. \tag{4.15}$$

When a fitness value scaled by the linear scaling method becomes negative, the negative value is replaced by a null value.

Selection: Step (i)
The roulette method selects N_g genes with large fitness values from N_g genes. First, a fan-shaped piece on a roulette wheel is assigned where its area is proportional to the fitness value of each gene. Then, the wheel is spun, a ball is thrown, and a gene is selected according to the fan-shaped piece where the ball stops. N_g genes are selected by repeating N_s spins of the roulette wheel.

Cross-Over and Mutation: Step (j)
Cross-overs are generated in the selected N_g genes with a probability P_{∞}. A pair of genotypes G_i and G_j is cut at an arbitrary point to get four fragments G_i^a, G_i^b, G_j^a, and G_j^b. Then G_i^a is linked to G_j^a, and G_j^a is linked to G_i^b. Afterward, mutations are generated for each gene. A mutation is defined as reversing a bit of a genotype with a probability P_{mut}.

Output Data: Step (k)
The optimal state vector is obtained as a result of this search procedure.

4.2 Dynamical System

4.2.1 Harmonic Oscillator

Mice are often used as experimental materials to clarify complex problems in the life sciences by reducing them to as simple a problem as possible. In the same sense, harmonic oscillators are used as experimental materials to understand the behavior of complex dynamical systems in a simple way. A harmonic oscillator in Fig. 4.2a is described by the second-order differential equation:

$$\frac{d^2x}{dt^2} = -kx, \tag{4.16}$$

where k is a spring constant. This equation is rewritten as a simultaneous first-order differential equation by introducing an additional variable y:

$$\begin{cases} \dfrac{dx}{dt} = \sqrt{k}y, \\ \dfrac{dy}{dt} = -\sqrt{k}x. \end{cases} \tag{4.17}$$

A dynamical system is a system whose state changes with time according to specific rules. We consider state variables, which are variables necessary to describe the system's state, and a phase space, which is a space with n state variables (basis vectors). The time variation of a point in the phase space determined by the state variables is called a trajectory.

We obtain a solution $x(t) = A\sin\sqrt{k}t$ and $y(t) = A\cos\sqrt{k}t$. If we give an initial condition $x(0) = A$ and $y(0) = 0$, we have $x(t_1) = 0$ and $y(t_1) = A$ at

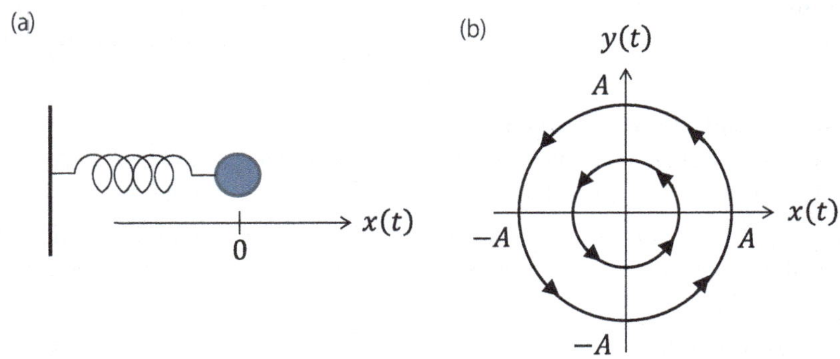

Fig. 4.2 Harmonic oscillator. Panel (**a**) shows mass point and coordinate, and panel (**b**) shows trajectories in the phase space

$t = t_1 = \frac{\pi}{2\sqrt{k}}$. This means that a trajectory $(x(t), y(t))$ becomes a circle with a given radius A, as depicted in Fig. 4.2b.

We introduce a damping term proportional to the velocity dx/dt in a harmonic oscillator:

$$\frac{d^2x}{dt^2} = -kx + \alpha\frac{dx}{dt}. \tag{4.18}$$

This equation is rewritten as a simultaneous first-order differential equation in a similar way:

$$\begin{cases} \dfrac{dx}{dt} = y, \\ \dfrac{dy}{dt} = -x + \alpha y, \end{cases} \tag{4.19}$$

where α is a model parameter.

A system described by Eq. (4.19) has a stable equilibrium point at origin $x = y = 0.0$ in phase space because we have $dx/dt = dy/dt = 0$ at this point. In the case of $\alpha = -0.3$, a trajectory starting from $x(0) = 2.0$, $y(0) = 2.0$ approaches toward a stable equilibrium point $x(\infty) = 0.0$, $y(\infty) = 0.0$, as shown in Fig. 4.3a. However, in the case of $\alpha = 0.3$, the origin is not a stable equilibrium point. For instance, a trajectory starting from $x(0) = 0.001$, $y(0) = 0.001$ gradually moves away from the origin as it rotates around it, as shown in Fig. 4.3b. The origin is called the unstable equilibrium point.

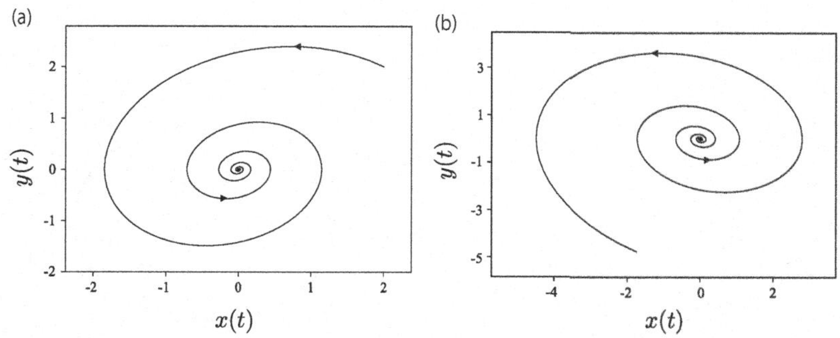

Fig. 4.3 Harmonic oscillator with a damping term. Panel (**a**) shows a stable equilibrium point, and panel (**b**) shows an unstable equilibrium point

4.2.2 Attractors

We consider a closed orbit in phase space. Such an orbit is called a periodic orbit. The attractor that attracts the neighboring orbit to the periodic orbit is called the periodic attractor. Such an attractor, which returns to the original state after a small perturbation, is unique to nonlinear systems. A dynamical system associated with the periodic attractor is called a limit cycle. A well-known example of the limit cycle oscillator is called Van der Pol oscillator, described by

$$\begin{cases} \dfrac{dx}{dt} = -y, \\ \dfrac{dy}{dt} = x + \alpha y \left(y^2 - \beta\right), \end{cases} \tag{4.20}$$

where α and β are model parameters.

A system described by Eq. (4.20) has a periodic orbit equilibrium. In the case of $\alpha = 0.3$ and $\beta = 1.0$, a trajectory starting from $x(0) = 2.0$, $y(0) = 2.0$ approaches toward the periodic attractor, as shown in Fig. 4.4a. However, with the same value of parameters, a trajectory starting from $x(0) = 0.001$, $y(0) = 0.001$ gradually moves away from the origin as it rotates around it and approaches the periodic attractor, as shown in Fig. 4.4b.

We consider the second-order differential equation with external force as follows:

$$\frac{d^2x}{dt^2} + \gamma \frac{dx}{dt} + \omega^2 x = \varepsilon f(t), \tag{4.21}$$

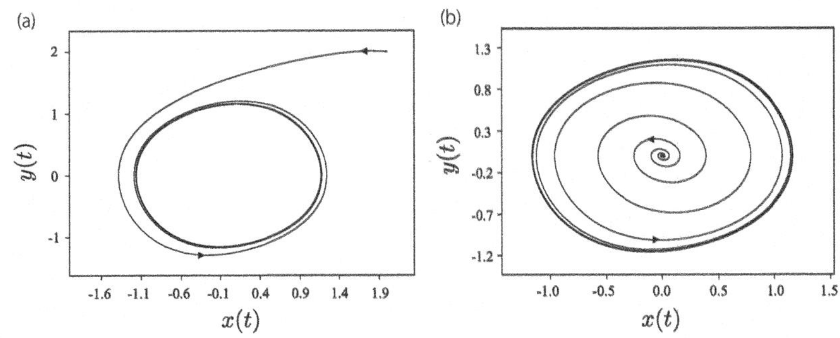

Fig. 4.4 Harmonic oscillator with a nonlinear damping term. Panels (**a**) and (**b**) show trajectories toward the periodic attractor from different initial conditions

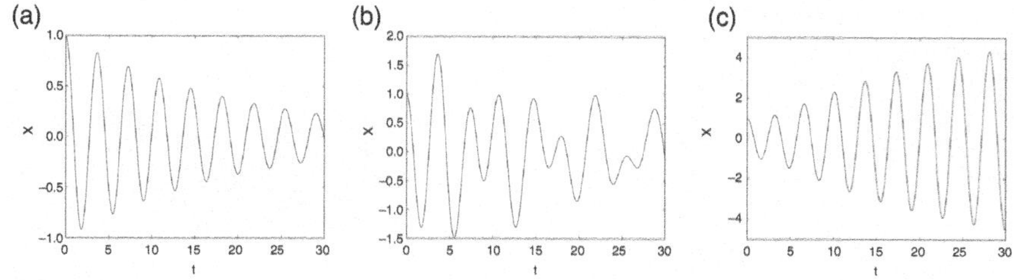

Fig. 4.5 Harmonic oscillator with a nonlinear damping term. Panel (**a**) shows a dumping oscillation, panel (**b**) shows a forced oscillation, and panel (**c**) shows a resonance oscillation

where γ, ω, and ε are model parameters, and $f(t) = \cos\omega' t$ is the external force. This equation is rewritten as a simultaneous first-order differential equation:

$$\begin{cases} \dfrac{dx}{dt} = y, \\ \dfrac{dy}{dt} = -\gamma y - \omega^2 x + \varepsilon f(t). \end{cases} \tag{4.22}$$

We solve the simultaneous first-order differential equation with three external forces: (a) dumping oscillation $\varepsilon = 0$, (b) forced oscillation $\varepsilon = 1$, and (c) resonance oscillation $\varepsilon = 1$ and $\omega = \omega'$ (Fig. 4.5).

The dynamical system described by Lorenz equation

$$\begin{cases} \dfrac{dx}{dt} = -px + py, \\ \dfrac{dy}{dt} = -xz + rx - y, \\ \dfrac{dz}{dt} = xy - bz \end{cases} \tag{4.23}$$

has an attractor in the three-dimensional phase space. Figure 4.6 shows an attractor obtained for parameters $p = 10$, $r = 28$, and $b = 8/3$. The attractor of Lorenz equation is called the strange attractor. The strange attractor is known to have self-similarity, i.e., fractal structure, as well as a sensitive dependence on initial conditions. This means that small differences in initial conditions gradually increase with time evolution. The dynamics associated with the sensitive dependence on initial conditions are called deterministic chaos.

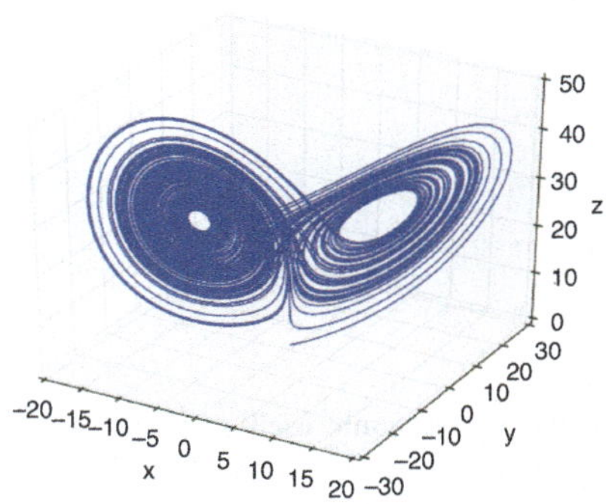

Fig. 4.6 Strange attractor is characterized by the self-similarity, i.e., fractal structure, and the sensitive dependence on initial conditions

4.3 Kuramoto Oscillator Model

4.3.1 Collective Motions

Collective motion is a phenomenon observed in a whole system caused by interaction between constituents. It has been known that various collective motions exist in natural phenomena. For instance, a heavy nucleus, consisting of a few hundred nucleons, is largely deformed at a highly excited state and proceeds to nuclear fission. This phenomenon is a well-known example of quantum mechanical collective motion due to strong nuclear force between nucleons.

The business cycle is observed in most industrialized economies. Economists have studied this phenomenon using mathematical models, including various linear, nonlinear, and coupled oscillator models. Interdependence, or coupling, between industries in the business cycle has been studied for more than half a century. A study of the linkages between markets and industries using nonlinear difference equations suggests a dynamical coupling among industries [8]. A nonlinear oscillator model of the business cycle was then developed using a nonlinear accelerator as the generation mechanism [9]. In this section, we stress the necessity of nonlinearity because linear models cannot reproduce sustained cyclical behavior and tend to die out or diverge to infinity.

However, it is noted that a simple linear economic model, based on ordinary economic principles, optimization behavior, and rational expectations, can produce cyclical behavior much like that found in business cycles [10]. Aside from synchronization in the business cycle, an important question is whether sectoral or aggregate shocks are responsible for the observed cycle. This question has been empirically examined, and it was clarified that business cycle fluctuations are caused by small sectoral shocks rather than by large common shocks [11].

As the third model category, coupled oscillators were developed to study noisy oscillating processes like national economies [12, 13]. Simulations and empirical analyses showed that synchronization between the business cycles of different countries is consistent with such mode-locking behavior. Along with this approach, a nonlinear mode-locking mechanism was further studied that described a synchronized business cycle between different industrial sectors [14].

4.3.2 *Synchronization in Physical and Biological Systems*

Many synchronization phenomena are known in physical and biological systems [16]. Physical examples include clocks hanging on a wall, an array of lasers, microwave oscillators, and Josephson junctions. Biological examples include synchronously flashing fireflies, networks of pacemaker cells in the heart, and metabolic synchrony in yeast cell suspensions. Certain species of fireflies living in Southeast Asia are known for flocking together on a single tree with their lights blinking in unison.

4.3.3 *Kuramoto Oscillator*

Kuramoto proposed a coupled oscillator model to explain this wide variety of synchronization phenomena [15–17]. In his model, the dynamics of the oscillators are governed by

$$\dot{\theta_i} = \omega_i + \sum_{j=1}^{N} k_{ji} \sin(\theta_j - \theta_i), \tag{4.24}$$

where θ_i, ω_i, and k_{ji} are the oscillator phase, the natural frequency, and the coupling strength, respectively. The second term of the RHS of Eq. (4.24) is identical to Eq. (1.18). The system exhibits synchronization if the coupling strength k_{ij} exceeds a certain threshold.

For two oscillator systems, the equation of motion to describe phases $\theta_1(t)$ and $\theta_2(t)$ is as follows:

$$\dot{\theta_1} = \omega_1 + k_{21} \sin(\theta_2 - \theta_1), \dot{\theta_2} = \omega_2 + k_{12} \sin(\theta_1 - \theta_2). \tag{4.25}$$

Here ω_1 and ω_2 are the natural frequencies of oscillators. Constant k is a constant of interaction strength. We consider phase difference $\delta(t) = \theta_1(t) - \theta_2(t)$ to understand the synchronization

$$\dot{\delta}(t) = \Delta\omega - k \sin \delta(t). \tag{4.26}$$

Next, we consider the stationary state: $d\delta(t)/dt = 0$. In Case I shown in Fig. 4.7a, $|\delta\omega| > k$. When the difference of natural frequencies $\Delta\omega$ is larger than interaction strength k, the phase difference is undetermined. This is the phase-slip. In Case II shown in Fig. 4.7b, $|\delta\omega| < k$. When the difference of natural frequencies $\Delta\omega$ is smaller than interaction strength k, we have constant phase difference δ_1^*. If the difference in natural frequencies is small, many systems are categorized as Case II.

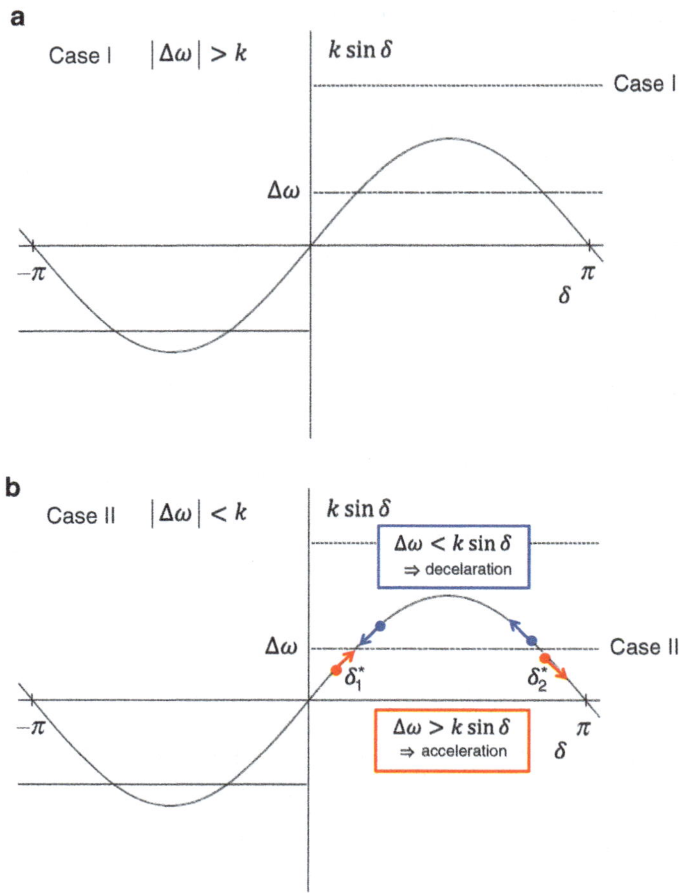

Fig. 4.7 Kuramoto oscillator. Panel (**a**) shows phase-slip, and panel (**b**) shows phase lock. δ_1^* is stable, whereas δ_2^* is unstable

4.3.4 *Concept of Phase Time Series*

The Kuramoto oscillator model is described using phase $\theta(t)$, though the observation is made for a time-varying physical quantity itself.

We calculate phase time series from the observed time series of physical quantity $x(t)$. The Hilbert transform of a continuous time series $x(t)$ is defined by

$$y(t) = H[x(t)] = \frac{1}{\pi} PV \int_{\infty}^{\infty} \frac{x(s)}{t-s} ds. \tag{4.27}$$

Using time series $x(t)$ and $y(t)$, we define a complex time series:

$$z(t) = x(t) + iy(t) = A(t)e^{i\theta(t)}, \tag{4.28}$$

where $\theta(t)$ is the phase time series calculated from the observed time series of physical quantity $x(t)$.

The following is an illustrative example to estimate the phase time series of $x(t) = \cos(\omega t)$. The Hilbert transform of $x(t)$ is calculated as $y(t) = H[\cos(\omega t)] = \sin(\omega t)$. The complex time series $z(t) = \cos(\omega t) + i\sin(\omega t) = A(t)e^{i\theta(t)}$ gives the phase time series $\theta(t)$. In addition, for $x(t) = \sin(\omega t)$ the Hilbert transform is given by $y(t) = H[\sin(\omega t)] = -\cos(\omega t)$. We calculate $\theta(t)$ in a similar way.

Next, we explain an actual calculation of the estimation of the phase time series for the observed time series of physical quantity $x(t)$. The time series $x(t)$ is expanded as a Fourier time series, shown in

$$x(t) = \frac{A_0}{2} + \sum_{n=1}^{\infty}\left(A_n \cos\frac{n\pi t}{T} + B_n \sin\frac{n\pi t}{T}\right). \tag{4.29}$$

The Hilbert transformation of a continuous time series $x(t)$ is calculated as

$$\begin{aligned} y(t) &= \frac{A_0}{2} + \sum_{n=1}^{\infty}\left(A_n H\left[\cos\frac{n\pi t}{T}\right] + B_n H\left[\sin\frac{n\pi t}{T}\right]\right) \\ &= \frac{A_0}{2} + \sum_{n=1}^{\infty}\left(A_n \sin\frac{n\pi t}{T} - B_n \cos\frac{n\pi t}{T}\right), \end{aligned} \tag{4.30}$$

where PV represents the Cauchy principal value, and then the time series y is calculated using the Fourier expansion in Eq. (4.29). A complex time series is obtained by adopting the time series $y(t)$ as an imaginary part. Consequently, a phase time series $\theta(t)$ is obtained using Eq. (4.28) for the observed time series of physical quantity $x(t)$.

4.3.5 *Measure of Synchronization*

Synchronization is defined as the phases locking $\theta_1(t) - \theta_2(t) = const.$ In the case of $const. \neq 0$, where the phase difference signifies a delay, a direct evaluation of the phase instead of the correlation coefficient is more adequate. This is because the correlation coefficient ρ varies depending on the delay δ. For example, in a trigonometric function with the period of oscillation equal to 2π, we have $\rho = 1$ for $\delta = 0$, $\rho = 0$ for $\delta = \pi/2$, and $\rho = -1$ for $\delta = \pi$. These examples imply that the estimation of phase time series is essential for the general synchronization phenomena, including phase locking.

For N-oscillators system, the collective rhythm produced by the whole population of oscillators is captured by a macroscopic quantity, such as the complex order parameter.

$$u(t) = r(t)e^{i\phi(t)} = \frac{1}{N}\sum_{j=1}^{N} e^{i\theta_j(t)}, \tag{4.31}$$

where the radius $r(t)$ measures the phase coherence, and $\phi(t)$ represents the average phase [16]. The radius $r(t)$ is often simply called the order parameter. $r(t)$ ranges from 0 to 1, with closer to 1 meaning a higher degree of synchronization.

4.4 Kuramoto Model with Inertia

We consider a power grid system comprising generators and an electricity market. Various types of shock may occur in the grid system of interest. If generation in a specific generator suddenly decreases significantly, the imbalance of demand and supply may destroy the synchronization among the generators. However, suppose the price of the electricity in the market is quickly adjusted. In that case, a sudden change of generation in the generator will be absorbed by consumers as a decrease in demand.

In this section, we first explain the formulation of our coupled oscillator model and then give an illustrative example to understand the fundamental behavior of the model.

4.4.1 Formulation

Our model is developed based on the Kuramoto model with inertia [18] by adding the electricity market [19]. We consider the system to consist of oscillator i and oscillator j. The oscillator represents the generator. The angle θ_i of oscillator i is written as

$$\theta_i = \omega t + \tilde{\theta}_i, \tag{4.32}$$

where ω and $\tilde{\theta}_i$ are the frequency and phase, respectively.

The energy dissipated as heat from oscillator i at a rate proportional to the square of the angular velocity is

$$P_d = K_D(\dot{\theta}_i)^2, \tag{4.33}$$

and the kinetic energy accumulated in oscillator i at a rate proportional to the square of the angular velocity is

$$P_a = \frac{1}{2} I \frac{d}{dt} (\dot{\theta}_i)^2, \tag{4.34}$$

where K_D and I are a dissipation constant and a moment of inertia, respectively. From Eq. (4.32), the angular difference $\Delta\theta_{ji}$ is written as the phase difference $\Delta\theta_{ji} = \theta_j - \theta_i = \tilde{\theta}_j - \tilde{\theta}_i$ using the phase $\tilde{\theta}_i$. The power transmitted from one oscillator to another is given by

$$P_t = -k_{ji} \sin \Delta\theta_{ji}, \tag{4.35}$$

using the phase difference $\Delta\theta_{ji}$. In Eq. (4.35), the negative sign indicates that power is lost from oscillator i.

By substituting relations (4.33), (4.34), and (4.35) into the power balance equation for oscillator i ($P_s = P_d + P_a + P_t$), we obtain an equation corresponding to Eq. (4.24)

$$I\omega\ddot{\tilde{\theta}}_i = P_s - K_D\omega^2 - 2K_D\omega\dot{\tilde{\theta}}_i + k_{ji} \sin \Delta\theta_{ji}, \tag{4.36}$$

using the approximate relation $\dot{\tilde{\theta}}_i \ll \omega$. From the above discussion, without loss of generality, we obtain an equation describing the dynamics of an N-oscillator system

$$\ddot{\tilde{\theta}}_i = P_i - \alpha\dot{\tilde{\theta}}_i + \sum_{j=1}^{N} k_{ji} \sin \Delta\theta_{ji}, \tag{4.37}$$

where $I\omega$, $P_s - K_D\omega^2$, and $2K_D\omega$ are replaced by 1, P_i, and α, respectively. P_i is regarded as the net input to oscillator i, which is equal to the difference between the input to oscillator i from outside the N-oscillator system and the output from oscillator i to outside the N-oscillator system. Hereafter $\tilde{\theta}_i$ is written as θ_i for simplicity.

It is noted that the synchronizing force is interpreted as being the electricity market. The meaning of the electricity market is clarified by adding the sectoral fluctuations of demand or supply, δ_{ji}, as

$$\ddot{\theta}_i = P_i - \alpha\dot{\theta}_i + \sum_{j=1}^{N} \{k_{ji} \sin \Delta\theta_{ji} + \delta_{ji}\}. \tag{4.38}$$

Figure 4.8 depicts a system with six generators. Figure 4.8a is the Nearest Neighbor (NN) graph, and Fig. 4.8b is the Complete (C) graph. Generators and the electricity market are indicated by circles and rectangles, respectively. In Fig. 4.8a, the number of the markets is six because they are only open to the neighboring generators. In

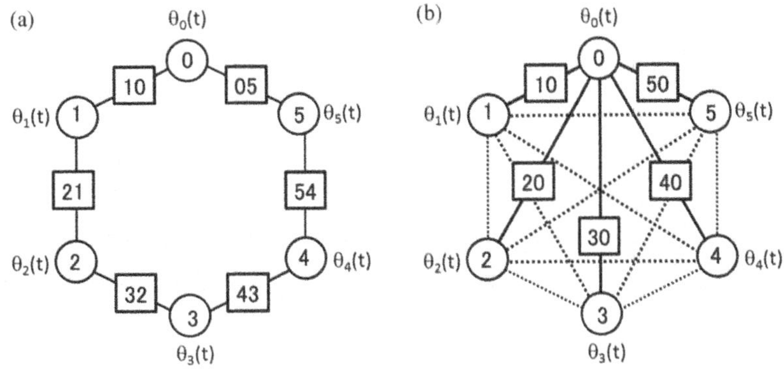

Fig. 4.8 Network topology. Panel (**a**) shows the nearest neighbor graph, and panel (**b**) shows a complete graph. Generators and the electricity market are indicated by circles and rectangles, respectively

Fig. 4.8b, however, the number of markets is five for each generator, and the total number of markets is equal to $(6 \times 5)/2 = 15$.

Demand d_{ij} for and supply s_{ji} of electricity i can be written as

$$d_{ij} = d_0 + \delta_{ij}, \tag{4.39}$$

$$s_{ji} = s_0 + \delta_{ji} \tag{4.40}$$

and are determined through the electricity market ij. d_0 and s_0 are the equilibrated demand and the equilibrated supply of electricity i in electricity market ji, and δ_{ij} and δ_{ji} are the fluctuations of demand and supply, respectively. In Eqs. (4.39) and (4.40) the equilibrated demand d_0 and the equilibrated supply s_0 are given by

$$d_0 = s_0 = k_{ji}|\sin(\theta_j^* - \theta_i^*)|, \tag{4.41}$$

where the equilibrated phase $\theta_i^*(i = 1, \cdots, N)$ is the solution of Eq. (4.38) with $\ddot{\theta}_i = \dot{\theta}_i = 0$. The relations between demand d or supply s and the price p of electricity i are written using the price elasticity of demand ϵ_d or the price elasticity of supply ϵ_s

$$\frac{d}{d_0} = \left(\frac{p}{p_0}\right)^{\epsilon_d}, \tag{4.42}$$

$$\frac{s}{s_0} = \left(\frac{p}{p_0}\right)^{\epsilon_s}. \tag{4.43}$$

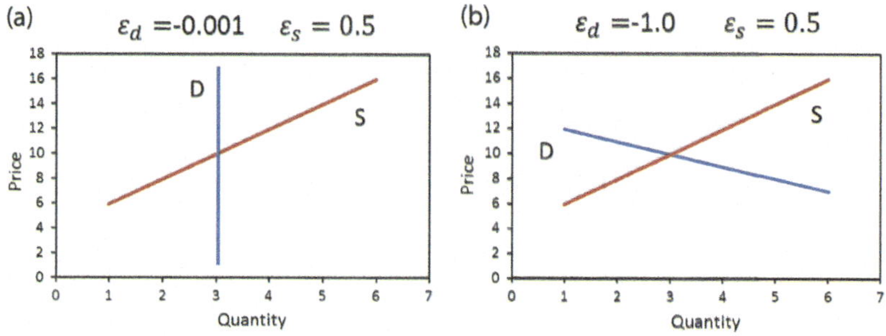

Fig. 4.9 Demand and supply. For smaller ϵ_d, the demand approaches to a vertical line

The price as a function of demand and supply is depicted in Fig. 4.9. Figure 4.9a is plotted for a small value of ϵ_d. In this case, the demand is shown as a vertical line. Oscillator j responds to the sectoral fluctuations of supply δ_{ji} according to price p. Depending on the market flexibility, generator j changes its demand by δ_{ij}, given by

$$\delta_{ij} = \begin{cases} -\delta ji & (\epsilon_d < 0) \\ 0 & (\epsilon_d = 0), \end{cases} \tag{4.44}$$

by responding to the fluctuation of supply δ_{ji}.

4.4.2 Illustrative Example

We consider a grid system in order to understand the fundamental behavior of the coupled oscillator model. For the NN graph, the model equations are written as

$$\begin{aligned} \ddot{\theta}_1 &= P_1 - \alpha\dot{\theta}_1 + \{k_{21}\sin(\theta_2 - \theta_1) + \delta_{21}\}, \\ \ddot{\theta}_2 &= P_2 - \alpha\dot{\theta}_2 + \{k_{21}\sin(\theta_1 - \theta_2) + \delta_{12}\} + \{k_{32}\sin(\theta_3 - \theta_2) + \delta_{32}\}, \\ &\cdots \\ \ddot{\theta}_{N-1} &= P_{N-1} - \alpha\dot{\theta}_{N-1} + \{k_{N-1N-2}\sin(\theta_{N-2} - \theta_{N-1}) + \delta_{N-2N-1}\}, \\ &+ \{k_{NN-1}\sin(\theta_N - \theta_{N-1}) + \delta_{NN-1}\}, \\ \ddot{\theta}_N &= P_N - \alpha\dot{\theta}_N + \{k_{NN-1}\sin(\theta_{N-1} - \theta_N) + \delta_{NN-1}\}. \end{aligned} \tag{4.45}$$

Analytic solutions of the stationary state will be obtained by solving the simultaneous equations of (4.45) with $\dot{\theta}_i = \ddot{\theta}_i = 0 (i = 1, \cdots, N)$. We consider the case, where $P_1 = P = 1$, $P_i = 0 (i = 2, \cdots, N-1)$, $P_N = -P = -1$, and $\theta_N = 0$.

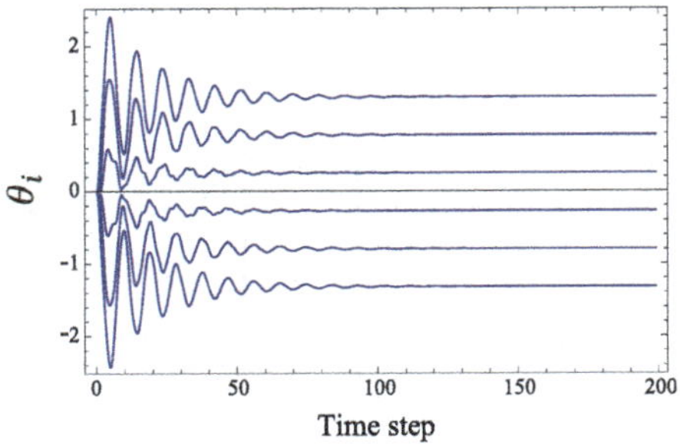

Fig. 4.10 Stationary solution. Synchronization was reproduced as an equilibrium solution

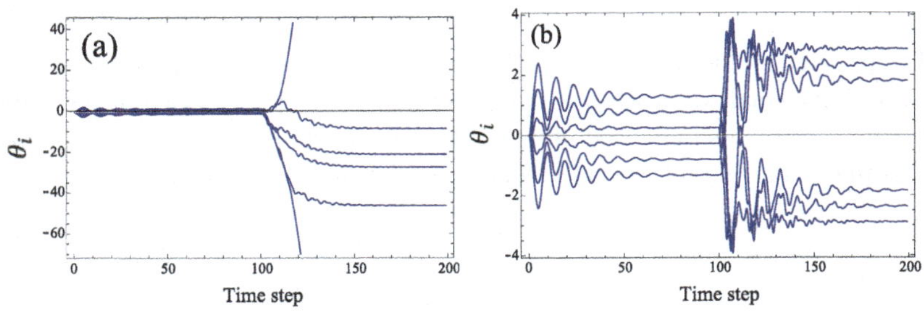

Fig. 4.11 Fluctuation and elasticity. Panel (**a**) is for zero elasticity ($\epsilon_d = 0$), and panel (**b**) is for finite elasticity ($\epsilon_d < 0$)

The analytic solutions obtained for the stationary state θ_i^* and the synchronizing coupling strengths k_{ji} are

$$\theta_i^* = \theta_{i+1}^* + \arcsin \frac{P}{k_{ji}}, \tag{4.46}$$

$$k_{ji} = \frac{P}{\sin(\theta_i^* - \theta_{i+1}^*)}. \tag{4.47}$$

We simulate the behavior of the system by solving Eq. (4.45) numerically with the initial condition $\dot{\theta}_i(0) = \ddot{\theta}_i(0) = 0 (i = 1, \cdots, N)$. Synchronization was reproduced as an equilibrium solution in a simple NN graph as shown in Fig. 4.10. The response to a sectoral fluctuation at $t = 100$ is also simulated in the NN graph. In the case of zero elasticity ($\epsilon_d = 0$), the synchronization was broken, as shown in Fig. 4.11a. In contrast, in the case of finite elasticity ($\epsilon_d < 0$), stability was restored after a shift of phase, as shown in Fig. 4.11b.

4.5 Analysis of Generator Synchronization

In this section, we analyze the synchronization observed in a power grid system using the coupled oscillator model described in Sect. 4.4. The effects on synchronization resulting from a shock are studied for systems with different price elasticities and coupling strengths.

4.5.1 *Synchronization*

The generators are rearranged in decreasing order of the observed phase of generated electricity. If we assume that the network is the NN graph, we can calibrate the coupling strengths k_{ij} using Eq. (4.47) with the observed phases θ_i^*. Analytical solutions of the stationary phases are then obtained using Eq. (4.46).

The analytical solutions of stationary phases are compared with the observed phases in Fig. 4.12a. The agreement between the analytical solutions and the observed phases is quite electricity. We simulate the behavior of the system by solving Eq. (4.45) numerically with the initial condition $\dot{\theta}_i(0) = \ddot{\theta}_i(0) = 0 (i = 1, \cdots, N)$. Synchronization was reproduced as an equilibrium solution in a simple NN graph, as shown in Fig. 4.12b.

4.5.2 *Fluctuating Electricity Market*

The complex order parameter

$$q(t) = \frac{1}{N} \sum_{j=1}^{N} e^{i\theta_j(t)} = r(\cos(\phi) + i \sin(\phi)) \tag{4.48}$$

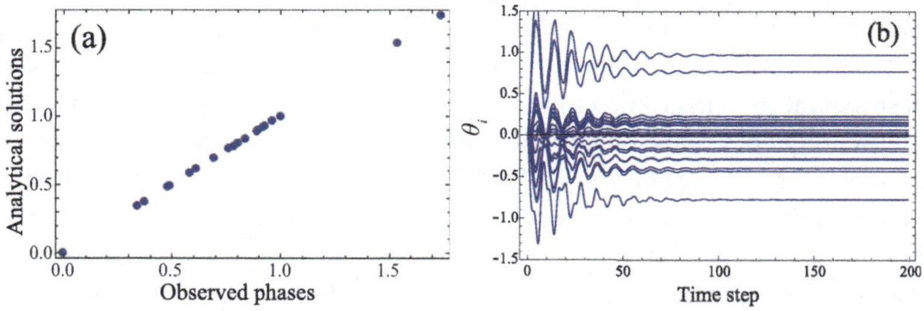

Fig. 4.12 Parameter calibration and synchronization. Panel (**a**) shows a comparison of analytical solutions and observed phases, and panel (**b**) numerical solutions of Eq. (4.45)

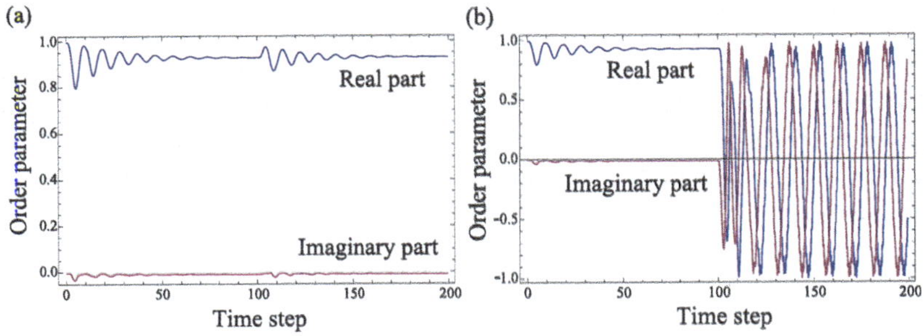

Fig. 4.13 Stability of synchronization. Panel (**a**) is for a finite elasticity ($\epsilon_d < 0$), and panel (**b**) is for zero elasticity ($\epsilon_d = 0$)

is defined as a macroscopic quantity that corresponds to the centroid of the phases of oscillators [16]. The radius r measures the coherence, and ϕ is the average phase. If $\mathrm{Re}(q(t)) \approx 1$ and $\mathrm{Im}(q(t)) \approx 0$, the oscillators remain in the synchronization region where the phase differences are relatively small. In contrast, if $\mathrm{Re}(q(t))$ and $\mathrm{Im}(q(t))$ oscillate between 1 and -1, the oscillators behave like a giant oscillator with a frequency additional to the common frequency ω.

First, we simulate the response to the fluctuation at $t = 100$ in the NN graph. The result for a finite elasticity ($\epsilon_d < 0$) is shown in Fig. 4.13a. It is seen that $\mathrm{Re}(q(t)) \approx 1$ and $\mathrm{Im}(q(t)) \approx 0$, even after the shock was applied in the middle of the network at $t = 100$. This means that the synchronization is stable, i.e., the oscillators remain in the region where phase differences are relatively small. For the case of zero elasticity ($\epsilon_d = 0$), $\mathrm{Re}(q(t))$ and $\mathrm{Im}(q(t))$ oscillate rapidly between 1 and -1 as shown in Fig. 4.13b. This means that synchronization is broken, and the generators behave like a giant oscillator with a high frequency additional to the common frequency ω.

Next, we simulate the power grid system with weaker coupling strengths k_{ij}. We expect the system to behave as if the generators were uncoupled for k_{ij} below a certain threshold k_{ij}^c. For the Kuramoto oscillator, the exact formula of the critical coupling strength k_{ij}^c has been derived and verified with the results of numerical simulations. Our simulation results with weaker coupling strengths of $0.3k_{ij}$ are shown in Fig. 4.14. The time evolution of the phases θ_i shown in Fig. 4.14a depicts a few generators separating from the main part of the coupled generator system, which still exhibits synchronization. The generator system acts like a giant oscillator with a low frequency additional to the common frequency ω, as seen in Fig. 4.14b. The phases θ_i at $t = 200$ in Fig. 4.14c show that the system disintegrated into four parts and that the main part moves with small phase differences.

This result is quite different from our expectation, but this is also reasonable because the system under consideration has $P_1 = 1$, $P_i = 0 (i = 2, \cdots, N-1)$, and $P_N = -1 (N = 21)$. Hence, one end of the system is pulled in the positive direction, and the other end is pulled in the opposite direction. Therefore, if the coupling strengths are weak enough, the oscillators at both ends of the system are

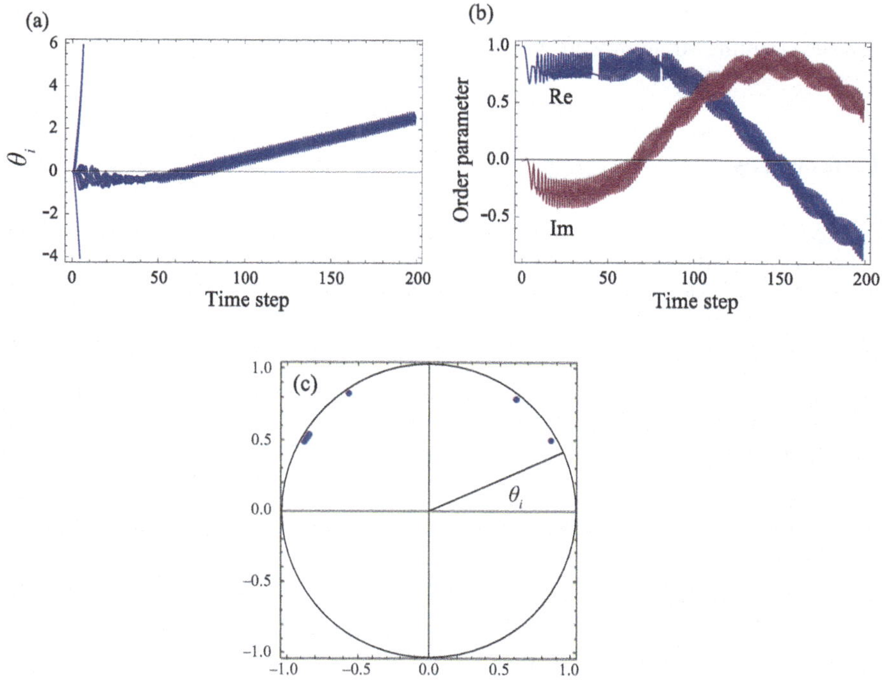

Fig. 4.14 Case with the coupling strength multiplied by 0.3. Panel (**a**) shows the time evolution of the phases, panel (**b**) shows the time evolution of the order parameter, and panel (**c**) shows the phases at $t = 200$

separated from the main part. The oscillators in the main part lose tension and shrink to the small phase difference. However, it is noted that the system in the C graph with $P_i \neq 0 (i = 1, \cdots, N)$ might behave similarly to the Kuramoto oscillator.

4.6 Summary

In this chapter, the fundamentals of typical optimization methods are described. These are linear programming methods, combinatorial optimization methods, and heuristic optimization methods. All of these are compelling methodologies when considering renewable energy problems. Furthermore, a model for understanding synchronization phenomena, which is directly crucial for the stability of AC power systems, was described, and an analytical example was given for a simple system. Existing theories for synchronization for economic, power, and physical and biological systems were reviewed. We then considered a power grid system consisting of generators and the electricity market. A coupled oscillator model exhibiting synchronization was developed based on the Kuramoto model with inertia by adding the electricity market, and analytic solutions of the stationary state and the coupling strength were obtained.

References

1. Stephen P. Bradley, "Applied Mathematical Programming", Addison-Wesley, February 1977, ISBN 978-0201004649
2. Wayne L. Winston, Munirpallam Venkataramanan, "Introduction to Mathematical Programming", Cengage Learning, 4th edition, June 2002, ISBN 978-0534359645
3. Robert J Vanderbei, "Linear Programming: Foundations and Extensions", Springer, 3rd ed., January 2008, ISBN 978-0387743875
4. Eugene Lawler, "Combinatorial Optimization: Networks and Matroids", Dover Publications; Reprint edition, November 2011, ISBN 978-0486414539
5. Hiroshi Morita, Naoki Shio, "Hybrid Branch and Bound Method with Genetic Algorithm for Flexible Flowshop Scheduling Problem", JSME International Journal, Series C, Vol 48, No 1, pp. 46–52, 2005.
6. John Holland, "Adaptation in Natural and Artificial Systems", Cambridge, MA: MIT Press, 1992, ISBN 978-0262581110.
7. John H. Holland, "Genetic Algorithms", Scientific American, Vol. 267, No. 1, pp. 66–73, JULY 1992.
8. R.M. Goodwin. "Dynamical coupling with especial reference to markets having production lags", Econometrica, 15 (1947), 181–204.
9. R. M. Goodwin, "The Nonlinear Accelerator and the Persistence of Business Cycles", Econometrica 19 (1951), 1–17.
10. J. B. Long and C. I. Plosser, "Real Business Cycles", J. of Political Economy 91 (1983), 39–69.
11. J .B. Long and C. I. Plosser, "Sectoral and aggregate shocks in the business cycle", American Economic Review 77, 333–336 (1987).
12. H. M. Anderson and J. B. Ramsey, Economic research reports PR # 99-01 (New York University, 1999).
13. D. D. Selover, R. V. Jensen and J. Kroll, "Industrial Sector Mode-Locking and Business Cycle Formation", Studies in Nonlinear Dynamics & Econometrics 7 (2003).
14. B. Sussmuth, "Business Cycles in the Contemporary World" (Springer, Berlin, Heidelberg, 2003).
15. Y. Kuramoto, in International Symposium on Mathematical Problems in Theoretical Physics, Lecture Notes Phys. vol. 39, ed. H. Araki, (Springer-Verlag, New York, 1975), p. 420 Physica 106A (1981), 128 "Chemical Oscillations, Waves, and Turbulence" (Springer, Heidelberg, New York, Tokyo, 1984)
16. S. H. Strogatz, "From Kuramoto to Crawford: exploring the onset of synchronization in populations of coupled oscillators", Physica D 143 (2000), 1–20.
17. J. A. Acebron, L. L. Bonilla, C. J. P. Vicente, F. Ritort and R. Spigler, "The Kuramoto model: A simple paradigm for synchronization phenomena", Rev. Mod. Phys. 77 (2005), 137.
18. G. Filatrella, A. H. Nielsen and N. F. Pedersen, "Analysis of a power grid using a Kuramoto-like model", Eur. Phys. J. B 61 (2008), 485–491.
19. Y. Ikeda, H. Aoyama, Y. Fujiwara, H. Iyetomi, K. Ogimoto, W. Souma, and H. Yoshikawa, "Coupled Oscillator Model of the Business Cycle with Fluctuating Goods Markets", Progress of Theoretical Physics Supplement, No.194, pp.111–121, June 2012

Chapter 5
Grid Flexibility

5.1 Technologies for Balancing Power

Gas turbine generation and electricity storage are essential as a balancing power source for renewable energy integration with large output fluctuations. Introducing large amounts of renewable energies, such as solar and wind power generation, is a global trend. However, the output of renewable energy is unstable, and in the case of solar power generation, power cannot be generated from sunset. In some regions, solar power generation has to be restricted in order to ensure the stable operation of the power system. Gas turbines are considered to be a fundamental technology for power generation in the future due to their ability to reduce greenhouse gas emissions to a lesser extent, for example, by co-firing natural gas and hydrogen. Electricity storage devices such as NaS batteries (sodium–sulfur batteries), RF batteries (redox flow batteries), and lithium-ion batteries are also very effective in regulating forces. In recent years, nickel, cobalt, copper, and rare earth elements used in gas turbines and electricity storage devices have attracted attention as strategic materials. This section describes material design techniques that lead to the reduction of the use of these strategic materials in gas turbine rotor blade materials and electricity storage device materials.

5.1.1 Gass Turbine Materials

Traditionally, combined cycle power generation using large gas turbines has been the mainstay of gas-fired power generation. The combined cycle consists of a gas turbine, a heat recovery boiler, a steam turbine, and a generator and can generate several hundred thousand to one million kW per line, achieving thermal efficiencies of up to 64%. In the future, further improving the efficiency of large gas turbines

Y. Ikeda, *Data Science of Renewable Energy Integration*,
Evolutionary Economics and Social Complexity Science 30,
https://doi.org/10.1007/978-981-99-8779-5_5

will be necessary. Specifically, the turbine inlet temperature must be increased. The maximum temperature at the gas turbine inlet is over 1,600 °C. Development to further increase this temperature to 1,700 °C is underway. Improving the heat resistance of the turbine materials will increase the thermal efficiency of the gas turbine itself and, at the same time, expand its application to hydrogen turbines. If green hydrogen can be used as a fuel, a large-scale power system with zero CO_2 emissions and high power generation efficiency will be realized.

To further expand the integration of renewable energy, installing thermal power sources or electric storage devices that compensate for load fluctuations in the power system is essential. One possibility for this is a small- to medium-sized gas turbine, such as an aircraft conversion-type gas turbine (Aero GT). Aero GT has a fast start-up time of about 10 minutes for a stand-alone gas turbine. While also improving thermal efficiency, a bottleneck for small gas turbines, the small output per series allows for more fine-tuning than larger combined cycle turbines. Such "spinning reserve power generation" by medium and small gas turbines is one of the best systems for regulating power sources such as renewable energy. With the transition to a fully liberalized electricity market through the separation of transmission and distribution, the capacity and balancing power market will be developed based on the aforementioned grid stabilization during the introduction of renewable energy. In Europe and the USA, in addition to the energy market (MWh), an ancillary market (capacity MW and regulating power $\Delta MW/\Delta h$ market) is operated to stabilize the frequency and voltage of the power grid following the separation of transmission and distribution. This ancillary market is closely related to stabilizing the power grid in response to the expansion of renewable energy integration.

The subject of this section is the design of heat-resistant alloys for gas turbines.

5.1.1.1 Alloy Design

Considerable expectations have been placed on computer simulation in developing new alloys. Computer simulations for a given alloy structure and composition calculate physical properties invaluable in developing new alloys. The molecular dynamics simulation method is the most suitable for calculating mechanical properties among various computer simulation methods. Alloy design is an optimization problem to search for an alloy's structure and composition to satisfy a given design rule. An alloy design is recognized as a kind of inverse problem. It is, however, impossible to optimize a composition for a multi-component alloy by a round-robin approach using molecular dynamics simulations because of the tremendous computational demand. A new method of alloy design was proposed by combining a genetic algorithm and molecular dynamics simulations [1]. The genetic algorithm is a biologically inspired optimization method mimicking the evolution of life. Living things have optimized properties under a given environment based on their self-organization and evolution. Evolution can also be recognized as an inverse problem. If an alloy and a design rule replace the living thing and its environment, it seems possible that alloy design could be accomplished by a method similar to biological self-organization and evolution; natural selection occurs according to each living thing's properties. In the case of alloy design, its properties are calculated by molecular dynamics simulations. Self-organization is effectively considered by assuming a specific form of an alloy gene in the genetic algorithm. The proposed method optimizes a composition for a multi-component alloy under a given design rule by evaluating physical quantities using molecular dynamics simulations and doing an efficient global search using the genetic algorithm. The design of several nickel-based superalloys is done to demonstrate the method's usefulness, and results are explained below.

5.1.1.2 Design Algorithm

The basic algorithm of a new alloy design method and each computation step of the algorithm are described in detail. The alloy design should be done based on the calculated values of the chemical potential μ, lattice misfit δ_a, elastic misfit $\delta_{c_{ij}}$, electron vacancy density $\bar{N}^v$, d-electron density $\bar{M}^d$ without using experimental values of the equilibrium composition, and the γ'/γ volume ratio. These physical quantities are calculated at a finite temperature using the first-principle molecular dynamics simulations for a given equilibrium composition. It is, however, impossible to optimize an equilibrium composition for a multi-component alloy by a round-robin approach using molecular dynamics simulations because of the substantial computational demand. This problem prohibits researchers from designing alloys using molecular dynamics simulations solely.

Various properties of alloys originate in a microscopic structure, an atomistic structure, and an alloy composition. Nickel-based superalloys have a precipitation hardening structure in which the γ phase matrix surrounds the γ' phase precipitates. The gamma phase is a nickel-based solid solution, and the γ' is based on a Ni_3Al intermetallic compound with the $L1_2$ structure. Once elements used in the nickel-based superalloy are specified, the next step of alloy designs is searching for alloy compositions. Thus, the alloy composition is encoded in the alloy gene. A binary number on the alloy gene expresses the compositions of each alloy. The alloy gene expressed by a binary number is defined as the nenotype gene. There are two ways to express alloy compositions, i.e., the fraction of each element or the number of each element. For the alloy gene expressed by atomic fraction, there is a problem in that the sum of each alloy fraction differs from 100% in the case of cross-over or mutation. The alloy gene, expressed by the number of each element, is used in this chapter.

The structure of an alloy gene of nickel-based superalloy is given in the following way: The alloy gene consists of three parts, i.e., the γ phase part, Ni sublattice of the γ' phase part, and Al sublattice of the γ' phase part. The number of Al, Ti, V, Cr, Co, Ni, Nb, Mo, Hf, Ta, W, and Re in the corresponding phase or sublattice is written in each part. The data length and offset for each element are used to translate a binary expression to a decimal one. An alloy gene expressed as a decimal number is called the phenotype gene. The data lengths of Ni in the γ phase, Ni in the γ' phase, and Al in the γ' phase are each 22 bits. Other elements have a 20-bit length. Offsets of Ni in the γ phase, Ni in the γ' phase, and Al in the γ' phase are $2^{20} = 1048576$, and offsets of the other elements are $2^0 = 1$. Transforming the genotype to the phenotype is accomplished by translating the gene's binary expression to a decimal expression and adding the offset value for each element.

Input Data: Step (a)

Several input parameters have to be read in step (a). They are N_g, k_{max}, Δn, P_∞, P_{mut}, and c_{ls}. N_g is the number of genes. k_{max} is the number of generations to be calculated and is hence equal to the number of iterations from steps (d) to (j). Δn is an increment of density used in calculations of the chemical potentials of each element. P_∞ and P_{mut} are probabilities of a cross-over and a mutation, respectively. P_∞ and P_{mut} are normalized for a pair of genes and each bit of a gene, respectively. c_{ls} is a parameter for linear scaling in the calculation of the fitness value.

Initialization of Physical Quantity Database: Step (b)

The physical quantity database is initialized in step (b). Details of the database are described in Sect. 5.1.1.3. The parameters related to the γ phase are a lattice constant a^{Ni}, elastic constant c_{11}^{Ni}, c_{12}^{Ni}, cohesive energy E^{Ni} of nickel, a slope of the lattice constant $da^\gamma/d\rho_i^\gamma$, slopes of the elastic constants $dc_{11}^\gamma/d\rho_i^\gamma$ and $dc_{12}^\gamma/d\rho_i^\gamma$, a slope of the cohesive constant $dE^\gamma/d\rho_i^\gamma$ when the element i is added to the γ phase. The parameters related to the γ' phase are lattice constant $a^{\mathrm{Ni_3Al}}$, elastic constants $c_11^{\mathrm{Ni_3Al}}$ and $c_12^{\mathrm{Ni_3Al}}$, cohesive energy $E^{\mathrm{Ni_3Al}}$ of $\mathrm{Ni_3Al}$ intermetallic compound, and a slope of the lattice constant $da^{\gamma'(\mathrm{Ni})}/d\rho_i^{\gamma'(\mathrm{Ni})}$, slopes of the elastic constants $dc_{11}^{\gamma'(\mathrm{Ni})}/d\rho_i^{\gamma'(\mathrm{Ni})}$ and $dc_{12}^{\gamma'(\mathrm{Ni})}/d\rho_i^{\gamma'(\mathrm{Ni})}$, a slope of the cohesive constant $dE^{\gamma'(\mathrm{Ni})}/d\rho_i^{\gamma'(\mathrm{Ni})}$ when the element i is added to the nickel sublattice of the γ' phase, and a slope of the lattice constant $da^{\gamma'(\mathrm{Al})}/d\rho_i^{\gamma'(\mathrm{Al})}$, slopes of the elastic constants $dc_{11}^{\gamma'(\mathrm{Al})}/d\rho_i^{\gamma'(\mathrm{Al})}$ and $dc_{12}^{\gamma'(\mathrm{Al})}/d\rho_i^{\gamma'(\mathrm{Al})}$, and a slope of the cohesive constant $dE^{\gamma'(\mathrm{Al})}/d\rho_i^{\gamma'(\mathrm{Al})}$ when the element i is added to the aluminum sublattice of the γ' phase. These parameters are obtained by molecular dynamics simulations for various alloys and intermetallic compounds.

Generation of Initial Alloy Gene: Step (c)

Compositions of alloys are written in alloy genes. Each alloy gene represents an alloy with a specific composition. Initial compositions are generated using a random number so that each alloy has a random composition. The composition is written as a genotype as described below. As a result, initial alloy compositions are distributed over the entire alloy composition space. Such a distribution allows a global search to be made among all possible alloy compositions. The structure of the alloy gene is described later in detail.

Conversion of Genotype to Phenotype: Step (d)

A binary expression of the alloy gene is called the genotype, and it is needed for a cross-over and a mutation. On the other hand, a decimal expression of the alloy gene is called the phenotype. The genotype is converted to the phenotype in this step.

Calculation of Physical Quantity: Step (e)

A lattice constant, elastic constants, free energy, and chemical potentials are calculated using the physical quantity database, which stores parameters computed using molecular dynamics simulation in Sect. 5.1.1.3, for the obtained alloy composition. Effects of multi-additive elements are assumed as linear superpositions of each additive element. Details are described in Sect. 5.1.1.3

Fitness: Step (f)

Fitness indicates the degree of satisfaction with a given design rule. The minimal design rule for a nickel-based superalloy is that the equal chemical potential in the two phases is to be satisfied for each additive element. The fitness function is determined to be equal to unity when the chemical potential in the two phases is satisfied precisely. The function form of the fitness is discussed in Sect. 5.1.1.4

Discrimination: Step (g)

The alloy gene with the maximum fitness value is selected from all alloy genes. If the selected value is larger than f_{max}, then the ally composition search is finished, and the process goes to step (k). Otherwise, the search continues to proceed to step (h).

Linear Scaling: Step (h)

An optimization using the genetic algorithm is very efficient if the fitness value is small. However, the efficiency goes down for larger fitness values, so many searching steps are needed for the optimum fitness value. In order to prevent an inefficient search, fitness values are scaled using the linear scaling method. The scaled fitness value f' is obtained by scaling the raw fitness value f as below:

$$f' = a_{ls} f + b_{ls}, \tag{5.1}$$

where coefficients a_{ls} and b_{ls} are evaluated by two conditions: (i) the average value of the scaled fitness f'_{av} is equal to the average value of the raw fitness f_{av} and (ii) the maximum value of the scaled fitness f'_{max} is c_{ls} times of the average value of the raw fitness f_{av}

$$a_{ls} = (c_{ls} - 1) \frac{f_{av}}{f_{max} - f_{av}} \tag{5.2}$$

$$b_{ls} = (1 - a_{ls}) f_{av}. \tag{5.3}$$

When a fitness value scaled by the linear scaling method becomes negative, the negative value is replaced by a null value.

Selection: Step (i)

The roulette method selects N_g genes with large fitness values from N_g genes. First, a fan-shaped piece on a roulette wheel is assigned where its area is proportional to the fitness value of each gene. Then, the wheel is spun, a ball is thrown, and a gene is selected according to the fan-shaped piece where the ball stops. N_g genes are selected by repeating N_s spins of the roulette wheel.

Cross-Over and Mutation: Step (j)

Cross-overs are generated in the selected N_g genes with a probability P_{∞}. A pair of genotype G_i and G_j is cut arbitrarily to get four fragments G_i^a, G_i^b, G_j^a, and G_j^b. Then G_i^a is linked to G_j^a, and G_j^a is linked to G_i^b. Afterward, mutations are generated for each gene. The mutation is defined as reversing a bit of a genotype with a probability P_{mut}.

Output Data: Step (k)

Bulk composition, composition of the γ phase, composition of the γ' phase, composition of the Ni sublattice in the γ' phase, and composition of the Al sublattice in the γ' phase are obtained as a result of the alloy design. Lattice constant of the γ phase a^{γ}, lattice constant of the γ' phase $a^{\gamma'}$, elastic constant of the γ phase c_{ij}^{γ}, elastic constant of the γ' phase $c_{ij}^{\gamma'}$, lattice misfit δ_a, and elastic misfits $\delta_{c_{ij}}$ are also obtained.

5.1.1.3 Evaluation of Physical Quantities

Molecular dynamics simulations were used to analyze lattice constant a, cohesive energy E_0, and elastic constants c_{11} and c_{12} of nickel-based superalloys for Al, Ti, V, Cr, Co, Ni, Nb, Mo, Hf, Ta, W, and Re additions. It was clarified that changes in all physical quantities evaluated were proportional to the changes in the concentration of additive elements. These results were fitted by the linear least square method, and the obtained parameters are saved as a database given below. A set of data that contains these parameters is called the physical quantities database.

(1) Lattice constant a^{Ni}, elastic constants c_{11}^{Ni} and c_{12}^{Ni}, and cohesive energy E^{Ni} for Ni
(2) Lattice constant $a^{\mathrm{Ni_3Al}}$, elastic constants $c_{11}^{\mathrm{Ni_3Al}}$ and $c_{12}^{\mathrm{Ni_3Al}}$, and cohesive energy $E^{\mathrm{Ni_3Al}}$ for Ni_3Al intermetallic compound
(3) Slope parameter of lattice constant $da^{\gamma}/d\rho_i^{\gamma}$, slope parameters of elastic constants $dc_{11}^{\gamma}/d\rho_i^{\gamma}$ and $dc_{12}^{\gamma}/d\rho_i^{\gamma}$, and slope parameter of cohesive energy $dE^{\gamma}/d\rho_i^{\gamma}$ for an addition of element i to Ni
(4) Slope parameter of lattice constant $da^{\gamma'(\mathrm{Ni})}/d\rho_i^{\gamma'(\mathrm{Ni})}$, slope parameters of elastic constants $dc_{11}^{\gamma'(\mathrm{Ni})}/d\rho_i^{\gamma'(\mathrm{Ni})}$ and $dc_{12}^{\gamma'(\mathrm{Ni})}/d\rho_i^{\gamma'(\mathrm{Ni})}$, and slope parameter of cohesive energy $dE^{\gamma'(\mathrm{Ni})}/d\rho_i^{\gamma'(\mathrm{Ni})}$ for addition of element i to the Ni sublattice for Ni_3Al
(5) Slope parameter of lattice constant $da^{\gamma'(\mathrm{Al})}/d\rho_i^{\gamma'(\mathrm{Al})}$, slope parameters of elastic constants $dc_{11}^{\gamma'(\mathrm{Al})}/d\rho_i^{\gamma'(\mathrm{Al})}$ and $dc_{12}^{\gamma'(\mathrm{Al})}/d\rho_i^{\gamma'(\mathrm{Al})}$, and slope parameter of cohesive energy $dE^{\gamma'(\mathrm{Al})}/d\rho_i^{\gamma'(\mathrm{Al})}$ for addition of element i to the Al sublattice for Ni_3Al

Suppose elements including Ni and Al among the 12 elements shown above are chosen, and a single-crystal nickel-based superalloy is cast. Lattice constant, cohesive energy, and elastic constants for such nickel-based superalloys are calculated with the physical quantities database using the following equations:

$$a^{\gamma} = a^{\mathrm{Ni}} + \sum_i \frac{da^{\gamma}}{d\rho_i^{\gamma}} \rho_i^{\gamma} \tag{5.4}$$

$$c_{11}^{\gamma} = c_{11}^{\mathrm{Ni}} + \sum_i \frac{dc_{11}^{\gamma}}{d\rho_i^{\gamma}} \rho_i^{\gamma} \tag{5.5}$$

$$c_{12}^{\gamma} = c_{12}^{\mathrm{Ni}} + \sum_i \frac{dc_{12}^{\gamma}}{d\rho_i^{\gamma}} \rho_i^{\gamma} \tag{5.6}$$

$$E^{\gamma} = E^{\mathrm{Ni}} + \sum_i \frac{dE^{\gamma}}{d\rho_i^{\gamma}} \rho_i^{\gamma} \tag{5.7}$$

$$a^{\gamma'} = a^{\mathrm{Ni_3Al}} + \sum_i \frac{da^{\gamma'(\mathrm{Ni})}}{d\rho_i^{\gamma'(\mathrm{Ni})}} \rho_i^{\gamma'(\mathrm{Ni})} + \sum_i \frac{da^{\gamma'(\mathrm{Al})}}{d\rho_i^{\gamma'(\mathrm{Al})}} \rho_i^{\gamma'(\mathrm{Al})} \tag{5.8}$$

$$c_{11}^{\gamma'} = c_{11}^{\mathrm{Ni_3Al}} + \sum_i \frac{dc_{11}^{\gamma'(\mathrm{Ni})}}{d\rho_i^{\gamma'(\mathrm{Ni})}} \rho_i^{\gamma'(\mathrm{Ni})} + \sum_i \frac{dc_{11}^{\gamma'(\mathrm{Al})}}{d\rho_i^{\gamma'(\mathrm{Al})}} \rho_i^{\gamma'(\mathrm{Al})} \tag{5.9}$$

$$c_{12}^{\gamma'} = c_{12}^{\mathrm{Ni_3Al}} + \sum_i \frac{dc_{12}^{\gamma'(\mathrm{Ni})}}{d\rho_i^{\gamma'(\mathrm{Ni})}} \rho_i^{\gamma'(\mathrm{Ni})} + \sum_i \frac{dc_{12}^{\gamma'(\mathrm{Al})}}{d\rho_i^{\gamma'(\mathrm{Al})}} \rho_i^{\gamma'(\mathrm{Al})} \tag{5.10}$$

$$E^{\gamma'} = E^{\mathrm{Ni_3Al}} + \sum_i \frac{dE^{\gamma'(\mathrm{Ni})}}{d\rho_i^{\gamma'(\mathrm{Ni})}} \rho_i^{\gamma'(\mathrm{Ni})} + \sum_i \frac{dE^{\gamma'(\mathrm{Al})}}{d\rho_i^{\gamma'(\mathrm{Al})}} \rho_i^{\gamma'(\mathrm{Al})}. \tag{5.11}$$

Here ρ_i^{γ}, $\rho_i^{\gamma'(\mathrm{Ni})}$, and $\rho_i^{\gamma'(\mathrm{Al})}$ are the concentrations of the additive element i in the γ phase, Ni sublattice of the γ' phase, and Al sublattice of the γ' phase, respectively. These concentrations are normalized by the total number of atoms of the γ phase, Ni sublattice of the γ' phase, and Al sublattice of the γ' phase, respectively. The summation terms in Eqs. (5.10) to (5.13) are made for all additive elements except Ni. In Eqs. (5.14) to (5.17), the first summation terms are made for all additive elements except Ni, and the second summation terms are made for all additive elements except Al. It should be noted that all slope parameters defined for each element are obtained by adding a single element to the Ni or $\mathrm{Ni_3Al}$ matrix. Thus, it is not rigorous to use Eqs. (5.14) to (5.17) in the case of multiple element additions. However, it may be a good approximation as long as the concentration of additive elements is low and they give a fair approximation, even if the concentration of additive elements is high, because the main concern is the relative difference of physical quantities between the γ and γ' phases.

Furthermore, Helmholtz free energy F of the superalloy is calculated using a cohesive energy E

$$F^{\gamma} = -\frac{N^{\gamma}}{2} E^{\gamma} - TS^{\gamma} \tag{5.12}$$

$$F^{\gamma'} = -\frac{N^{\gamma'(\mathrm{Ni})} + N^{\gamma'(\mathrm{Al})}}{2} E^{\gamma'} - TS^{\gamma'}, \tag{5.13}$$

where T, S, N^{γ}, $N^{\gamma'(\mathrm{Ni})}$, and $N^{\gamma'(\mathrm{Al})}$ are temperature, entropy, the total number of atoms in the γ phase, the total number of Ni sublattices in the γ' phase, and the total number of Al sublattices in the γ' phase, respectively. For each alloy gene, the total number of atoms in the γ phase N^{γ}, the total number of Ni sublattices in the γ' phase $N^{\gamma'(\mathrm{Ni})}$, and the total number of Al sublattices in the γ' phase $N^{\gamma'(\mathrm{Al})}$ are varied within a range determined by bit length and offset value of the alloy gene. Thus, before calculating free energy F, these numbers have to be normalized so that the ratios $N^{\gamma'(\mathrm{Ni}))} : N^{\gamma'(\mathrm{Al}))} = 3 : 1$ and $N^{\gamma} : N^{\gamma'(\mathrm{Ni}))} + N^{\gamma'(\mathrm{Al}))} = 1 : 1$ are kept.

Entropy S is calculated by counting the number of states of atomic configuration for additive elements as follows:

$$S^{\gamma} = k_B \ln\left(\binom{N^{\gamma}}{n_1^{\gamma}} \cdot \binom{N^{\gamma} - n_1^{\gamma}}{n_2^{\gamma}} \cdots \binom{n_f^{\gamma}}{n_f^{\gamma}}\right) \tag{5.14}$$

$$S^{\gamma'} = k_B \ln\left(\begin{array}{c} \binom{N^{\gamma'(\mathrm{Ni})}}{n_1^{\gamma'(\mathrm{Ni})}} \cdot \binom{N^{\gamma'(\mathrm{Ni})} - n_1^{\gamma'(\mathrm{Ni})}}{n_2^{\gamma'(\mathrm{Ni})}} \cdots \binom{n_f^{\gamma'(\mathrm{Ni})}}{n_f^{\gamma'(\mathrm{Ni})}} \cdot \\ \binom{N^{\gamma'(\mathrm{Al})}}{n_1^{\gamma'(\mathrm{Al})}} \cdot \binom{N^{\gamma'(\mathrm{Al})} - n_1^{\gamma'(\mathrm{Al})}}{n_2^{\gamma'(\mathrm{Al})}} \cdots \binom{n_f^{\gamma'(\mathrm{Al})}}{n_f^{\gamma'(\mathrm{Al})}} \end{array}\right), \tag{5.15}$$

where k_B, n_i^{γ}, $n_i^{\gamma'(\mathrm{Ni})}$, $n_i^{\gamma'(\mathrm{Al})}$, $\binom{n}{m}$, and f are Boltzmann constant, the number of atoms of the additive element i in the γ phase, the number of atoms of the additive element i in the Ni sublattice of the γ' phase, the number of atoms of the additive element i in the Al sublattice of the γ' phase, the number of combinations choosing n atoms among m atoms, and the number of elements in the superalloy, respectively.

When the total number of atoms is visible, a thermodynamic equilibrium state is the minimum point of grand potential Ω

$$\Omega^{\gamma} = F^{\gamma} - \sum_i \mu_i^{\gamma} n_i^{\gamma} \tag{5.16}$$

$$\Omega^{\gamma'} = F^{\gamma'} - \sum_i \mu_i^{\gamma'} \left(n_i^{\gamma'(\mathrm{Ni})} + n_i^{\gamma'(\mathrm{Al})}\right), \tag{5.17}$$

where μ_i^{γ} and $\mu_i^{\gamma'}$ are chemical potentials of the element i in the γ phase and the γ' phase, respectively. Then the chemical potential of the element i μ_i is calculated using the following equations:

$$\mu_i^{\gamma} = \frac{\partial F^{\gamma}}{\partial n_i^{\gamma}} \cong \frac{F^{\gamma}(n_i^{\gamma} + \Delta n_i^{\gamma}) - F^{\gamma}(n_i^{\gamma})}{\Delta n_i^{\gamma}} \tag{5.18}$$

$$\begin{aligned} \mu_i^{\gamma'} &= \frac{\partial F^{\gamma'}}{\partial\left(n_i^{\gamma'(\mathrm{Ni})} + n_i^{\gamma'(\mathrm{Al})}\right)} \\ &\cong \frac{F^{\gamma'}\left(n_i^{\gamma'(\mathrm{Ni})} + n_i^{\gamma'(\mathrm{Al})} + \Delta n_i^{\gamma'}\right) - F^{\gamma'}\left(n_i^{\gamma'(\mathrm{Ni})} + n_i^{\gamma'(\mathrm{Al})}\right)}{\Delta n_i^{\gamma'}}, \end{aligned} \tag{5.19}$$

where Δn_i is an increment of the number of atoms of element i, and the number of atoms of all elements is kept constant.

5.1.1.4 Design Rule and Fitness

For an alloy consisting of many elements, the concentration of each element at an equilibrium state is determined so that the chemical potential of each element is balanced in each phase. The design rule for nickel-based superalloy is that the equal chemical potential is satisfied in the γ and γ' phases.

First, a tangent line is drawn using free energy F^{γ} of the γ phase and its slope s_i^{γ} at $\rho_i = \rho_i^{\gamma}$, toward the γ' phase in order to obtain the estimated free energy of the γ' phase $\tilde{F}^{\gamma'}$ at $\rho_i = \rho_i^{\gamma'}$:

$$\tilde{F}^{\gamma'}(\rho_i^{\gamma'}) = s_i^{\gamma}\left(\rho_i^{\gamma'} - \rho_i^{\gamma}\right) + F^{\gamma}(\rho_i^{\gamma}), \tag{5.20}$$

where the slope s_i^{γ} is given by the chemical potential of the γ phase

$$\rho_i^{\gamma'} = N^{\gamma}\mu_i^{\gamma}. \tag{5.21}$$

Next, another tangent line is drawn using free energy of the γ' phase $F^{\gamma'}$ and its slope $s_i^{\gamma'}$ at $\rho_i = \rho_i^{\gamma'}$, toward the γ phase in order to obtain an estimated free energy of the γ phase $\tilde{F}^{\gamma}$ at $\rho_i = \rho_i^{\gamma'}$:

$$\tilde{F}^{\gamma}(\rho_i^{\gamma}) = s_i^{\gamma'}\left(\rho_i^{\gamma} - \rho_i^{\gamma'}\right) + F^{\gamma'}(\rho_i^{\gamma'}), \tag{5.22}$$

where the slope $s_i^{\gamma'}$ is given by the chemical potential of the γ' phase

$$s_i^{\gamma'} = \left(N^{\gamma'(\mathrm{Ni})} + N^{\gamma'(\mathrm{Al})}\right)\mu_i^{\gamma'}. \tag{5.23}$$

The tie line of the element i is obtained when both the difference between the estimated free energy of the γ phase $\tilde{F}^{\gamma}$ and the free energy of the γ phase F^{γ} at $\rho_i = \rho_i^{\gamma}$ and the difference between the estimated free energy of the γ' phase $\tilde{F}^{\gamma'}$ and the free energy of the γ' phase $F^{\gamma'}$ are minimized. Thus a fitness f is determined as the quantity to include the degree of satisfaction of the design rule.

$$f = \left[\begin{array}{l} 1 + \dfrac{1}{M-1}\sum_i \left|\dfrac{\tilde{F}^{\gamma}(\rho_i^{\gamma}) - F^{\gamma}(\rho_i^{\gamma})}{F^{\gamma}(\rho_i^{\gamma})} \times 100\right| + \\ \dfrac{1}{M-1}\sum_i \left|\dfrac{\tilde{F}^{\gamma'}(\rho_i^{\gamma'}) - F^{\gamma'}(\rho_i^{\gamma'})}{F^{\gamma'}(\rho_i^{\gamma'})} \times 100\right| \end{array}\right]^{-1}, \tag{5.24}$$

where M is the number of elements in the alloy, and the summation is made for all elements except nickel.

5.1.1.5 Results

Alloy design using GA was performed for the Ni-Al-Cr-Mo-Ta element system and Ni-Al-Co-Cr-W-Ti-Ta element system. The input parameters are shown in Table 5.1. For the two systems, all parameters were set to be identical, except that the gene lengths differed depending on the number of elements constituting the system.

The results for the Ni-Al-Cr-Mo-Ta element system are shown in Tables 5.2, 5.3, 5.4, 5.5, and 5.6. Tables 5.2, 5.3, 5.4, and 5.5 show the elemental compositions of each phase, comparing calculated results with measured values. These tables show that the agreement between the calculated and measured values is relatively good for the Ni-Al-Cr-Mo-Ta element system. Furthermore, Table 5.6 confirms that the misfit of the lattice constants in the two phases is relatively small. On the other hand, the misfit of elastic constants in the two phases was found to be on the order of a few 10%.

The results for the Ni-Al-Co-Cr-W-Ti-Ta element system are shown in Tables 5.7, 5.8, 5.9, 5.10, and 5.11. Tables 5.7, 5.8, 5.9, and 5.10 show the elemental compositions of each phase, comparing calculated results with measured values. From these tables, it is noticeable that some elements in the Ni-Al-Co-Cr-W-Ti-Ta element system do not agree with the calculated and measured values compared to the Ni-Al-Cr-Mo-Ta element system. Table 5.11 shows that the misfit of the lattice constants in the two phases is relatively small, but larger than in the Ni-Al-Cr-Mo-Ta element system. Similarly, the misfit of elastic constants in the two phases is larger than in the Ni-Al-Cr-Mo-Ta element system.

As can be seen from this example, the alloy design method using molecular dynamics and GA is still not at a practical level, but the overall characteristics can be said to be reproduced. By combining first-principle molecular dynamics, which calculates electronic states, with metaheuristic optimization methods, including GA

Table 5.1 Input parameters

Parameter	Ni–Al–Cr–Mo–Ta element system	Ni–Al–Co–Cr–W–Ti–Ta element system
N_g	100	100
k_{max}	5000	5000
Δn	1.0×10^{-2}	1.0×10^{-2}
P_∞	0.90	0.90
P_{mut}	3.0×10^{-3}	3.0×10^{-3}
C_{ls}	2.0	2.0

Table 5.2 Ni–Al–Cr–Mo–Ta element system: the γ phase

	Ni	Al	Cr	Mo	Ta
Exp	71.0%	7.9%	11.3%	8.5%	1.1%
Cal	62.0%	3.9%	11.8%	9.6%	7.3%

Table 5.3 Ni–Al–Cr–Mo–Ta element system: the γ' phase

	Ni	Al	Cr	Mo	Ta
Exp	73.2%	19.2%	1.7%	2.3%	4.5%
Cal	50.7%	18.6%	14.6%	16.9%	0.0%

Table 5.4 Ni–Al–Cr–Mo–Ta element system: the Ni sublattice of the γ' phase

	Ni	Al	Cr	Mo	Ta
Exp	89.4%	0.0%	6.4%	1.9%	3.2%
Cal	65.1%	3.2%	16.0%	16.0%	0.0%

Table 5.5 Ni–Al–Cr–Mo–Ta element system: the Al sublattice of the γ' phase

	Ni	Al	Cr	Mo	Ta
Exp	11.0%	65.0%	0.0%	10.0%	14.0%
Cal	10.0%	60.5%	10.5%	19.5%	0.0%

Table 5.6 Physical quantities of the Ni–Al–Cr–Mo–Ta element system

	γ phase	γ' phase	Misfit
Lattice constant, $a/10^{-1}$ nm	3.773	3.741	−0.852 %
Elastic constant, $c_{11}/10^{11}$ Pa	1.677	1.934	15.3 %
Elastic constant, $c_{12}/10^{11}$ Pa	0.8654	1.074	24.1 %

Table 5.7 Ni–Al–Co–Cr–W–Ti–Ta element system: the γ phase

	Ni	Al	Co	Cr	W	Ti	Ta
Exp	59.0%	3.4%	8.5%	25.0%	2.8%	1.1%	0.0%
Cal	50.5%	0.0%	12.5%	10.2%	7.9%	10.2%	7.9%

Table 5.8 Ni–Al–Co–Cr–W–Ti–Ta element system: the γ' phase

	Ni	Al	Co	Cr	W	Ti	Ta
Exp	70.4%	15.9%	2.8%	4.0%	2.3%	1.7%	1.1%
Cal	57.3%	7.9%	11.3%	9.1%	2.8%	8.5%	1.7%

Table 5.9 Ni–Al–Co–Cr–W–Ti–Ta element system: the Ni sublattice of the γ' phase

	Ni	Al	Co	Cr	W	Ti	Ta
Exp	84.3%	7.0%	5.7%	6.4%	0.0%	1.9%	0.0%
Cal	72.1%	1.9%	12.1%	7.7%	1.3%	3.8%	0.6%

Table 5.10 Ni–Al–Co–Cr–W–Ti–Ta element system: the Al sublattice of the γ' phase

	Ni	Al	Co	Cr	W	Ti	Ta
Exp	18.0%	59.0%	1.5%	0.0%	11.0%	2.5%	6.0%
Cal	16.5%	28.0%	11.0%	10.0%	9.0%	22.0%	3.5%

Table 5.11 Physical quantities of the Ni–Al–Co–Cr–W–Ti–Ta element system

	γ phase	γ' phase	Misfit
Lattice constant, $a/10^{-1}$ nm	3.790	3.666	−3.28 %
Elastic constant, $c_{11}/10^{11}$ Pa	1.535	2.038	32.8 %
Elastic constant, $c_{12}/10^{11}$ Pa	0.8247	1.276	52.4 %

and other machine learning methods, it will be possible to use this method not only for the design of heat-resistant superalloys for gas turbines but also for the design of materials for energy storage devices for electric vehicles.

5.1.2 Electricity Storage

We describe the operation of conventional electric power systems briefly. The supply and demand imbalance in electric power may cause not only a failure in production due to a power frequency problem but also a blackout due to the sequential parallel-off of thermal power plants. The usual operation of an electric power system requires centralized energy management to avoid the above failures and maintain a stable power supply.

Facilities such as power plants, power substations, and distribution and transmission lines must effectively balance supply and demand. For this purpose, operating procedures are rigorously determined, and the central load dispatching office monitors the overall system and orders various load dispatch instructions, such as parallel, parallel-off, power control, and operating switch.

In recent years, the integration of renewable energy, including wind power, solar power, hydropower, biomass, and geothermal, into the power system has been ongoing to reduce fossil fuel consumption. We note that wind and solar power have a characteristic that is not possessed by other renewables. That characteristic is output fluctuation, as depicted in Fig. 5.1. Panels (a) and (c) show an output time series for solar PV and wind power, respectively. Panels (b) and (d) are the corresponding growth rate time series. The output fluctuation makes integrating them into the conventional power system difficult (Fig. 5.2).

In the conventional system, load fluctuations are caused by fluctuations in demand. Thermal and hydropower plants restore load balance (see Fig. 5.3a). When we integrate wind power and roof-top solar PV power, load fluctuations increase as the characteristic of wind and solar PV power combines with demand fluctuations. If thermal and hydropower plants lack a sufficient balancing capability, we need large electricity storage such as batteries (see Fig. 5.3b). However, if demand side management is introduced, electricity storage on a moderate scale suffices to restore load balance. This implies that managing demand introduces additional balancing capability in the supply side of the system (see Fig. 5.3c).

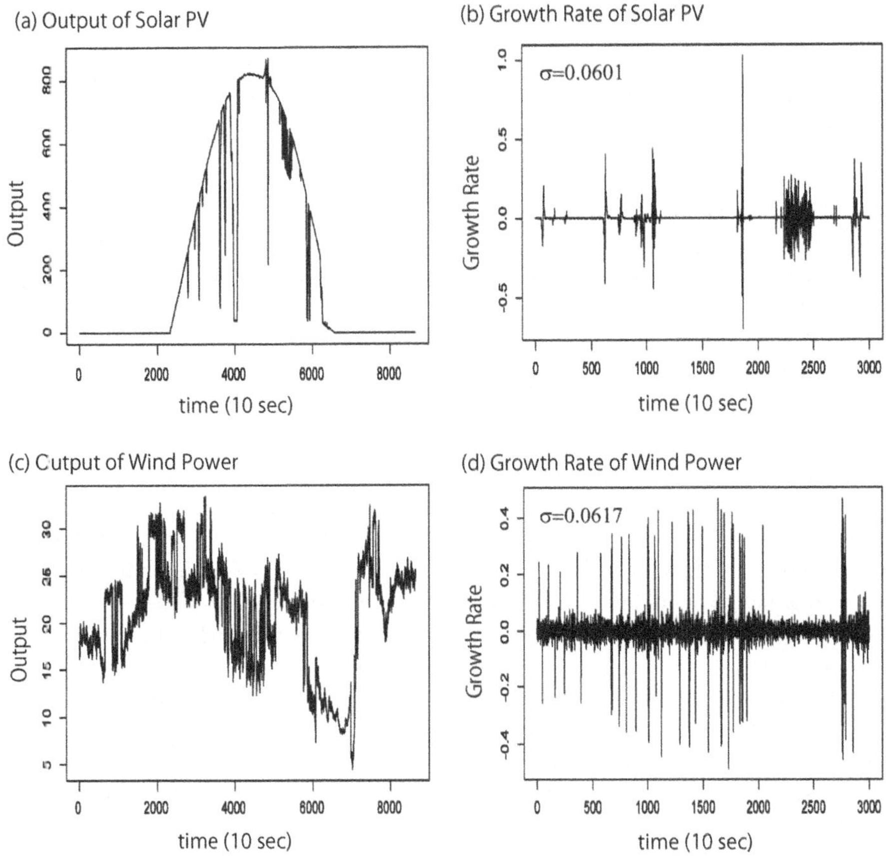

Fig. 5.1 Characteristic of output fluctuation in renewable energy: Panels (**a**) and (**c**) are the original time series for solar PV and wind power, respectively. Panels (**b**) and (**d**) are the corresponding growth rate time series

Renewable energy will be integrated as much as possible to reduce fossil fuel consumption in our power grid to establish a sustainable society.

Long-term bilateral contracts mark the first stage of an energy market. Generating companies and retailers conclude long-term bilateral contracts based on their long-term demand forecasting. Long-term bilateral contracts between generating companies and retailers will take up the largest share of the energy market.

The second stage of the energy market is the day-ahead spot market (see Fig. 5.3a). First, the meteorological bureau announces the weather forecast for the next day. Mega-solar plants place their output electricity, estimated using the weather forecast, on the market. HEMS in smart houses estimates the house-to-house load by considering the output from roof-top solar PVs. Retailers estimate their regional load in eco-towns by aggregating the load for all eco-towns and bidding on the market. Companies operating thermal power plants forecast demand,

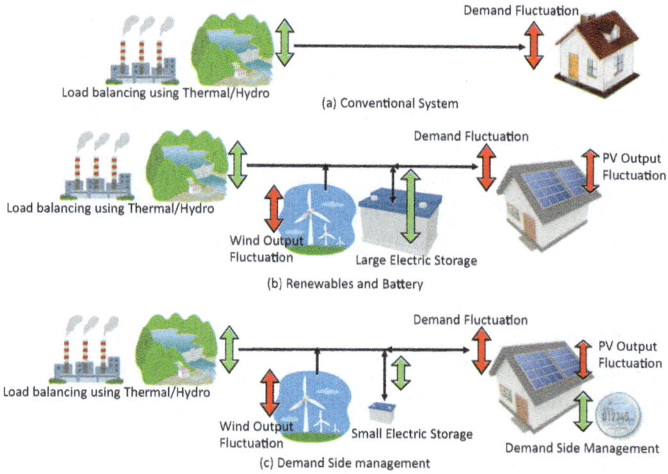

Fig. 5.2 Grid integration of wind power and solar PV: Demand management increases balancing capability and requires smaller electricity storage

Fig. 5.3 operation of microgrid before and after gate closure: Panel (**a**) is the day-ahead spot market, and panel (**b**) is the balancing market

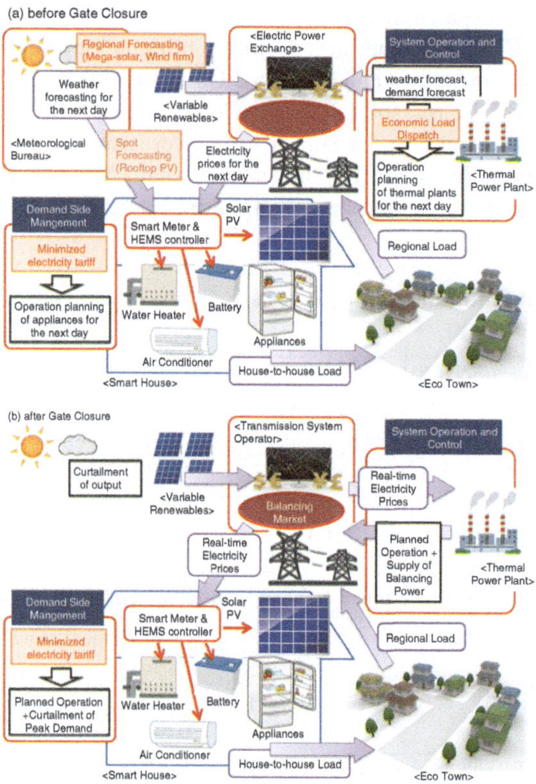

make operation plans based on economic load dispatch, and place their supply on the market. Electric power exchange is responsible for operating the day-ahead spot market. The gate of the day-ahead spot market is closed before a specific time.

After the gate closure, the balancing market is opened by the transmission system operator (TSO) (see Fig. 5.3b). It is critical to making the balancing market possible, measuring demand in real-time using smart meters, and capturing the system-wide demand–supply imbalance through the SCADA (Supervisory Control and Data Acquisition) system. HEMS located in a smart house curtails peak demand to minimize the electricity tariff in each house if the real-time electricity price rises. Companies operating thermal power plants place their balancing power on the market if they have extra generation capacity. TSO is responsible for operating the overall power system in the eco-town, keeping demand and supply in balance. If PV output is too much to maintain a balance, TSO can order mega-solar plants to curtail their output.

As explained above, distributed generation, market mechanism, demand response, and IT technology are the keys to integrating more renewable energy to emit less CO_2. Note that the power system in an eco-town is clearly at the opposite end of the system operated by an oligopolistic utility company's central load dispatching office.

Cost is another crucial factor in deploying renewable energy on a large scale. Investment cost and Levelized Cost of Electricity (LCOE) are shown in Tables 5.12 and 5.13 for solar PV and Concentrated Solar Power (CSP), respectively. The investment cost for Solar PV is lower than that for CSP, but LCOE for solar PV is higher than CSP. This means that solar PV is economically easy to implement, but recouping the investment takes longer. Note that the LCOE for CSP with storage is

Table 5.12 Investment cost and levelized cost of electricity for solar PV in 2015

		Investment cost (USD/kW)	LCOE (USD/MWh)
Mega-Solar PV	Min.	1522	–
	Max.	2913	–
	Ave.	2043	–
Rooftop PV	Min.	1609	200
	Max.	4739	316
	Ave.	2130	–

Table 5.13 Investment cost and levelized cost of electricity for CSP in 2015

		Investment cost (USD/kW)	LCOE (USD/MWh)
CSP w/o storage	Min.	3739	158
	Max.	6348	263
	Ave.	4609	191
CSP with 6 hour storage	Min.	6348	146
	Max.	9130	213
	Ave.	7304	168

lower than that for CSP without storage, although the investment cost is higher. This is because storage separates heat acquisition during the day and power generation after the sun sets. This capability has a considerable economic advantage in a country where peak demand is in the early evening.

5.1.3 Demand Side Management

The heart of demand side management is the Home Energy Management System (HEMS), consisting of an energy measurement unit and an information acquisition unit (see Fig. 5.4). HEMS currently assumes time-based pricing but is ideal for dynamic pricing in the balancing market. This system allows it to manage power-saving operations during the day (high price) and automatic operation at night (low price). In this system, smart appliances, such as refrigerators, washing machines, air conditioners, TV sets, heat pumps, water heaters, household fuel cell cogeneration systems, and induction heating cooking devices, are controlled through a home gateway and a cloud energy management server via mobile phone while outside the home. In addition to these appliances, DC air conditioners are efficient for residences or offices with a roof-top solar PV panel. This is because solar PV generates DC power, and air conditioners use a DC brushless motor.

When we modernize power systems using the market mechanism, demand response, and IT technology, but the system needs more flexibility, we need technological innovations using various electricity storage technologies. One of the most promising storage technologies is hydrogen production using extra wind/solar

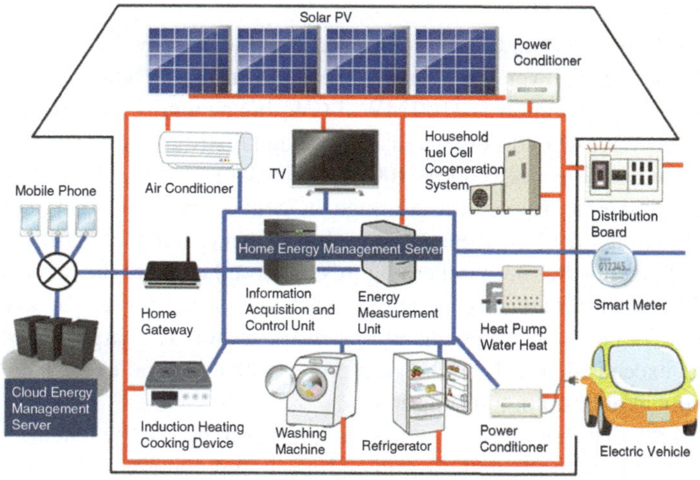

Fig. 5.4 Demand side management: The key technologies are energy measurement and information acquisition units

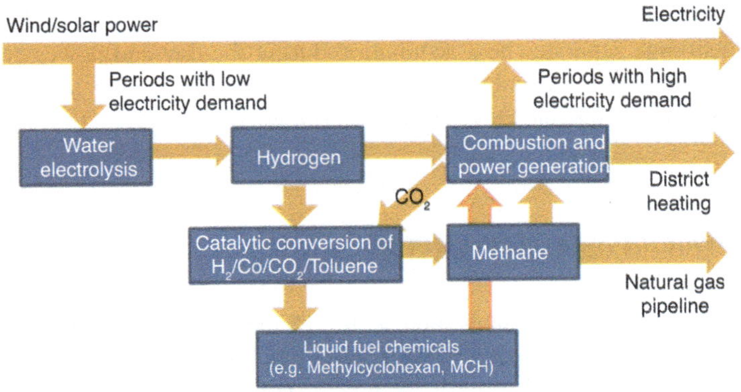

Fig. 5.5 Hydrogen production using renewable energy: Hydrogen production using extra wind/solar power is energy storage

power (see Fig. 5.5). In periods with low electricity demand, extra wind/solar power is used for water electrolysis to produce hydrogen instead of curtailing output power. The produced hydrogen can be stored in a high-pressure tank or as liquefied hydrogen. The stored hydrogen is combusted in periods with high electricity demand. Alternatively, methane is produced from a catalytic reaction of hydrogen and CO_2 in the exhaust gas of thermal power plants. Methane is liquefied at low temperatures and is stored in a tank in the same manner as storing natural gas. Another promising technology is a chemical reaction of toluene and hydrogen to synthesize methylcyclohexane (MCH), which is a liquid state in an ambient environment (normal temperature and atmospheric pressure). Therefore, MCH is easy to store and transport. This means that MCH could be exported, just like oil and natural gas.

Storage applications and technologies are characterized by the two-dimensional space of discharge power (MW) and discharge duration (hour) . For instance, application in a generation is (100 MW–1 GW, hour-month), system operation is (100 kW–100 MW, second-hour), transmission and distribution is (1 MW–100 MW, hour-day), and end-use is (100 W–100 kW, min-day) (see Fig. 5.6). Applications with large economic values are arbitrage in generation (80), load following in system operation (150), investment deferral in transmission and distribution (100), and off-grid in end-use (330), where figures in parentheses are the economic value in USD/MWh. Arbitrage in generation stores low-priced power for later sale at higher peak price; load following in system operation is charging power when generation exceeds demand, discharging power when demand exceeds generation, and investment deferral in transmission.

Meanwhile, distribution is relieving congestion on the grid by placing storage at the connection bottleneck, and off-grid end-use is supplying power using solar PV with storage for small-scale users.

The appropriate technologies for the applications are shown in Fig. 5.6 by colored box, based on their characteristics of discharge power (MW) and discharge duration

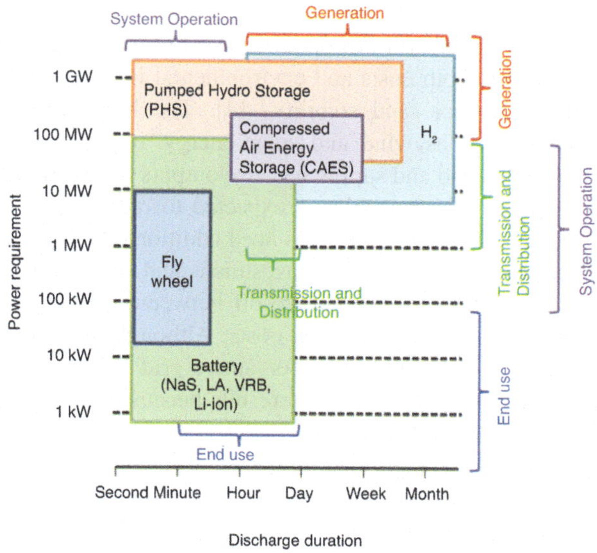

Fig. 5.6 Electricity Storage: Storage technologies are characterized by discharge power (MW) and discharge duration (hour)

(hour). The investment cost for Pumped Hydro Storage (PHS), H_2, and Compressed Air Energy Storage (CAES) are low in comparison to other storage technologies. These three technologies are suitable for arbitrage application in generation due to those most competitive levelized costs. PHS, CAES, and batteries (NaS) are suitable for load following application in system operation. PHS, H_2, and CAES are suitable for investment deferral application in transmission and distribution, and batteries (LA, VRB, and Li ion) are best for off-grid in end-use. LA, VRB, and Li-ion batteries are relatively expensive but easy for small-scale users to implement.

5.2 Simple Model of Smart Grid

Reducing CO_2 emissions is a pressing issue for our highly industrialized economies. There are several studies about future energy scenarios to reduce energy consumption [2–4]. One recent study shows that reducing energy consumption and CO_2 emissions is possible if sufficient investment is made immediately [4]. In this study, extrapolation of the current trend is a business-as-usual scenario, which is also called the Baseline (BASE) scenario, and a scenario halving CO_2 emissions at 2050, the BLUE MAP (BLUE) scenario [4].

A smart grid is an electricity network that uses digital technology to monitor and manage the transport of electricity from all generation sources to meet the varying electricity demands of end users. Such girds will be able to coordinate the needs and capabilities of all generators, grid operators, end users, and electricity market stakeholders in such a way that it can optimize asset utilization and operation and, in the process, minimize both costs and environmental impacts while maintaining system reliability, resilience, and stability [4]. The large-scale integration of variable renewables, such as wind and solar energy, requires demand response programs to balance demand and supply because outputs from renewables fluctuate in time. The integration of renewables is expected to reduce CO_2 emissions. On the other hand, demand response programs need additional hardware and software, as described later, which will increase investment and operational costs. In this sense, the smart grid faces a difficult trade-off between reducing CO_2 emissions and increasing demand response program costs. Although there are many studies about the estimation of cost and CO_2 for smart grid [1–3, 5–7], quantitative estimations have not provided satisfactory results because of inexplicit definitions and appropriate modeling for smart grids.

The smart grid is defined as the electric power system consistent with the BASE or BLUE scenarios with large-scale renewable integration and a demand response program's capability as a working hypothesis to avoid the difficulties mentioned above. We assume that transmission lines are enhanced in both scenarios, and accordingly we do not consider the cost related to transmission lines. Quantitative estimations of cost and CO_2 emissions are made using a linear programming approach for smart grids in the USA. Variants of scenarios are discussed, an estimation of electricity storage required for the grid integration of renewables, and a break-even point analysis is made to get the picture of a smart grid shortly [8].

We used the Integrated MARKAL-EFOM System (TIMES) as a linear programming tool to analyze energy technology systems [9–11]. The estimations of cost and CO_2 for smart grids were made using TIMES under the assumptions described below.

5.2.1 Assumptions

The smart-grid cost is categorized by the cost for electricity generation, operation and enhancement of transmission lines, and demand response programs. In this study, we prioritize the cost of electricity generation because electricity generation is the primary source of CO_2 emissions in the smart grid. We consider a demand response program, e.g., the load shifting, where the total electricity load is maintained but timing is shifted.

The following assumptions were made in our estimations.

1. The United States (US) is divided into ten regions: CALifornia (CAL), East South Central (ESC), MiD Atlantic (MDA), MounTaiN (MTN), East North Central (NEC), New England (NEE), PACific (PAC), South ATlantic (SAT), West North Central (WNC), and West South Central (WSC).
2. Fourteen technologies for electricity generation are considered: Coal Steam, Integrated Gasification Combined Cycle (IGCC), IGCC with Carbon Capture and Storage (CCS), Petro Steam, Combustion Turbine (natural gas), Natural Gas Combined Cycle (NGCC), NGCC with CCS, Nuclear, Hydro, Geothermal, Biomass, Solar (mostly Photo Voltaic, PV), and Wind farm.
3. Petroleum is produced in CAL and WSC and is traded to other regions. Natural gas is produced in ESC, MTN, and WSC and is traded to other regions. Coal is produced in ESC, MTN, NEC, and SAT and is traded to other regions. Biomass is produced in all regions.
4. Geothermal generation is used only in CAL, MTN, and PAC.
5. The installed capacities of generation technologies in 2005 are given as initial conditions for each region.
6. The BASE scenario and the BLUE scenario of ETP2010 are used as future scenarios for electricity demand, solar generation, wind generation, fuel price, and CO_2 price for each region from 2005 to 2050.
7. Various investment and operational costs as well as CO_2 emissions are estimated in order to satisfy the electricity demand scenario from 2005 to 2050 in each region for the US electricity sector.

5.2.2 Formulation of Linear Programming

The outline of our formulation of linear programming is described in the following. The objective function NPV is defined as the net present value of the annual cost.

$$NPV = \sum_{r=1}^{R} \sum_{t=1}^{T} \frac{C(r,t)}{(1+d)^{t-1}}, \tag{5.25}$$

$$\begin{aligned} C(r,t) = \sum_{k=1}^{K} [&i(r,t,k)I(r,t,k) + m_f(r,t,k)P(r,t,k) \\ &+ m_v(r,t,k)\sum_{s=1}^{S} A(r,t,k,s)] + p(r,t)E(r.t) \\ +\sum_{s=1}^{S} [&n(r,t)M(r,t) + d(r,t)T(r,t,s,i/e) + q(r,t)O(r,t)], \end{aligned} \tag{5.26}$$

where r, t, k, s, and d are region, time (year), generation technology (process), intra-day time slice, and discount rate, respectively. The annual cost $C(r, t)$ is defined using the decision variables and the coefficients. The decision variables $I(r, t, k)$, $P(r, t, k)$, $A(r, t, k, s)$, $E(r, t)$, $M(r, t)$, $T(r, t, s, i/e)$, and $O(r, t)$ are additional capacity, installed capacity, activity level, CO_2 emissions, mining, inter-regional trade, and import from overseas, respectively. The variables i and e in inter-regional trade $T(r, t, s, i/e)$ stand for inter-regional import and export, respectively. The coefficients $i(r, t, k)$, $m_f(r, t, k)$, $m_v(r, t, k)$, $p(r, t)$, $n(r, t)$, $d(r, t)$, and $q(r, t)$ are unit investment cost, unit fixed cost of operation and maintenance, unit variable cost of operation and maintenance, CO_2 price, mining cost, trade cost, and import cost, respectively.

The objective function NPV in Eq. (5.25) is minimized under the following constraints:

1. Satisfaction of demands

$$\sum_{k_e=1}^{K_e} A(r, t, k_e, s) \geq D(r, t, k, s), \tag{5.27}$$

   where k_e and $D(r, t)$ are end-use technology and electricity demand.
2. Capacity transfer

$$P(r, t, k) = \sum_{t'<t} I(r, t', k) + R(r, t, k), \tag{5.28}$$

   where $R(r, t, k)$ is the residual capacity due to investment before $t = 1$.
3. Use of capacity

$$A(r, t, k, s) \leq f(r, t, k, s) P(r, t, k), \tag{5.29}$$

   where $f(r, t, k, s)$ is the load factor. The estimation of the load factors for the variable renewables will be described later.
4. Energy balance

$$\sum_{k=1}^{K} cns(r, t, k, c) A(r, t, k, s) + T(r, t, s, e) < \sum_{k=1}^{K} prd(r, t, k, c) A(r, t, k, s) + M(r, t) + O(r, t) + T(r, t, s, i), \tag{5.30}$$

   where $cns(r, t, k, c)$ and $prd(r, t, k, c)$ are the amount of commodity c consumed to operate one unit of technology k and the amount of commodity c produced by one unit of technology k in region r and time t, respectively.

5.2.3 Parameters

Electricity demand for each region in the USA in the BASE and the BLUE scenarios is shown in Fig. 5.7a,b, respectively [4]. These demand profiles were calculated using various data for the electricity sector and all sectors. The price of CO_2 is not taken into account in the BASE scenario as shown in Fig. 5.7c. Figure 5.7d depicts the future scenario for fuel prices, which is identical for both scenarios. The technology parameters for various generation technologies were a standard set of parameters used in the Energy Technology Perspectives model [4]. Some cost parameters are decreased due to technological improvement and diffusion in markets.

The output power from a wind farm P_{wf} is calculated using the power curve for a given wind velocity v

$$P_{wf} = \begin{cases} P^r_{wf} v/v^r & (v < v^r) \\ P^r_{wf} & (v \geq v^r), \end{cases} \tag{5.31}$$

where the rated output P^r_{wf} and the rated velocity v^r were assumed as $P^r_{wf} = 2\,\text{MW}$ and $v^r = 16m/s$ m/s, respectively.

The output power from a PV module P_{pv} is calculated for a given solar irradiance I_s

$$P_{pv} = \eta I_s S_m, \tag{5.32}$$

where the efficiency η and module size S_m are assumed as $\eta = 0.2$ and $S_m = 1.0m^2$, respectively. This specification of PV panel is equivalent to the rated output $P^r_{pv} = 200\,\text{W/m}^2$ at solar irradiance $1\,\text{MW/m}^2$ with the size of $1\,\text{m}^2$.

We used time series of wind velocity and solar irradiance from the global atmospheric data ERA-40, the reanalysis dataset distributed by the European Center for Medium-Range Weather Forecasts. The dataset's features are (1) period: 1957-2002, (2) temporal resolution: 6 hours, and (3) spatial resolution: 2.5 degrees in latitude and longitude. The calculated time series of output for wind firm and PV at 12:00 pm in July are shown in Fig. 5.8 from 1992 to 2001 for MDA. The first 31 data points correspond to 12:00 pm of each day in July in 1992, the next 31 data points correspond to 12:00 pm of each day in July in 1993, and the final 31 data points correspond to 12:00 pm of each day in July 2001.

The average power μ_{wf} and μ_{pv} and those standard deviations σ_{wf} and σ_{pv} were estimated for each region. Then the load factors $f(r, t, wf, s) = \mu_{wf}/P^r_{wf}$ and $f(r, t, pv, s) = \mu_{pv}/P^r_{pv}$ and those standard deviations σ_{wf}/P^r_{wf} and σ_{pv}/P^r_{pv} were obtained for each region.

Fig. 5.7 Model parameters: (**a**) Electricity demands in the BASE scenario, (**b**) electricity demands in the BLUE scenario, (**c**) CO_2 price, and (**d**) fuel prices in the BASE and BLUE scenarios

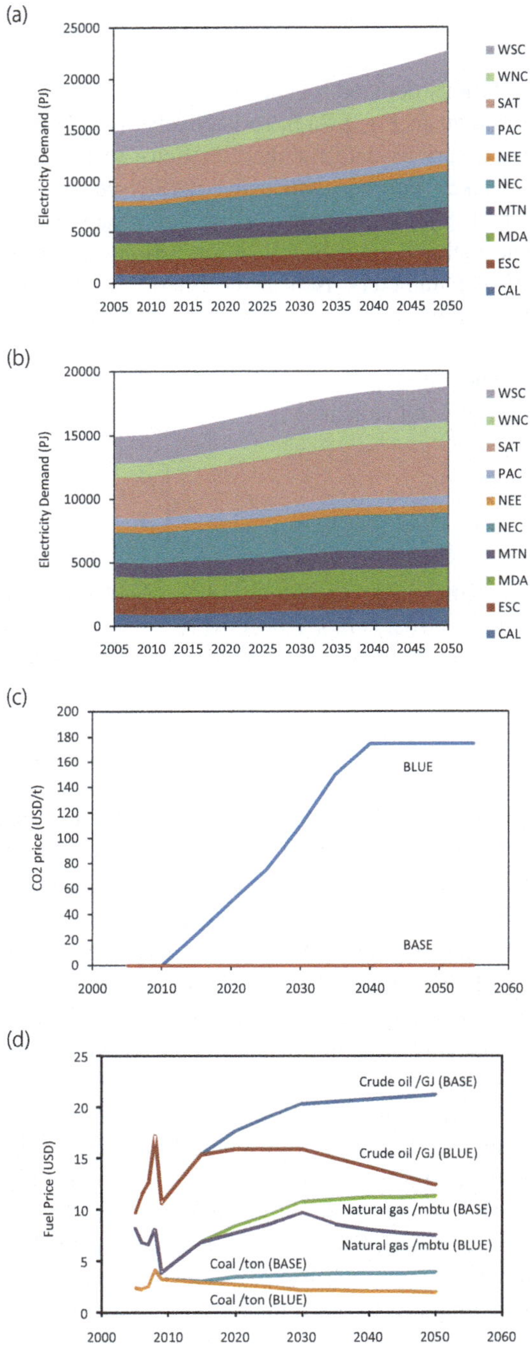

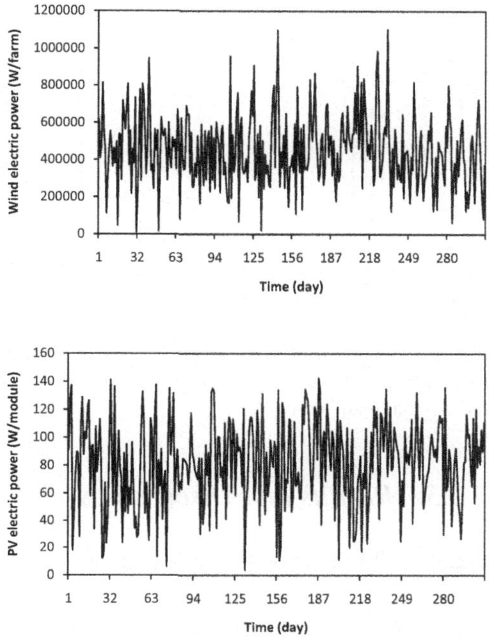

Fig. 5.8 Time series of wind and solar generation at 12:00 pm in July from 1992 to 2001 in MDA: ERA-40 data of the European Center for Medium-Range Weather Forecasts was used for the calculation

5.2.4 Load Curve and Load Shifting

The demand response program, including demand side management and load shifting, consists of the following technologies:

1. Advanced metering infrastructure with smart meters
2. Home energy management systems, in-house network and display, smart appliances, and electricity storage
3. Distribution management systems equipped with ubiquitous sensors
4. Data center, i.e., software and hardware for advanced analysis

The cost for the load shifting could be clearer because we cannot estimate the volume efficiency in producing various equipment and appliances. Thus, we do not consider the cost of the load shifting in the estimation. The cost for the load shifting will be discussed using the break-even point analysis in the last section.

Normalized hourly load curves for summer and winter in NEC are shown in Fig. 5.9. Blue and red lines show the load curves without and with load shifting, respectively. The load curve with load shifting was generated by reducing the standard deviation by 80% with the same average load from the original load curve. Peak load decreases from 10% to 15% for all regions after the load shifting. In this study, we used three seasons, i.e., summer, winter, and intermediate, and the intraday time slices S equal to 24.

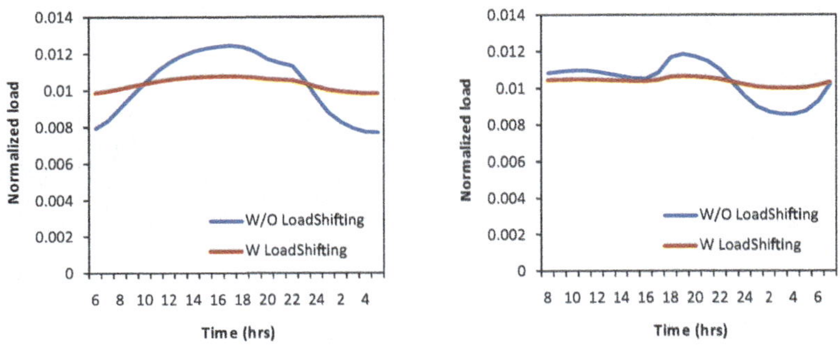

Fig. 5.9 Normalized load curves for summer and winter in NEC: The load curve with load shifting was generated by reducing standard deviation by 80% with the same average load from the original load curve

5.2.5 Results and Discussions

5.2.5.1 BASE and BLUE Scenarios

The electricity generation for each technology, which is required to satisfy the electricity demand in the BASE and the BLUE scenarios, is shown in Fig. 5.7a,b [4]. The obtained estimations for cost and CO_2 emissions for the BASE and BLUE scenarios will be used as reference values to quantify the effect of the smart grid in the next subsections.

5.2.5.2 BASE Smart Grids

We define the BASE smart grid as the BASE scenario with high renewables and load shifting through the demand response program. In Table 5.14, the conditions of the BASE smart grid are shown. The high share of renewables in the BASE smart grid is assumed to be equivalent to the electricity generation using renewables in the BLUE scenario. The shares of wind and solar are 13% and 5%, respectively, in 2050. It is expected that no significant cost is required to integrate such shares of the renewables into the grid.

Electricity generation for each technology is summarized in Table 5.15 for the BASE smart grid. Hydro, nuclear, coal-fired, and IGCC are widely used as baseload

Table 5.14 The condition of the BASE smart grids

	Without load shifting	With load shifting
Normal renewables	Base	SG1
High renewables	SG2	SG3

Table 5.15 Electricity generations for each technology in the BASE smart grids in 2050

	BASE	SG1	SG2	SG3
Hydro	5%	6%	5%	6%
Geothermal	1%	1%	1%	1%
Nuclear	10%	11%	10%	11%
Coal steam	49%	51%	42%	43%
IGCC	14%	15%	12%	14%
Combustion turbine	2%	1%	3%	1%
NGCC	2%	1%	2%	1%
Petro steam	0%	0%	0%	0%
Biomass	7%	6%	7%	6%
Solar	3%	3%	5%	5%
Wind	6%	6%	13%	13%

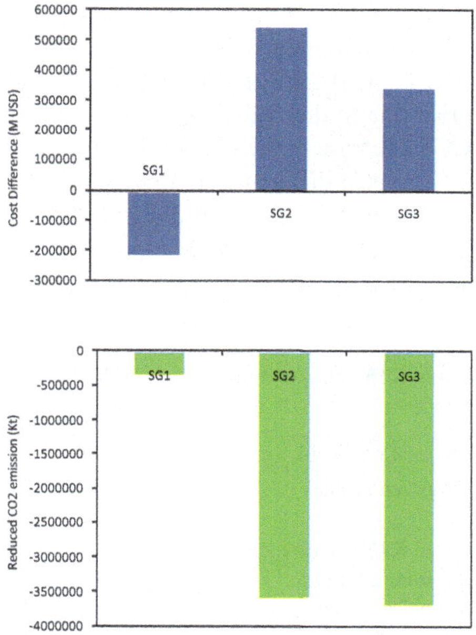

Fig. 5.10 Cost and CO_2 emissions in the BASE smart grids relative to the BASE scenario: CO_2 emissions decreased but keep increasing trend even in 2050

technologies due to the load shifting in SG1 and SG3. CCS is not penetrating the market because the CO_2 price is zero in the BASE smart grid.

The cost difference and the reduction of accumulated CO_2 emissions compared to the BASE scenario from 2005 to 2050 are shown in Fig. 5.10 for the BASE smart grids. SG2 costs are higher than the BASE scenario when only high renewables are considered. The increased cost is due to expensive renewables. SG3, where high renewables and load shifting are applied, shows least CO_2 emissions. It is noted that SG1 and SG3 are the least costly due to their load shifting. Because the widely used baseload technologies in the smart grid are relatively low cost.

The three BASE smart grids show less CO_2 emissions compared to the BASE scenario in Fig. 5.10, but temporal change of CO_2 emissions keeps increasing trend even in 2050.

5.2.5.3 BLUE Smart Grids

We define the BLUE smart grid as the BLUE scenario with high renewables and load shifting through demand response. In Table 5.16, the conditions of the BLUE smart grids are shown. The high renewables in the BLUE smart-grid case are assumed to be twice as high as the electricity generation using the renewables in the BLUE scenario. The share of wind and solar is 30% and 11%, respectively, in 2050. Some electricity storage would be required to integrate this share of renewables [12, 13]. An estimation of the electricity storage is described in the Sect. 5.2.5.4.

Electricity generation for each technology is summarized in Table 5.17 for the BLUE smart grids. IGCC and NGCC without CCS are not used in the BLUE smart grids due to the high CO_2 price. It is noted that IGCC with CCS and NGCC with CCS decrease in the case of the high renewables in SG5 and SG6.

The cost difference and reduction of accumulated CO_2 emissions compared to the BLUE scenario from 2005 to 2050 are shown in Fig. 5.11 for the BLUE smart grids. Both cost and CO_2 emissions are decreased in SG4 and SG6, where load shifting is applied.

Table 5.16 The condition of the BLUE smart grids

	Without load shifting	With load shifting
Normal renewables	Blue	SG4
High renewables	SG5	SG6

Table 5.17 Electricity generation for each technology in the BLUE smart grids at 2050

	BLUE	SG4	SG5	SG6
Hydro	6%	7%	5%	5%
Geothermal	3%	3%	2%	2%
Nuclear	33%	33%	32%	32%
Coal steam	0%	0%	0%	0%
IGCC with CCS	17%	17%	10%	11%
Combustion Turbine	2%	1%	2%	1%
NGCC	0%	0%	0%	0%
NGCC with CCS	9%	10%	2%	3%
Petro steam	0%	0%	0%	0%
Biomass	9%	8%	6%	6%
Solar	6%	6%	11%	11%
Wind	15%	15%	30%	30%

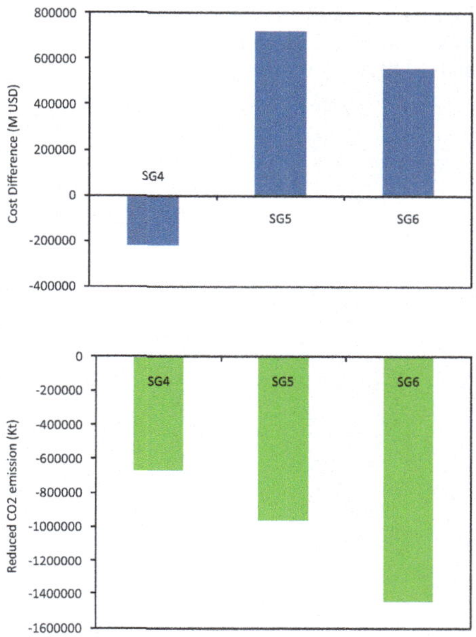

Fig. 5.11 Cost and CO_2 emissions in the BLUE smart grids relative to the BLUE scenario: The power sector becomes essentially decarbonized with the lowest emissions occurring in SG6 by 2050

The three BLUE smart grids emit less CO_2 emissions compared to the BLUE scenario as shown in Fig. 5.11, and by 2050, the power sector will become essentially decarbonized, with the lowest emissions occurring in SG6.

The load curve and electric energy supply for various technologies in WSC are shown in Fig. 5.12. When load shifting is applied, the baseload technologies, such as hydro, nuclear, and coal fire plants, including those with CCS, increase. At the same time, combustion turbine and biomass decrease due to the flatter profile. It is noted that petro steam and NGCC are almost zero with and without load shifting. This suggests that electricity storage is required for large-scale integration of renewables.

5.2.5.4 Electricity Storage for Renewable Integration

The short-term power fluctuation due to the variable renewables, such as wind farms and PV, will be absorbed by the peak load plants, such as combustion turbines, NGCC, and Petro steam plants. If the peak load plants' reserve capacity is insufficient, electricity storage, such as battery storage, could be used for renewable energy integration.

We define a wind site with a capacity of C_{wf} and a PV site with a capacity of C_{pv}. The wind site consists of many wind farms, and the PV site consists of many PV modules. A perfect correlation of the output is assumed inside these sites. Therefore, we call C_{wf} and C_{pv} the correlated capacity for wind farm and PV, respectively. The smoothing effect is expected between the sites because no correlation is assumed between the sites. The correlated capacities are essential parameters for the estimation of the capacity of electricity storage.

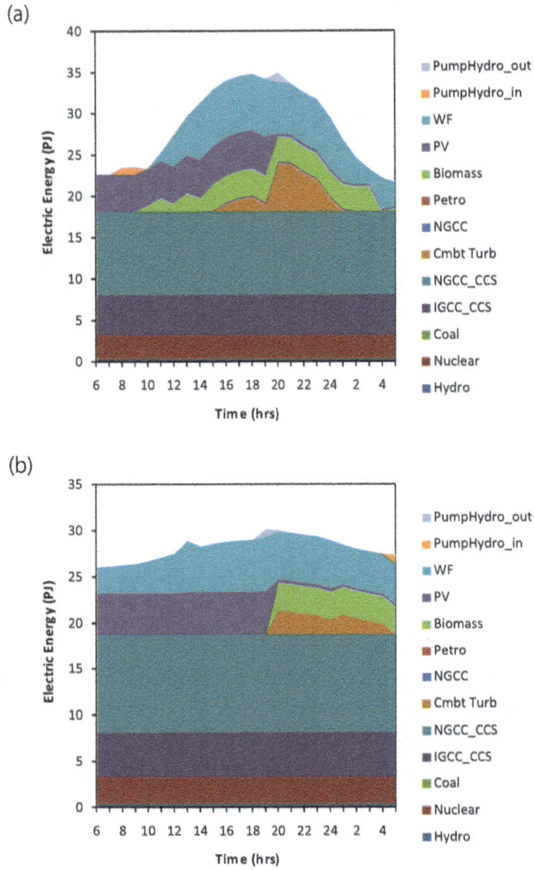

Fig. 5.12 Load Curve in WSC: (**a**) BLUE scenario and (**b**) SG4

The number of wind sites $n_{wf}(r, t; C_{wf})$ and the number of PV sites $n_{pv}(r, t; C_{pv})$ are calculated by

$$n_{wf}(r, t; C_{wf}) = P(r, t, wf)/C_{wf}, \tag{5.33}$$

$$n_{pv}(r, t; C_{pv}) = P(r, t, pv)/C_{pv}, \tag{5.34}$$

where $P(r, t, k)$ is installed capacity. The fluctuation of power for the wind site $\sigma_{wf}/P_{wf}^{r}C_{wf}$ and the PV site $\sigma_{pv}/P_{pv}^{r}C_{pv}$ are obtained using the standard deviation of load factor σ_{wf}/P_{wf}^{r} and σ_{pv}/P_{pv}^{r}.

Considering the smoothing effect between the sites, the aggregated fluctuation of power for the variable renewables is obtained:

$$\begin{aligned} F(r, t; C_{wf}, C_{pv}) = 3[n_{wf}(r, t; C_{wf}) \\ (\sigma_{wf}/P_{wf}^{r}C_{wf})^2 + n_{pv}(r, t; C_{pv})(\sigma_{pv}/P_{pv}^{r}C_{pv})^2]^{1/2}, \end{aligned} \tag{5.35}$$

where the factor 3 was chosen in order to set the confidence interval for the normal distribution to 0.9973.

Then, the reserve capacity of the peak load plants is calculated by

$$
\begin{aligned}
Q(r, t, peakplant) = {} & P(r, t, peakplant) - \\
& A(r, t, peakplant, peaktime)\frac{10^3 4S}{3.6 n_h},
\end{aligned}
\tag{5.36}
$$

where S is the number of the intra-day time slices, and n_h is the length of year in hours. Units for activity and capacity are PJ and GW, respectively. 10^3 and 3.6 are conversion factors between these units.

If the aggregated fluctuation of power from the variable renewables is smaller than the reserve capacity of the peak load plants, no electricity storage is required for the renewable integration. If this is not the case, the capacity of electricity storage for renewable integration is equal to $F(r, t; C_{wf}, C_{pv}) - Q(r, t, peakplant)$.

The capacity of electricity storage required for renewable integration for the BASE and BLUE smart grids in 2050 is shown as a function of the correlated capacity in Fig. 5.13. For instance, the estimation shows the electricity storage capacity is 9.0 GW for SG6 with a correlated capacity 0.6 GW.

The number of sales of electric vehicles in the USA is estimated to be about 3.3×10^6 vehicles in 2040 and about 5.0×10^6 vehicles in 2050 in the BLUE scenario

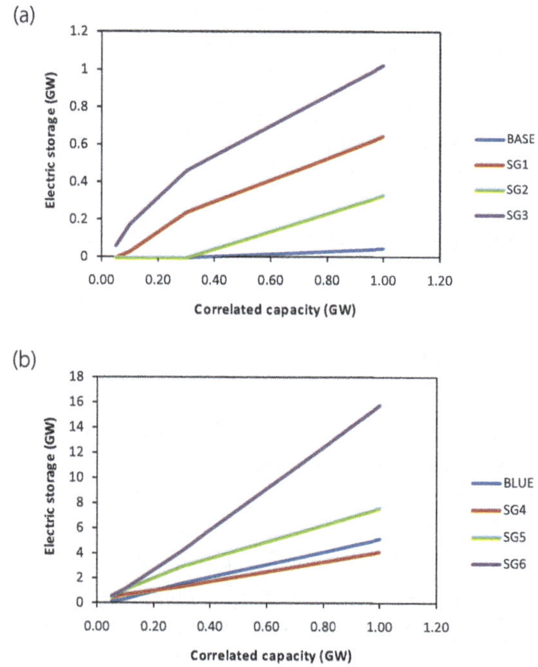

Fig. 5.13 Electricity storage at 2050: (**a**) BASE smart girds and (**b**) BLUE smart girds

[4]. Suppose we adopt the scenario that the lifetime of an electric vehicle is about 10 years, and all batteries used in electric vehicles sold in 2040 are used as storage in the power grid in 2050. In that case, an estimated capacity of electricity storage of 9.0 GW is obtained at the lowest cost. Here, we assumed that the battery's charging and discharging power is equal to 3 kW. Therefore, the cost of the electricity storage for the renewable integration is considered to be small.

5.2.5.5 Break-Even Point Analysis

The relation between the cost of reducing CO_2, the reduced generation cost, and the additional cost for load shifting is discussed in this section. The break-even point of the smart grid is obtained by

$$\Delta NPV = P_r(\Delta CO_2) + \Delta C_G = C_{LS}, \tag{5.37}$$

where ΔNPV is the difference in NPV of the smart grid and the NPV of the BASE or BLUE scenarios. This difference ΔNPV is divided into two parts: $P_r(\Delta CO_2)$ and ΔC_G, which represent the cost of reducing CO_2 and the reduced generation cost using the smart grid, respectively. Equation (5.37) shows that the difference ΔNPV has to be equal to C_{LS}, which is the additional cost due to the investment and operation required for load shifting. The higher the unit price of CO_2, the larger the margin for the investment cost C_{LS}.

The cost saving for load shifting C_{LS} is estimated to be $1.64 \times 10^{11} USD$ using Eq. (5.37) for SG5 and SG6 in Fig. 5.11. Assuming the number of households is kept constant between 2005 and 2050 at 10^9, the cost saving of $1.64 \times 10^{11} USD$ is equivalent to the yearly saving of $3.64 USD/household/year$ for 45 years. If the saving of $3.64 USD/household/year$ is not enough for utilities to cover the cost increase due to the load shifting equipments, the utilities and customers have to share the extra cost. The breakout of the cost to the four items in Sect. 5.2.4 is arbitrary in this study.

5.3 Summary

In this chapter, typical mathematical programming and heuristic optimization methods are described, followed by brief examples of these methods. Cost and CO_2 emissions were quantitatively estimated using linear programming for smart grids in the USA. Although this model has various limitations, it can easily estimate each power generation facility's installed capacity, cost, and greenhouse gas emissions in a smart-grid integrating renewable energy. The next issue to be considered is to calculate the optimal operation schedule of each power generation facility, considering the time variation of renewable energy output based on the estimated installed capacity of each power generation technology. This requires a unit commitment model, explained in the next chapter.

References

1. Y. Ikeda, "A New Method of Alloy Design Using a Genetic Algorithm and Molecular Dynamics Simulation and Its Application to Nickel-Based Supperalloys", Materials Transaction, JIM, Vol. 38, No. 9 (1997) pp.771–779
2. International Energy Outlook 2010, U.S. Energy Information Administration (2010).
3. European Climate Change Policy Beyond 2012, World Energy Council (2009).
4. Energy Technology Perspectives 2010, International Energy Agency (2010).
5. Power Delivery System of the Future: A Preliminary Estimate of Costs and Benefits, Electric Power Research Institute (2006).
6. Methodological Approach for Estimating the Benefits and Costs of Smart Grid Demonstration Projects, Electric Power Research Institute (2010).
7. The Smart Grid: an estimation of the Energy and CO_2 Benefits, Pacific Northwest national Laboratory (2010).
8. Y.Ikeda, U.Remme, D.Elzinga, S.Heinen, and P.Taylor, "Estimations of cost and CO2 emission for smart grids using a linear programing approach" IEEE Trondheim PowerTech pp.19–23, June (2011).
9. R. Loulou, U. Remme, A. Kanudia, A. Lehtila, and G. Goldstein, "Documentation for the TIMES Model", International Energy Agency (2005).
10. L.G. Fishbone, and H. Abilock, "MARKAL, A Linear Programming Model for Energy Systems Analysis: Technical Description of the BNL Version", International Journal of Energy Research, Vol. 5, 353–375, (1981).
11. E. van der Voort, E. Donni, C. Thonet, E. Bois d'Enghien , C. Dechamps, and J.F. Guilmot, "Energy Supply Modelling Package EFOM-12C Mark I, Mathematical description", Louvain-la-Neuve, Cabay: Commission of the European Communities, EUR-8896 (1984).
12. B. S. Borowy and Z. M. Salameh, "Dynamic Response of a Stand-Alone Wind Energy Conversion System with Battery Energy Storage to a Wind Gust", IEEE Transactions on Energy Conversion, 12 (1997) 73–78.
13. A. Cavallo, "Controllable and affordable utility-scale electricity from intermittent wind resources and compressed air energy storage", Energy, 32 (2007) 120–127.

Chapter 6
Grid Integration of Renewable Energy

6.1 Unit Commitment Model

Electricity demand will change from time to time. Many generating units are needed during heavy loads. However, keeping all of the generating units needed for heavy loads in operation during light loads is unnecessary. It is necessary to decide from an economic standpoint which generating units should start up and which should shut down in response to changes in load. This problem is called the unit commitment problem.

If the start-up and shutdown decisions are made only to minimize the fuel cost, obtaining an optimal solution is not difficult since each time can be treated as an independent problem. In practice, however, starting up a generating unit requires additional costs. In addition, once the unit shuts down, a certain amount of time is required to restart it. Since these time-related constraints must be considered, the start-up and shutdown cannot be decided independently for each hour but must serve the entire period under consideration. Such a problem requires a combinatorial optimization rather than a linear programming formulation.

6.1.1 Power Grid with Fluctuated Renewables

In order to mitigate climate change, it is essential to reduce the CO_2 emissions in various industry sectors by using renewable energies. The deployment of renewable energies, such as wind and photovoltaic (PV) systems, has begun in the power industry sector. The share of renewable energies in the total electricity generation will exceed 15% shortly.

Y. Ikeda, *Data Science of Renewable Energy Integration*,
Evolutionary Economics and Social Complexity Science 30,
https://doi.org/10.1007/978-981-99-8779-5_6

It is, however, noted that the output of renewable energies fluctuates significantly depending on weather conditions. With an increasing share of renewable energies, there is an increasing fluctuation in the electric power supply. At the same time, the share of energy contributed by various types of thermal power plants is expected to decrease. This implies that it will take more work to ensure the supply–demand balance of electric power in the currently used conventional power grids.

Smart grids have the potential to increase the capability to attain the supply–demand balance of electric power by using new technologies, such as smart meters, home energy management systems, building, and energy management systems, control of charge and discharge of electric vehicles, and various types of electric storages [1–4]. A power utility develops a plan to operate a power grid based on the forecast output of renewable energies for 24 hours. The forecast includes a time series of output scenarios and its error for wind power and PV systems in the power grid under consideration. An accurate forecast makes it possible to operate the power grid economically by avoiding overestimating the required adjustments for the supply–demand balance.

From our studies on the operation cost, the CO_2 emissions, and the peak load of the smart grid [5, 6], we identified the need to have a detailed understanding of the operation of the power grid. In this section, we describe a unit commitment model [7–12] with demand response for the integration of renewable energies to analyze requirements of the forecast output and its error.

6.1.2 Model: Objective Function

The purpose of our unit commitment model [13–15] is to obtain the time series for the operational state of thermal power plants that would maximize the profit of an electric power utility by taking into account both the forecast of output and its error for renewable energies [16–20] and the demand response of consumers on the change of electricity prices [21–26]. The model is formulated as a mixed integer linear programming problem.

The time series of the operational state of thermal power plant $i (i = 1, \cdots, N)$ is obtained by maximizing the objective function:

$$\begin{aligned} F(p_t^i, u_t^i, z_t^i, w_t^l) = &\sum_{t=1}^{T} d_t^{(f)} \sum_{l=1}^{L} w_t^l r^l \left(\frac{r^l}{\bar{r}}\right)^{\epsilon_d} \\ &- \sum_{t=1}^{T} \sum_{i=1}^{N} [b_i p_t^i + S_i z_t^i]. \end{aligned} \tag{6.1}$$

This objective function represents the profit of an electric power utility. The first term of the r.h.s. in Eq. (6.1) is the sales revenue, and the second is the operation cost. Here N, T, L are the number of thermal power plants, time horizon, and the

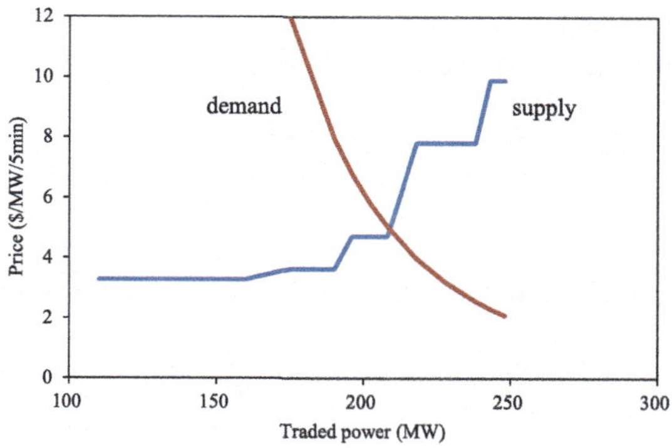

Fig. 6.1 Demand and supply: The price of electricity is determined at the intersection of the demand and supply curves

number of price levels, respectively. Continuous variable p_t^i is the output power variable of thermal power plant i, and integer variables u_t^i, z_t^i, w_t^l are the status production variable of thermal power plant i (1 = committed, 0 = decommitted), the start-up variable of thermal power plant i (1 = start up, 0 = others), and the demand response variable (1 = selected, 0 = not selected), respectively. Parameters S_i and b_i are the start-up cost of thermal power plant i and the fuel cost of the thermal power plant i, respectively. The forecasted demand and its error are indicated by $d_t^{(f)}$ and σ_d, respectively. Here (f) stands for forecasting.

Other parameters related to the demand response $\bar{r}$, r^l, ϵ_d are the average electricity price, the price level, and the price elasticity of demand, respectively. Figure 6.1 shows the qualitative relationship between demand and supply. If the electricity price r deviates from the average price $\bar{r}$, the demand d is changed from the average demand $\bar{d}$ as follows:

$$\frac{d}{\bar{d}} = \left(\frac{r}{\bar{r}}\right)^{\epsilon_d}. \tag{6.2}$$

The dependence of the demand d on the price r is depicted in Fig. 6.1 [27–31].

6.1.3 Model: Global Constraints

The sum of the demand response variable w_t^l has to satisfy the constraint

$$\sum_{l=1}^{L} w_t^l = 1 \tag{6.3}$$

to ensure that only a single price level r^l is selected. In addition to this constraint, the average of the selected price r^l has to be equal to the average price $\bar{r}$

$$\frac{1}{T}\sum_{t=1}^{T}\sum_{l=1}^{L} w_t^l r^l \leq \bar{r}. \tag{6.4}$$

Constraint that the daily aggregate demand does not change with demand response

$$\sum_{t=1}^{T} d_t^{(f)} = \sum_{t=1}^{T} \widetilde{d_t^{(f)}}, \tag{6.5}$$

$$\widetilde{d_t^{(f)}} = d_t^{(f)} \sum_{l=1}^{L} w_t^l \left(\frac{r^l}{\bar{r}}\right)^{\epsilon_d}, \tag{6.6}$$

must be satisfied. The constraint that the total supply is greater than the demand

$$\sum_{i=1}^{N} p_t^i + wd_t^{(f)} + pv_t^{(f)} + g_t - h_t \geq \widetilde{d_t^{(f)}} \tag{6.7}$$

must be satisfied. Here $wd_t^{(f)}$, $pv_t^{(f)}$, g_t, and h_t are the predicted wind power output, predicted solar power output, and discharged and charged power of the storage device, respectively. Note that an energy storage device means a battery or a pumped storage system.

If we consider the forecast error of demand σ_d, the forecast error of wind power σ_w, and the forecast error of PV σ_p, the constraint in Eq. (6.7) can be rewritten as

$$\frac{\sum_{i=1}^{N} p_t^i + wd_t^{(f)} + pv_t^{(f)} - d_t^{(f)} \sum_{l=1}^{L} w_t^l \left(\frac{r^l}{\bar{r}}\right)^{\epsilon_d}}{\sqrt{\sigma_d^2 + \sigma_w^2 + \sigma_p^2}} \geq \phi^{-1}(\alpha), \tag{6.8}$$

where α and $\phi(\cdot)$ are the probability to ensure the supply–demand balance and the cumulative distribution function, respectively. If the forecast error is distributed according to the normal distribution, the probability density function is

$$p(x) = \frac{1}{\sqrt{2\pi\sigma^2}} \exp\left[-\frac{(x-\mu)^2}{2\sigma^2}\right], \tag{6.9}$$

and the cumulative distribution function is written using the error function $erf[\cdot]$ as

$$\phi(x) = \frac{1}{2}\left(1 + \text{erf}\left[\frac{x-\mu}{\sqrt{2\sigma^2}}\right]\right), \tag{6.10}$$

Fig. 6.2 Distribution of forecast error: The Laplace distribution has a longer tail than the normal distribution

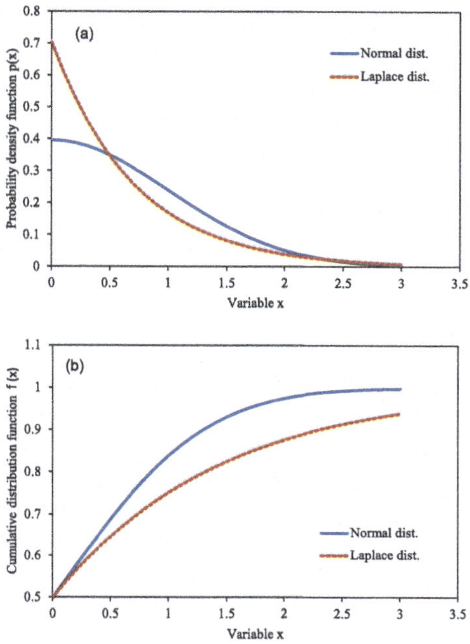

where μ and σ are a mean and a standard deviation, respectively. However, if the probability density function $p(x)$ is a Laplace distribution

$$p(x) = \frac{1}{2b} \exp\Big[-\frac{|x-\mu|}{b}\Big], \tag{6.11}$$

then the cumulative distribution function $\phi(x)$ is

$$\phi(x) = \frac{1}{2}\Big(1 + \text{sgn}(x-\mu)\Big(1 - \exp\Big[-\frac{|x-\mu|}{b}\Big]\Big)\Big). \tag{6.12}$$

Here a standard deviation is given by $\sigma = \sqrt{2}b$ and $\text{sgn}(x-\mu) = +(x \geq \mu), -(x < \mu)$. The functional forms for these distributions are depicted for $\mu = 0$ and $\sigma = 1$ in Fig. 6.2. It is noted that here, the Laplace distribution shows a distribution tail longer than the normal distribution. This implies that the probability of ensuring supply–demand balance differs for these two distributions. For the actual wind data distribution, see Sect. 3.3.

In addition, the energy storage device is subject to the constraint.

$$v_t c^{min} \leq g_t \leq v_t c^{max}, \tag{6.13}$$

$$(1 - v_t)c^{min} \leq h_t \leq (1 - v_t)c^{max}, \tag{6.14}$$

$$R^{min} \leq \sum_{s=1}^{t} (h_s \eta - g_s)\Delta_t \leq R^{max}, \tag{6.15}$$

must be satisfied. Here v_t, c^{min}, c^{max}, R^{min}, R^{max}, η, and Δ_t are the state variables of the storage device (1 = discharge, 0 = charge), minimum charge/discharge power, maximum charge/discharge power, minimum storage power, maximum storage power, efficiency, and time step.

Furthermore, the constraints on the reserve are

$$\sqrt{\varsigma_d^2 + \varsigma_w^2 + \varsigma_p^2} \leq \sum_{i=1}^{N} 0.05 u_t^i \bar{p}_{max}^i + 0.05 c^{max}. \tag{6.16}$$

Here, it is assumed that the maximum output $\bar{p}_{max}^i$ of thermal power unit i and 5% of the maximum charge/discharge power c^{max} of the storage unit can be used as reserve power. The ς_d, ς_w, and ς_p are the standard deviations of the short-term variations of demand, wind power, and solar power, respectively. However, we assume that 5% of the maximum output $\bar{p}_{max}^i$ is used as reserve power, and at the same time, we assume that the thermal power unit i is able to operate at its maximum output $\bar{p}_{max}^i$ as in the facility capacity constraint (6.17). Note that these assumptions imply coarse modeling.

6.1.4 Model: Local Constraints for Thermal Power Plants

The following constraints are used for each thermal power plant as typical constraints in a unit commitment model.

Generation Capacity
The output power p_t^i has to be between the maximum output power $\bar{p}_{max}^i$ and the minimum output power $\bar{p}_{min}^i$ when the operation is in steady state:

$$\bar{p}_{min}^i u_t^i \leq p_t^i \leq \bar{p}_{max}^i u_t^i. \tag{6.17}$$

Ramp-up Limit
The increase in the output of thermal power plant i has to be smaller than the maximum ramp-up speed Δ_+ when the unit is up at the previous time step and is smaller than the minimum output power $\bar{p}_{min}^i$ when the unit is down at the previous time step:

$$p_t^i - p_{t-1}^i \leq u_{t-1}^i \Delta_+^i + (1 - u_{t-1}^i) \bar{p}_{min}^i. \tag{6.18}$$

Ramp-down Limit
The decrease in the output of thermal power plant i has to be smaller than the maximum ramp-down speed Δ_- when the unit is up at time step t and is smaller

than the maximum output power $\bar{p}^i_{max}$ when the unit is down at time step t:

$$p^i_t - p^i_{t-1} \geq -u^i_t \Delta^i_- - (1 - u^i_t)\bar{p}^i_{max}. \tag{6.19}$$

Minimum Up-time Constraint

Thermal power plant i has to be operated longer than the minimum up-time requirement τ^i_+, once the unit is up:

$$\begin{aligned} u^i_t &\geq u^i_s - u^i_{s-1}, \\ s &\in [t - \tau^i_+, t - 1]. \end{aligned} \tag{6.20}$$

Minimum Downtime Constraint

Thermal power plant i has to be stopped longer than the minimum downtime requirement τ^i_-, once the unit is down:

$$\begin{aligned} u^i_t &\leq 1 + u^i_s - u^i_{s-1}, \\ s &\in [t - \tau^i_-, t - 1]. \end{aligned} \tag{6.21}$$

Constraint on the Start-up Variable

The start-up variable z^i_t has to satisfy the following constraints by definition:

$$\begin{aligned} z^i_1 &\geq u^i_1, \\ z^i_t &\geq u^i_t - u^i_{t-1} (t > 2). \end{aligned} \tag{6.22}$$

6.2 Application 1: A Small Power Grid

6.2.1 Analyzed Power Grid

We analyzed the operational state of thermal power plants by considering both forecast output and its error for renewable energies and the demand response of consumers, using the unit commitment model described in Sect. 6.1. In this chapter, we consider a small power grid consisting of thermal power plants, PV systems, and wind farms, although the model is easily extendible to a larger power grid. The number of thermal power plants is 12, and parameters are given in Table 6.1. The installed capacity for each PV system and wind farm is 30 MW. The intra-day peak demand is about 170 MW. We assume the forecast errors σ_w and σ_p are both $10\%(3MW)$ of the installed capacity in the reference case. In addition to the reference case, we analyze two more cases $\sigma_w = 6$ MW and $\sigma_w = 9$ MW, while σ_p remains unchanged. The forecast error for the demand is $\sigma_d = 0$ for all cases. Scenarios for the demand, the PV output power, and the wind output

Table 6.1 The parameters of thermal power plants

i	b_i	$\bar{p}^i_{max}$	$\bar{p}^i_{min}$	Δ_+	Δ_-	τ^i_+	τ^i_-	S_i
1	3.0	50.0	25.0	0.5	0.5	3.0	3.0	1000.0
2	3.0	50.0	25.0	0.5	0.5	3.0	3.0	1000.0
3	3.3	15.0	7.5	0.5	0.5	3.0	3.0	200.0
4	3.3	15.0	7.5	0.5	0.5	3.0	3.0	200.0
5	4.3	6.0	2.0	5.0	5.0	3.0	3.0	100.0
6	4.3	6.0	2.0	5.0	5.0	3.0	3.0	100.0
7	4.3	6.0	2.0	5.0	5.0	3.0	3.0	100.0
8	7.1	10.0	4.0	5.0	5.0	3.0	3.0	200.0
9	7.1	10.0	4.0	5.0	5.0	3.0	3.0	200.0
10	7.1	10.0	4.0	5.0	5.0	3.0	3.0	200.0
11	9.0	5.0	2.5	0.5	0.5	3.0	3.0	100.0
12	9.0	5.0	2.5	0.5	0.5	3.0	3.0	100.0

power in the reference case are shown in Fig. 6.3. While wind scenario 1 is used in the reference case, wind scenario 2, where wind power suddenly decreases in the evening, discusses the fluctuation of wind power and the effect on the demand response.

The marginal cost C_t^M, the spinning reserve R_t^S, and the load frequency control (LFC) margin R_t^L are calculated using the obtained time series $p_t^i (i = 1, \cdots, N)$. The marginal cost C_t^M is defined by

$$C_t^M = \max\{b_i | u_t^i = 1\}. \tag{6.23}$$

The spinning reserve R_t^S and the LFC margin R_t^L are defined as proxy quantities in this chapter as follows:

$$R_t^S = \sum_{i=1}^{N} p_t^i - \sum_{i=1}^{N} \hat{p}_t^i, \tag{6.24}$$

$$R_t^L = \sum_{i=1}^{N} \min\{\bar{p}^i_{max} - p_t^i, 0.05\bar{p}^i_{max} | u_t^i = 1\}. \tag{6.25}$$

Here, $\hat{p}_t^i$ is the output power of thermal power plant i obtained in the optimization with the constraint of Eq. (6.7). The spinning reserve R_t^S is ready to generate power to absorb the fluctuation instantaneously. On the other hand, the LFC margin R_t^L is the remaining capacity that is able to increase output with the constraints of Eqs. (6.18) and (6.19).

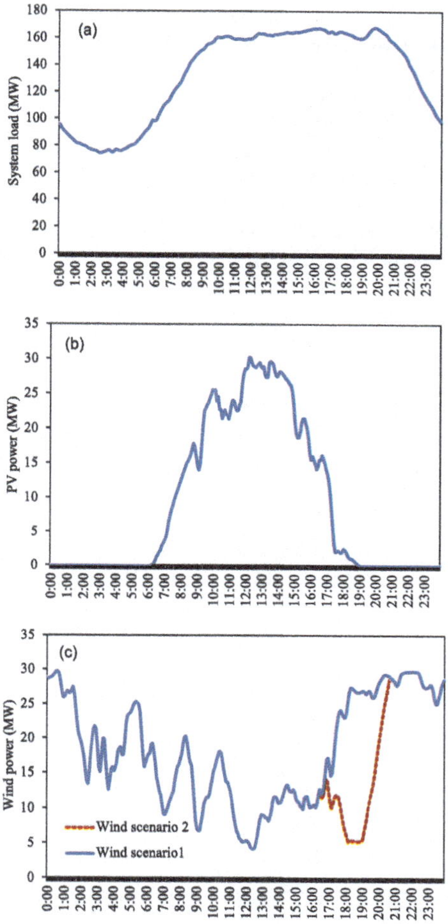

Fig. 6.3 Scenario in the reference case: (**a**) Load, (**b**) solar PV output, and (**c**) wind power output

6.2.2 Results and Discussions

We discuss the requirements for the forecast output and its error for the integration of renewable energies in this section. The unit commitment model formulated as a mixed integer linear programming problem in Eqs. (5.25) to (6.22) was solved to analyze the power grid described in Sect. 6.2.1 using a commercial solver [32, 33].

6.2.2.1 Reference Case

The results of the reference case are shown in Fig. 6.4. Figure 6.4a depicts the share of the thermal power plants, the wind power, and the PV systems to satisfy the given demand. The thermal power plants are numbered in the order of increasing operation

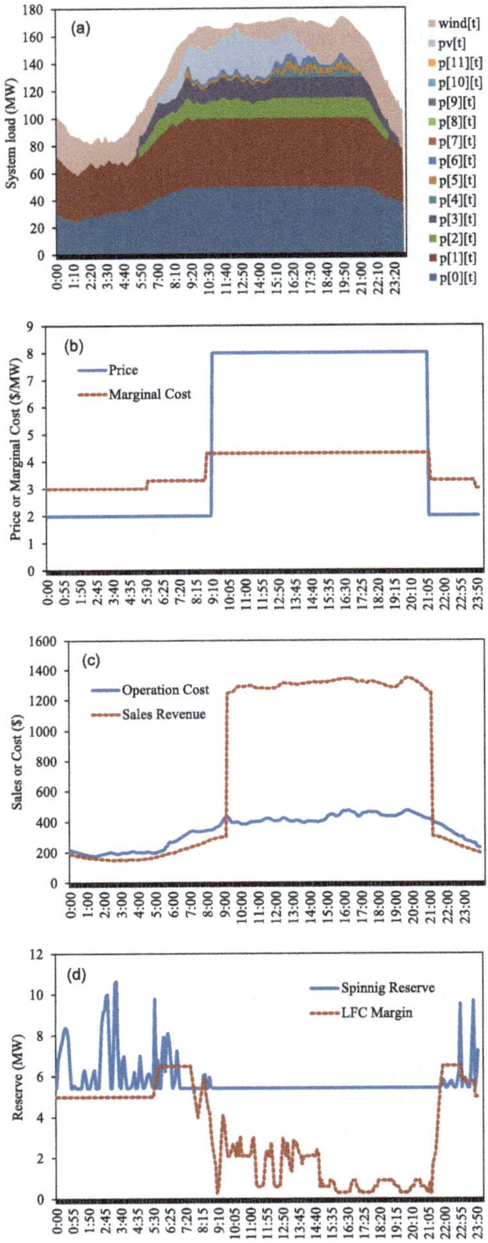

Fig. 6.4 Results of the reference case: (**a**) the share of the thermal power plants, the wind power, and the PV systems to satisfy the given demand, (**b**) the marginal cost C_t^M, (**c**) sales revenue and cost, and (**d**) the spinning reserve R_t^S and the LFC margin R_t^L

cost, i.e., the merit order. The fluctuation of the wind and the PV outputs was absorbed by starting the thermal power plants serially in the merit order. Figure 6.4b shows that the marginal cost C_t^M is high from 9:00 to 20:00. This is because some additional thermal power plants are operating in this period. The price structure

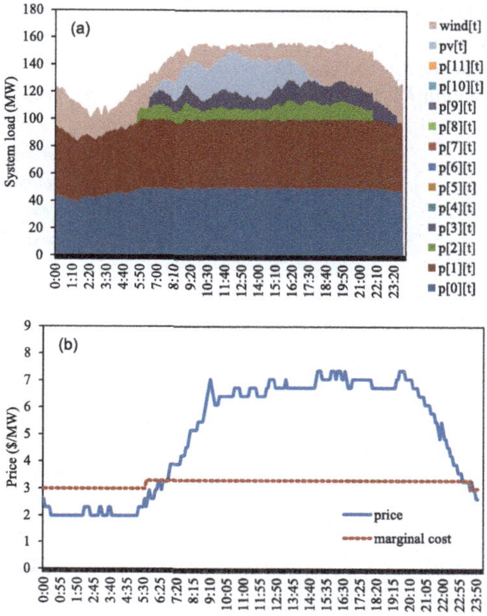

Fig. 6.5 Demand response: The load profile and the marginal cost become flatter due to the demand response. See Fig. 6.4 for comparison

where the daytime rate is higher than the nighttime rate is obtained as a result of maximizing the profits of an electric power utility. Reflecting the price structure, the sales revenue is high during the daytime, as shown in Fig. 6.4c. Figure 6.4d depicts that the spinning reserve R_t^S has a flat structure during the daytime, while the LFC margin R_t^L is high during nighttime.

6.2.2.2 Demand Response

It is well known that the price elasticity of demand ϵ_d is small because the share of expenditure for electricity in a household budget is low [34]. The effects of the demand response with $\epsilon_d = -0.30$ are shown in Fig. 6.5. Figure 6.5a depicts that the load profile becomes flat, and the peak load is reduced by about 20 MW. The profile of the marginal cost becomes flat, and the price structure is significantly changed as shown in Fig. 6.5b. A larger effect to the demand response is expected for a larger ϵ_d.

6.2.2.3 Effects of the Forecast Error

The effects of the forecast error on the operation cost and reserves are analyzed here. The forecast errors σ_w and σ_p are 10%(3) MW of the installed capacity in the reference case. In addition to the reference case, we analyzed two more cases: $\sigma_w = \sigma_p = 6$ MW and $\sigma_w = \sigma_p = 9$ MW. Figure 6.6a depicts that

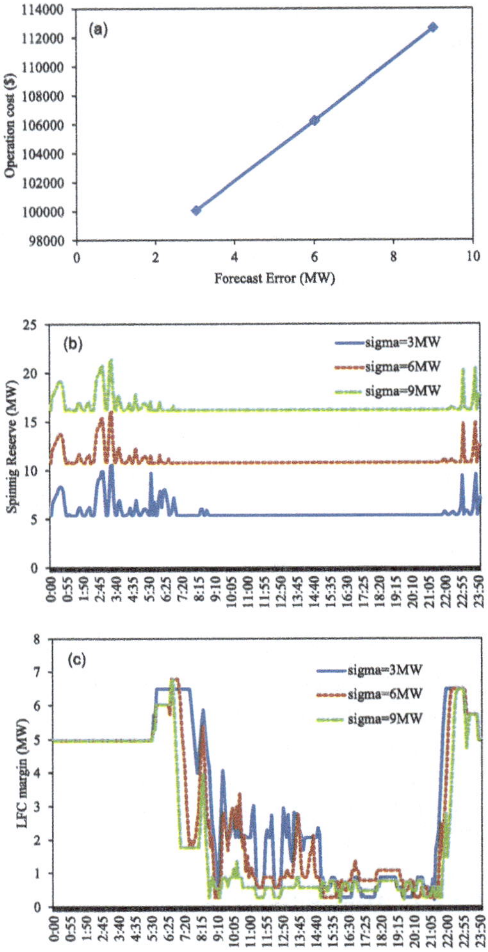

Fig. 6.6 Effect of forecast error: (**a**) cost, (**b**) spinning reserve R_t^S, and (**c**) LFC margin R_t^L

the operation cost increases with the forecast error σ_w. This tendency is also true for the spinning reserve R_t^S as shown in Fig. 6.6b. The intra-day structure is the same for the three cases, but the level increases by 5.4 MW as σ_w increases by 3 MW. This is reasonable because a larger reserve is required to absorb the larger fluctuation. Figure 6.6c shows that the LFC margin R_t^L decreases during daytime, while the margin for all three was relatively high during nighttime, independently of the forecast error. This implies that the demand during nighttime is relatively low, and therefore, the remaining capacity is large in the operating thermal power plants.

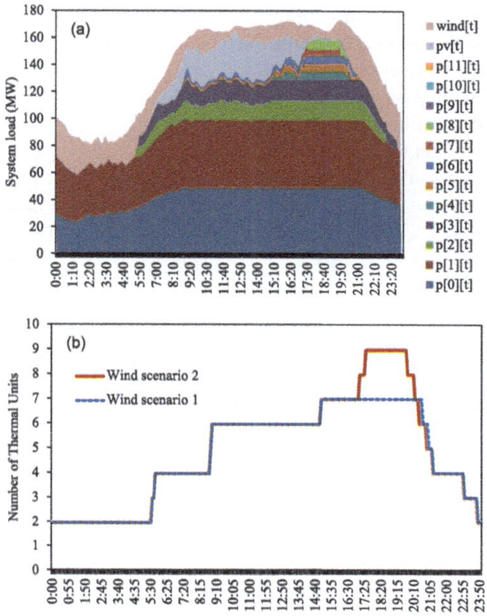

Fig. 6.7 Wind scenario 2: The wind power suddenly decreases in the evening

6.2.2.4 Effects of Sudden Decrease in Wind Power

In wind scenario 2, the wind power suddenly decreased in the evening compared to wind scenario 1. We expect that more thermal power plants will be operated to generate power to absorb this sudden decrease in wind power. The effects of the sudden decrease in wind power were analyzed, and the results are shown in Fig. 6.7. Figure 6.7a depicts an increase in operating thermal power plants between 17:00 and 20:00. This increase is clearly shown in Fig. 6.7b. Nine thermal power plants are in operation during 17:00 and 20:00 for scenario 2, whereas just seven plants operate in the same period for scenario 1. The increase in the number of operating thermal power plants in such a short period does not significantly affect the total operation cost. However, it is noted that the power utility must continue to use thermal power plants to ensure supply–demand balance even after installing many wind farms or PV systems. Therefore, the substitution of thermal power plants for wind farms or PV systems is expected to be low, although this issue has to be studied quantitatively using actual data.

6.2.2.5 Demand Response and a Sudden Decrease in Wind Power

Finally, the effects of the demand response with $\epsilon_d = -0.30$ in wind scenario 2 were analyzed. Suppose the number of operating thermal power plants is reduced by the demand response. In that case, the economic value of the wind farms will increase due to the high number of thermal power plants substituted by wind farms.

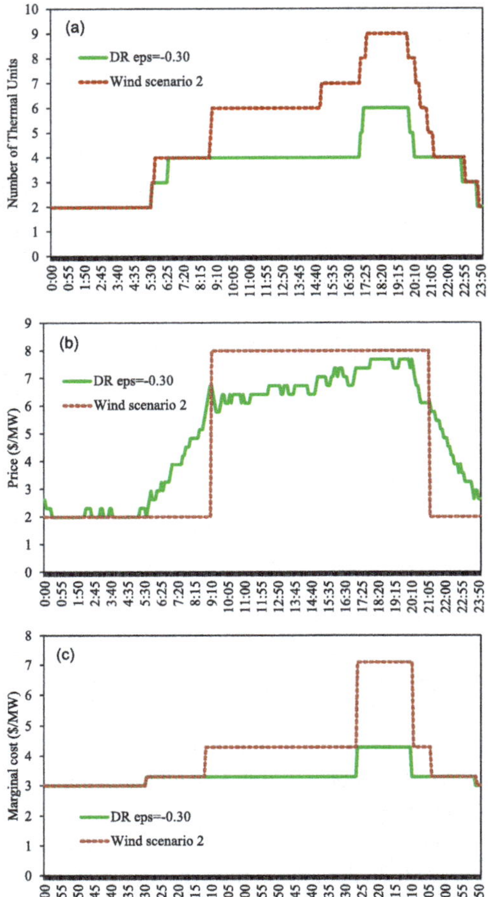

Fig. 6.8 Wind scenario 2 with demand response: (**a**) the number of operating thermal power plants, (**b**) the price structure with and without the demand response, and (**c**) the marginal cost

Figure 6.8a shows that the demand response reduces the number of operating thermal power plants during 9:00 and 20:00. This implies that the effects of the demand response can be clearly observed in the number of operating thermal plants. Figure 6.8b depicts the price structure with and without the demand response. The demand response smoothens the rectangular structure for wind scenario 2. The price between 17:00 and 20:00 is kept at a high level. It is recognized that the marginal cost is high during 17:00 and 20:00 in Fig. 6.8c. Therefore, the high price during 17:00 and 20:00 is a consequence of the high marginal cost during the same period. Because the operation cost was not increased in scenario 2, we cannot expect the demand response to cause a reduction in the operating cost in this case. However, a power utility has to continue to use thermal power plants to ensure supply–demand balance; some of these plants can be decommissioned after installing a large number of wind farms or PV systems, if the demand response is applied with an appropriate price structure.

6.2.3 Summary 1

The output of renewable energy fluctuates significantly depending on weather conditions. Therefore, it will not be easy to ensure the supply–demand balance of electric power using currently used conventional power grids. We have developed a unit commitment model to analyze the requirements of the forecast output and its error for renewable energies. Our model obtains the time series for the operational state of thermal power plants that would maximize the profits of an electric power utility by considering both the forecast error for renewable energies and the demand response of consumers. The model is formulated as a mixed integer linear programming problem.

We considered a power grid consisting of thermal power plants, PV systems, and wind farms. The unit commitment model was solved using a commercial solver to analyze the power grid.

First, the essential property of the model was discussed using the results of the reference case. The fluctuation of the wind and the PV outputs was absorbed by starting the thermal power plants serially in the merit order. The price structure where the daytime rate was higher than the nighttime rate was obtained as a result of maximizing the profits of an electric power utility.

Next, the effects of the forecast error on the operation cost and reserves were analyzed. In addition to the reference case ($\sigma_w = 3\,\text{MW}$), we analyzed two more cases, namely, $\sigma_w = \sigma_p = 6\,\text{MW}$ and $\sigma_w = \sigma_p = 9\,\text{MW}$. We confirmed that the operation cost increased with the forecast error. The intra-day structure of the spinning reserve was the same for the three cases, but the level increased with the forecast error. The LFC margin decreased during the daytime, while it remained relatively high during nighttime, independently of the forecast error.

Then, the effects of a sudden decrease in wind power were analyzed. More thermal power plants will have to be operated to generate power to absorb this sudden decrease in wind power. The increase in the number of operating thermal power plants within a short period did not significantly affect the total operation cost.

Finally, the effects of the demand response in the case of a sudden decrease in wind power were analyzed. We confirmed that the demand response reduced the number of operating thermal power plants. This implies that the demand response with an appropriate price structure controls the number of operating thermal power plants. A power utility must continue to use thermal power plants to ensure supply–demand balance; some of these plants can be decommissioned after installing many wind farms or PV systems if the demand response is applied with an appropriate price structure.

In future work, we intend to study the demand response technology and the forecast output of wind farms and PV systems using the developed unit commitment model for maximizing the renewable energy integration in an actual power grid. We also plan to quantitatively study the issue of substituting existing thermal power plants with renewable energy using a large set of relevant data [35].

6.3 Application 2: Two Large Power Grids

6.3.1 *Integrated Operation of Power Grids*

In order to reduce greenhouse gas emissions, renewable energies such as wind and solar power are being introduced in Europe and other parts of the world. However, since the output of these renewable energies fluctuates depending on weather conditions, grid interconnection is expected to become increasingly difficult as their large-scale introduction progresses. Therefore, there are growing expectations for smart grids that improve the power grid's ability to adjust supply and demand through active demand [5, 6]. Here, demand activation refers to the response of consumer behavior to electricity prices and is referred to as demand response in the following.

In this section, the power grids in the service areas of Tokyo Electric Power Company will be referred to as "the Tokyo power grid" and the service areas of Tohoku Electric Power Company as "the Tohoku power grid". In Japan, the massive introduction of renewable energy is being considered after the Great East Japan Earthquake of 2011. In particular, photovoltaic power generation installed on building roofs is attracting attention in the Tokyo power grid, and wind power generation utilizing favorable wind conditions is attracting attention in the Tohoku power grid.

In this new power grid, thermal power generation is to be operated in such a way as to achieve a balance between supply and demand while taking into account the output fluctuation of renewable energy sources [20] and demand response [21–23]. The power supplier is required to operate thermal power units to achieve a supply–demand equilibrium while taking into account the following factors. Therefore, based on the unit commitment model [7] of thermal power generators, a model that takes into account the forecasted output of renewable energies [16–19] and demand response [24, 25, 27, 28] has been developed and a model that considers demand response has been actively studied in recent years.

Using the unit commitment model described in Sect. 6.1 [14, 15], we analyzed the effects of the operation of energy storage devices and demand response during the lowest demand period and the relationship between the suppression of renewable energy output and grid interconnection throughout the year under the situation in which a large amount of renewable energy have been introduced in FY2030. We analyzed the operation plans of the Tokyo power grid and the Tohoku power grid to study the economics of the integrated operation of the two power grids.

We note that the lamp-up limit in Eq. (6.18) and the ramp-down limit in Eq. (6.19) were not used in this section's analysis.

6.3.2 Analysis of Tokyo and Tohoku Power Grids

6.3.2.1 Analysis Condition

The parameters for the entire power grid in the Tokyo power grid and the Tohoku power grid are shown in Table 6.2. The planning period T is assumed to be one day in the analysis of Sect. 6.3.2.2. In the analysis of Sects. 6.3.3 and 6.3.4, the planning period T is assumed to be one month, and each month is analyzed separately. For nuclear power, the same installed capacity as before the 2011 earthquake was assumed. Nuclear power and hydropower other than pumped storage are assumed to operate at constant output as the baseload. The parameters b_i, S_i, $\bar{p}^i_{max}$, $\bar{p}^i_{min}$, τ^i_+, and τ^i_- of thermal power units are taken from ESPRIT [36]. The elasticity ϵ_d was set to obtain the equivalent peak demand reduction rate as in the demand response experiment.

The planned installation capacity of photovoltaic power generation in each electric power grid in FY2030 is calculated as follows. The planned installation capacity of PV power generation in each power grid in FY2030 was calculated by multiplying the installed capacity of PV power generation in the 2010 Basic Energy Plan by 1.5 and then by proportionally dividing the electricity demand of each system. The calculated installed capacity of PV power generation is 26,481 MW in the Tokyo power grid and 7,741 MW in the Tohoku power grid. The forecasted value $pv_t^{(f)}$ was assumed to be the actual value of PV power output (1-hour value) [4]. The parameters assumed for the forecast error and short-term variability of PV power generation are shown in Table 6.3.

Next, the planned installed capacity of wind power generation in each power grid in FY2030 is calculated as follows. The installed capacity of wind power generation was calculated by dividing the installed capacity of the wind power generation target of the Japan Wind Power Association by the amount of wind power resources in each power grid. The calculated installed capacity of wind power generation is 1,009 MW in the Tokyo power grid and 9,081 MW in the Tohoku power grid. The forecast value $wd_t^{(f)}$ was assumed to be the actual output of wind power (1-hour

Table 6.2 Parameters of the whole system

Symbol	Parameters
N	91 (Tokyo), 43 (Tohoku)
T	48 (1 day), 1440 (30 days)
Δ_t	30 min.
L	20
$\bar{r}$	16 JPY/kWh
r^1	14 JPY/kWh
r^L	18 JPY/kWh
ϵ_d	0 (without DR), −0.3 (with DR)
α	1.28 (90 % CL for normal distribution)
Baseload	21000.0 MW (Tokyo), 4748.0 MW (Tohoku)

Table 6.3 Parameters of forecast error and fluctuation

Symbol	Parameters
σ_p	20% of PV output power
σ_w	20% of wind output power
ς_p	10% of PV output power
ς_w	10% of wind output power
σ_d	10% of system load
ς_d	3% of system load

Table 6.4 Parameters of electric storage device

Symbol	Tokyo	Tohoku
c^{min}	0.0 MW	0.0 MW
c^{max}	11800.0 MW	712.0 MW
R^{min}	0.0 MWh	0.0 MWh
R^{max}	118000.0 MWh	7120.0 MWh
η	0.7	0.7

value) [36]. The parameters assumed concerning the forecast error and short-term variability of wind power are shown in Table 6.3.

Electricity demand in the Tokyo power grid and the Tohoku power grid in FY2030 is assumed to be the forecasted value $d_t^{(f)}$ based on the actual FY2010 results; however, the decrease in demand after the 2011 March earthquake was adjusted. The historical electricity demand for FY 2010 for 10 electric power grids in Japan, including those served by Tokyo Electric Power Company and Tohoku Electric Power Company, is publicly available. The parameters assumed for the forecast errors and short-term fluctuations in electricity demand are shown in Table 6.3.

Furthermore, for multiple pumped storage power plants in each power grid, the capacities of the individual pumped storage power plants [36] were added together and treated as a single storage device. The parameters of the energy storage devices are shown in Table 6.4. The amount of stored electricity at time $t = 1$ and at time $t = T$ was assumed to be equal to half of the maximum stored electricity R^{max}.

The forecasted scenarios of solar power generation, wind power generation, and electricity demand in the Tokyo power grid on May 1, 2030, are shown in Fig. 6.9 and R^{max} in Fig. 6.9. The forecasted scenario for May 1, 2030, in the Tohoku power grid is shown in Fig. 6.10. In the Tokyo power grid, photovoltaic power output is large due to the widespread use of photovoltaic panels installed on building roofs. However, the amount of wind power installed is small. On the other hand, in the Tohoku power grid, wind power generation has been introduced on a large scale, taking advantage of favorable wind conditions, and its output is larger than that of solar power generation, which is a characteristic feature.

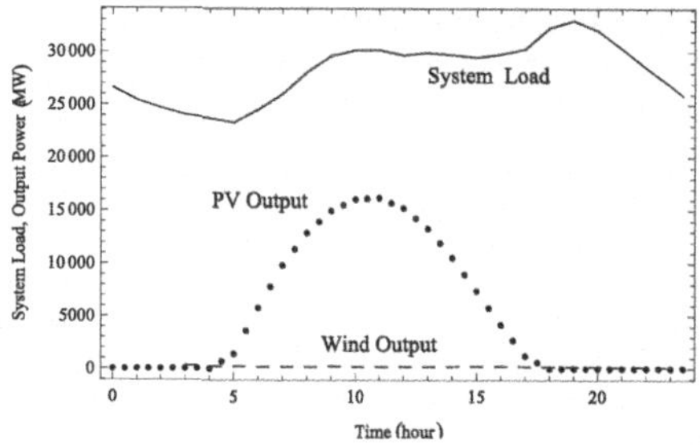

Fig. 6.9 Forecasted scenario: Solar PV output, wind power output, and system load on May 1, 2030, for the Tokyo power grid

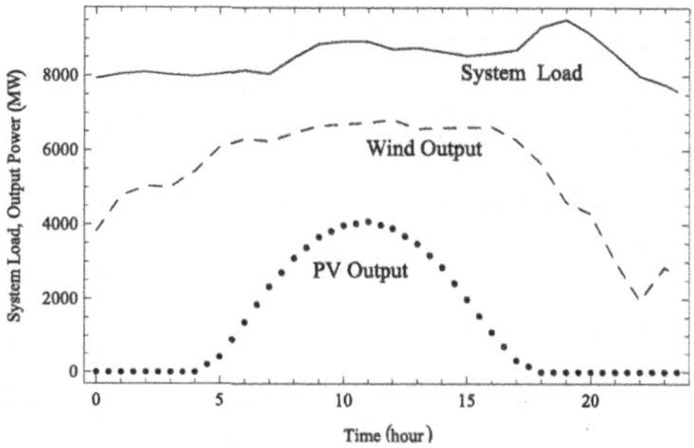

Fig. 6.10 Forecasted scenario: Solar PV output, wind power output, and system load on May 1, 2030, for the Tohoku power grid

6.3.2.2 Thermal Power Unit Operation Plan for the Period of Least Demand

Since the May holiday period has the lowest demand throughout the year, the conditions from the perspective of supply–demand coordination of PV and wind power grid interconnection are the most severe. Using the model formulated in Eqs. (5.25)–(6.22), the operation plans of thermal power units on May 1, 2030, were examined, assuming separate operation of the Tokyo power grid and the Tohoku power grid. The analysis results for the Tokyo power grid are shown in Figs. 6.11 and 6.13. The analysis results for the Tohoku power grid are shown in Figs. 6.12 and 6.14. However, the installed wind power generation capacity in the Tohoku power grid is 9,081 MW. In the actual analysis, the optimal solution can only be obtained if the wind power generation output is reasonable. Therefore, in this analysis, the wind power generation output was multiplied by 0.3 (output suppression factor $C = 0.3$). The specific setting of the output suppression factor C is explained in Sect. 6.3.3.

Figures 6.11 and 6.12 show the daily power generation of various power sources in the Tokyo and Tohoku power grid, respectively. As is clear from Eqs. (6.8) and (6.16), the model operates thermal power units to guarantee the supply–demand balance even when the output of solar or wind power is less than the forecasted value. Therefore, in Figs. 6.11 and 6.12, the output of solar or wind power generation over demand is used to charge storage devices (pumping operation), and the remainder is used to suppress output.

Figures 6.13 and 6.14 show the equivalent demand and the daily transition of pumped storage operation and generation operation in the Tokyo and Tohoku

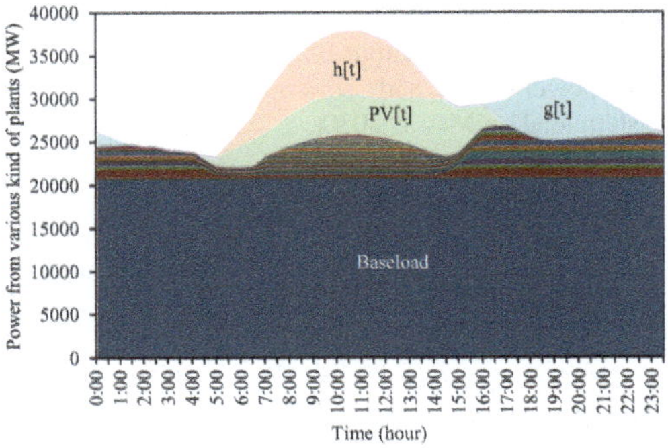

Fig. 6.11 Output power: the Tokyo power grid on May 1, 2030

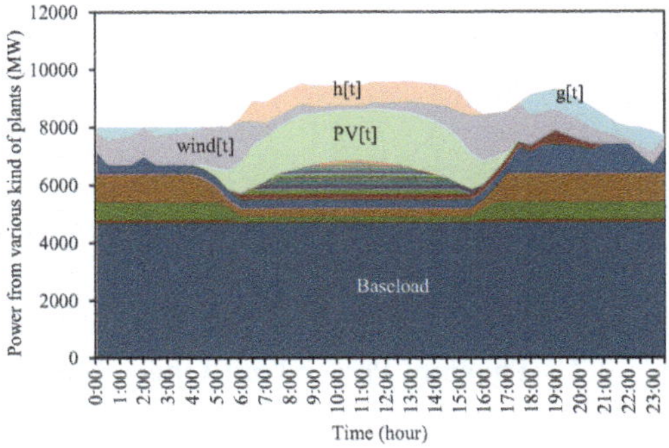

Fig. 6.12 Output power: the Tohoku power grid on May 1, 2030

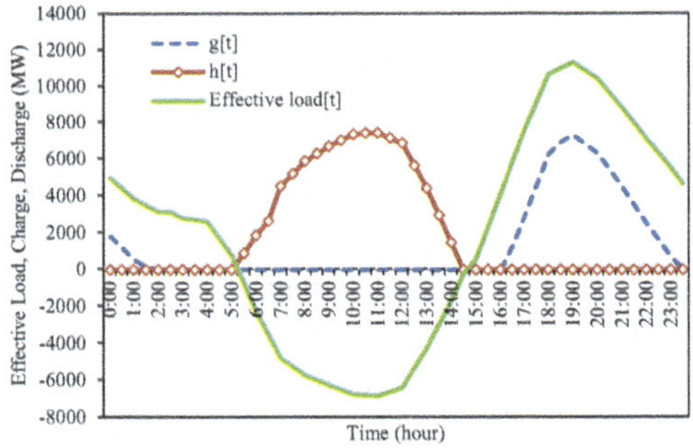

Fig. 6.13 Effective load and pumped hydro operation: the Tokyo power grid on May 1, 2030

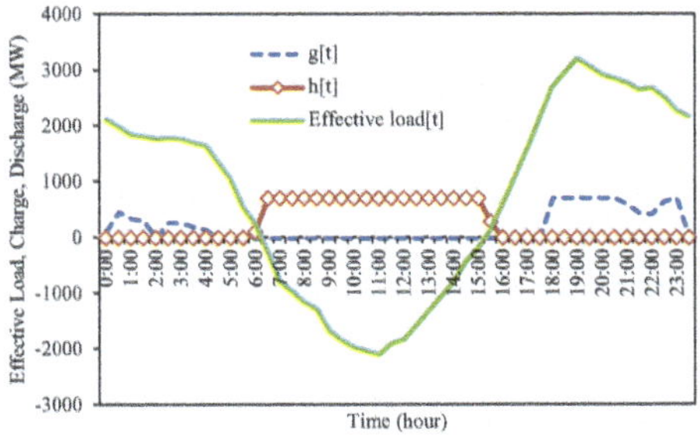

Fig. 6.14 Effective load and pumped hydro operation: the Tohoku power grid on May 1, 2030

power grid, respectively. The equivalent demand is defined as electricity demand minus baseload and output from solar and wind power generation. In both the Tokyo power grid and the Tohoku power grid, there are times when the daytime equivalent demand is negative. This negative equivalent demand does not directly mean that the PV and wind power output cannot be grid-connected. This is because, as mentioned earlier, the excess output from PV and wind power is first used for pumping operations, and the remaining excess output is curtailed within the forecast error. If the equivalent demand after this output curtailment is negative, the grid connection will not be possible. It was found that when a large amount of PV and wind power generation is introduced in FY2030, pumping operation will be conducted during the daytime, and power generation operation will be conducted during the nighttime during the lowest demand period in May.

Furthermore, in order to clarify the effect of demand response, Figs. 6.15 and 6.16 show the daily changes in electricity demand, electricity prices, and sales with demand response divided by those without demand response in the Tokyo power grid and the Tohoku power grid, respectively. In this model, a discount structure of electricity prices is spontaneously obtained such that electricity prices are lower during the nighttime hours when demand is low for most of the year. Setting electricity prices high during the daytime when demand is high and low during the nighttime when demand is low maximizes the profit of the utility company operating the power grid in Eq. (5.25). However, during the lowest demand

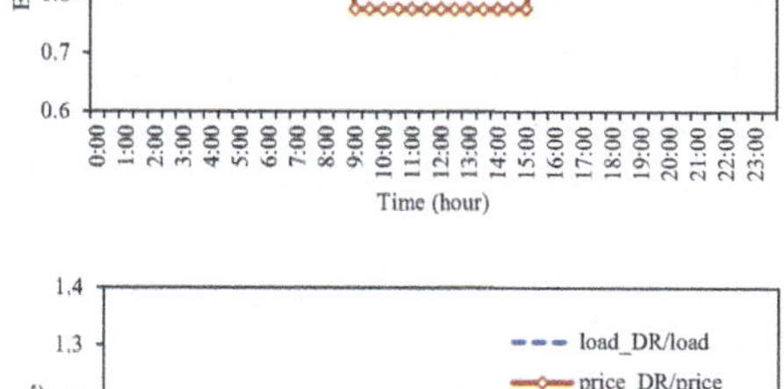

Fig. 6.15 Effect of Demand Response: the Tokyo power grid on May 1, 2030

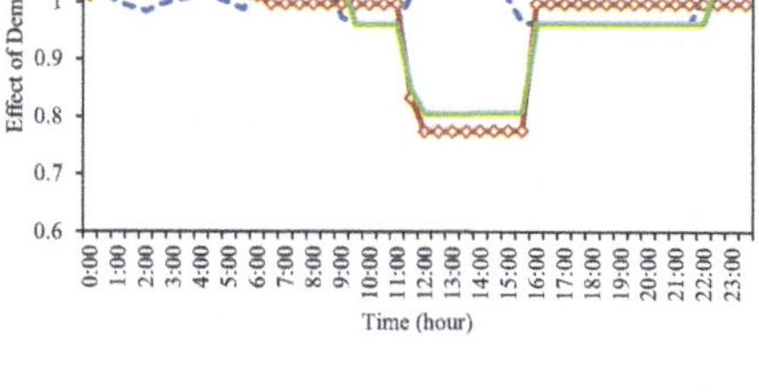

Fig. 6.16 Effect of Demand Response: the Tohoku power grid on May 1, 2030

period in May, a price structure is obtained such that daytime prices are lower than nighttime prices. In the presence of demand response, both in the Tokyo power grid and the Tohoku power grid, the daytime and nighttime electricity demand increases. Due to the decrease in daytime electricity prices, the amount of sales of utility companies corresponding to the first term on the right side of Eq. (5.25) also decreases. This trend can be reversed by increasing the demand for electricity during the daytime as much as possible. This trend occurs because the utility company increases its profits by increasing daytime electricity demand as much as possible, thereby reducing pumping operations with losses corresponding to the efficiency $\eta = 0.7$. This result can be interpreted as reproducing the behavior of consumers who reduce electricity prices by shifting demand from nighttime, when electricity prices are high, to daytime, when electricity prices are low, in the present model.

6.3.3 *Wind Power Curtailment in Tohoku System*

Fluctuations in the output of photovoltaic and wind power generation must be balanced by output adjustment, including start-up and shutdown of hydroelectric and thermal power generation, for grid interconnection. Especially in Japan, output adjustment by thermal power units is mainly used. If the proportion of hydroelectric and nuclear power generation remains unchanged, the proportion of thermal power units will decrease as the amount of solar and wind power generation increases. Therefore, as the proportion of these power sources with fluctuating output increases, it will become challenging to interconnect PV and wind power generation to the grid due to supply–demand balancing problems. If such a situation occurs and there are restrictions on energy storage, it will be necessary to curtail the output of photovoltaic or wind power generation.

This study assumed that 26,481 MW of photovoltaic power generation will be installed in the Tokyo power grid and 7,741 MW in the Tohoku power grid in FY2030. For wind power generation, we assumed that 1,009 MW and 9,081 MW will be installed in the Tokyo power grid and the Tohoku power grid in FY2030. Figures 6.17 and 6.18 show monthly changes in the ratio of PV and wind power

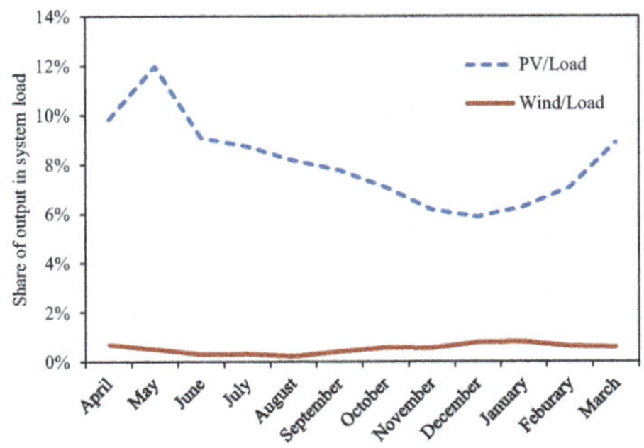

Fig. 6.17 Share of PV or wind output in system load: the Tokyo power grid in 2030

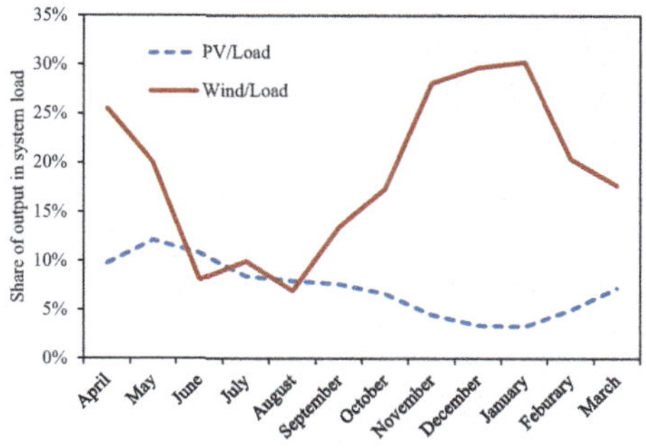

Fig. 6.18 Share of PV or wind output in system load: the Tohoku power grid in 2030

output to electricity demand in the Tokyo power grid and the Tohoku power grid in FY2030, respectively. In the Tokyo power grid, photovoltaic power generation is expected to spread, but even in May, when demand is at its lowest, its share is only about 12%. The share of wind power generation will remain below 1% throughout the year. On the other hand, in the Tohoku power grid, the share of photovoltaic power generation ranges from 5% to 10%. In contrast, the share of wind power generation continues to exceed 20% in May when electricity demand is low and in winter when the wind is strong. Using the model formulated in Eqs. (5.25)–(6.22), we analyzed whether PV or wind power generation can be integrated into the grid for each month in FY2030 without output curtailment under the analysis conditions described in Sect. 6.3.2.1. Note that demand response is not considered in the analysis in Sects. 6.3.3 and 6.3.4.

For the Tokyo power grid, we confirmed that PV and wind power generation can be integrated into the grid without output curtailment in all months. However, in the Tohoku power grid, the optimal solution could not be obtained in May, when electricity demand is low, and in winter, when the wind is strong. Therefore, we introduced a wind power output suppression factor C in Eq. (6.8):

$$\frac{\sum_{i=1}^{N} p_t^i + C \cdot wd_t^{(f)} + pv_t^{(f)} + g_t - h_t - \widetilde{d_t^{(f)}}}{\sqrt{\sigma_d^2 + \sigma_w^2 + \sigma_p^2}} \geq \phi^{-1}(\alpha). \tag{6.26}$$

Using the new constraint Eq. (6.26), we reanalyzed the monthly grid integration in the Tohoku power grid by varying the output suppression factor of wind power generation, C, in 0.1 increments as a parameter. Figure 6.19 shows the monthly change in the output suppression factor of wind power generation in the Tohoku power grid. It was found that output suppression is not necessary in summer with $C = 1.0$ but is necessary in May, when electricity demand is low, with $C = 0.3$, and in winter, when wind is strong, with $C = 0.6$ or $C = 0.7$. The conventional form of operating each power grid independently cannot take advantage of the favorable wind conditions in the Tohoku region.

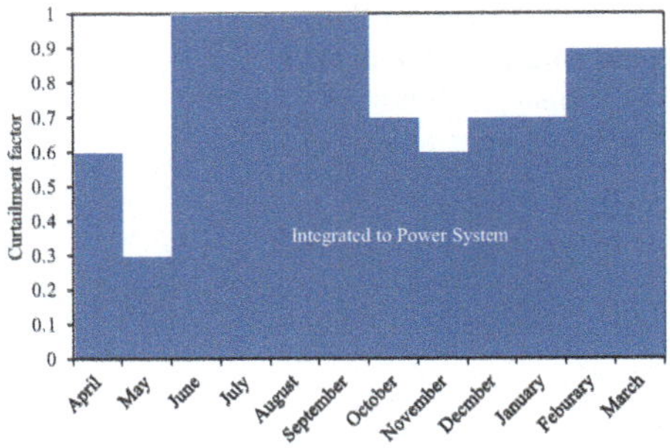

Fig. 6.19 Curtailment of wind power for the Tohoku power grid in 2030

6.3.4 *Tokyo–Tohoku Grid Interconnection*

Currently, in preparation for the installation of wind power generation, studies are being conducted on the integrated operation of the Tokyo power grid and the Tohoku power grid. In this section, we discuss the economics of integrated operation to avoid output curtailment of wind power generation in the Tohoku power grid, assuming the following operation method as an integrated operation. Suppose the wind power generation output is too large to be interconnected with the Tohoku power grid. In that case, the wind power output is transmitted to the Tokyo power grid at a price lower than the cost of generating power from thermal power units in the Tokyo power grid. The Tokyo power grid purchases power from the Tohoku power grid, and the thermal power units shut down by the amount of power purchased to reduce costs. In the following sections, we discuss the economics of the integrated operation, assuming that there are no transmission costs between Tokyo and Tohoku and that the two systems are simply treated as a single system.

The monthly trends of the share of PV and wind power output in the electricity demand of the interconnected system between the Tokyo power grid and the Tohoku power grid are shown in Fig. 6.20. The share of photovoltaic power generation in electricity demand reaches its maximum in May when electricity demand is low, but the value is about 12%. The share of wind power generation in the electricity demand is about 8% even in winter when the wind is vital due to the large electricity

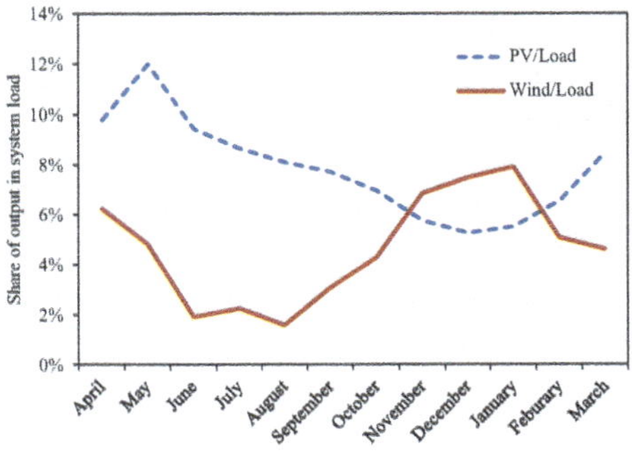

Fig. 6.20 Share of PV or wind output in system load in 2030 for Tokyo–Tohoku grid interconnection

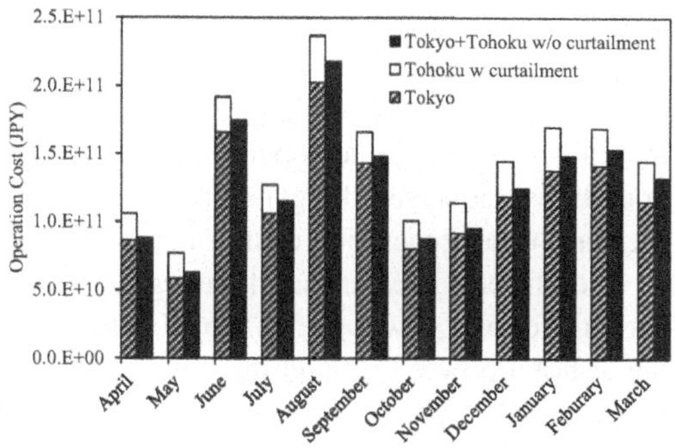

Fig. 6.21 Cost for Tokyo–Tohoku grid interconnection in 2030

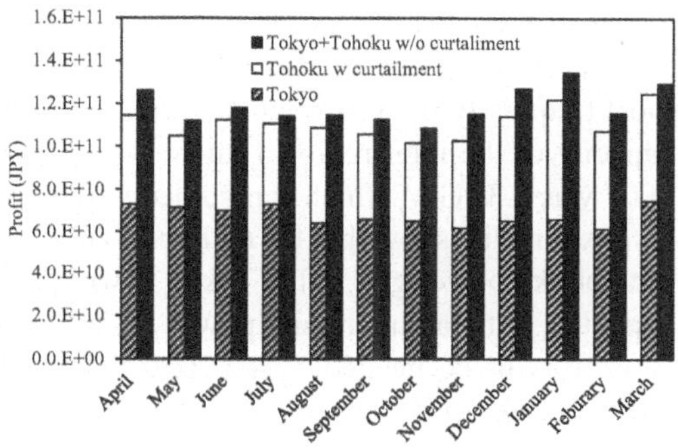

Fig. 6.22 Profit for Tokyo–Tohoku grid interconnection in 2030

demand in the Tokyo power grid, which is considerably lower than that of the individual operation in the Tohoku power grid. Therefore, it is expected that the integrated operation of the power grids in the Tokyo power grid and the Tohoku power grid will reduce the amount of wind power generation in the Tohoku power grid and effectively use its favorable wind conditions. In fact, in analyzing the interconnection of PV and wind power generation in the Tokyo power grid and the Tohoku power grid, $C = 1.0$ in all months, indicating that output suppression was not necessary.

Figures 6.21 and 6.22 show the monthly analysis for the operation only in the Tokyo power grid, the operation only in the Tohoku power grid, and the integrated operation in the Tokyo power grid and the Tohoku power grid, respectively. Here, sales price refers to the price of electricity sold to consumers, and generation cost refers to the cost of generating thermal power units without considering the trading value of power transactions between power grids. In operation only in the Tohoku power grid, the wind power generation output was suppressed by the suppression factor shown in Fig. 6.19.

Figure 6.21shows that the cost of power generation in the integrated operation of the Tokyo power grid and the Tohoku power grid is less than the sum of the cost in the Tokyo power grid and the Tohoku power grid throughout a year. This is because the integrated operation eliminates the need for wind power output curtailment, allowing more wind power generation and reducing the cost of power generation.

Figure 6.22 shows that the profit of the integrated operation in the Tokyo power grid and the Tohoku power grid is larger than the sum of the profit in the Tokyo power grid and the Tohoku power grid. This can be understood from the fact that the sales of the integrated operation are equal to the sum of the sales of the individual operations. However, the generation cost of the integrated operation is smaller than the sum of the generation cost of the individual operations.

Therefore, it is clear that the integrated operation between the Tokyo power grid and the Tohoku power grid is more economical than the separate operation between the Tohoku power grid and the Tokyo power grid because it can effectively utilize the favorable wind conditions in Tohoku. However, it is necessary to confirm that the power transmitted from the Tohoku power grid to the Tokyo power grid is less than the capacity of the interconnection lines between the Tokyo power grid and the Tohoku power grid. The month wind power output needs to be curbed most frequently in individual operations within the Tohoku power grid in May. The capacity of the inter-regional interconnection line between the two power grids in 2007 was 6.31 GW, which is assumed to remain unchanged until 2030. The analysis results for May show that for 65 hours out of 744 hours, the power transmitted from the Tohoku power grid to the Tokyo power grid exceeded the 6.31 GW transmission capacity, with a maximum power of 8.58 GW. Therefore, at least 2.27 GW of capacity will be needed by FY2030 to eliminate wind power output curtailment in the Tohoku power grid.

6.3.5 Summary 2

After the Great East Japan Earthquake in 2011, photovoltaic power generation installed on building roofs has been attracting attention in the Tokyo power grid, and wind power generation utilizing favorable wind conditions has been attracting attention in the Tohoku power grid. In this chapter, we formulated a unit commitment model for thermal power units in the power grid using demand response and energy storage devices, considering the forecasted output of wind power and solar power and its forecasting error. We analyzed the operation plans of power grids in the Tokyo power grid and the Tohoku power grid in 2030 and examined the economics of the integrated operation of the two power grids.

First, the daily operation plans of thermal power units were examined for the month of May, when the conditions for grid interconnection of solar and wind power generation were most severe. We found that a discount system of electricity prices can be obtained spontaneously so that electricity prices are lower during the low-demand period and that the price of electricity is lower during the low-demand period in May. In May, when demand is at its lowest, pumping operations are conducted during the daytime, and power generation operations are conducted at night.

Next, we analyzed whether PV or wind power generation can be integrated into the grid without output control for each month in FY2030. In the Tokyo power grid, it was confirmed that grid integration was possible in all months without output curtailment. On the other hand, in the Tohoku power grid, output curtailment is not necessary in summer but is necessary in May, when electricity demand is low, and in winter, when the wind is strong.

Furthermore, the study of the integrated operation of power grids within the Tokyo power grid and the Tohoku power grid showed that power generation costs can be reduced compared to individual operations because wind power output within the Tohoku power grid can be connected to the grid without suppression. However, there are cases where the amount of power transmitted from the Tohoku power grid to the Tokyo power grid exceeds the transmission capacity, and FY2030 will require an increase in transmission capacity of 2.27 GW to eliminate wind power output curtailment. The evaluation of the market price of electricity and the amount of installed capacity to increase transmission capacity remains a future research topic.

References

1. The Green Grid: Energy Saving and Carbon Emissions Reductions Enabled by a Smart Grid, Electric Power Research Institute (2008).
2. Smart Grid System Report, U.S. Department of Energy (2009).
3. European Electricity Grid Initiative Roadmap and Implementation Plan, European Electricity Grid Initiative (2010).
4. Technology Roadmap: Smart Grids, International Energy Agency (2011).
5. Y. Ikeda, U. Remme, D. Elzinga, S. Heinen, P. Taylor, and K. Ogimoto, "Estimations of cost and CO2 emissions for smart grids using a linear programing approach," PowerTech, 2011 IEEE Trondheim, pp.1–8, 19–23 June 2011.
6. S. Heinen, D. Elzinga, S. Kim, and Y. Ikeda, "Impact of Smart Grid Technologies on Peak Load to 2050", International Energy Agency Working Paper, August, 2011.
7. A. J. Wood, and B.F. Wollenberg, "Power Generation, Operation, and Control", Wiley-Interscience; 2nd edition (1996).
8. R.H. Kerr, J.L. Scheidt, A.J. Fontanna, and J.K. Wiley, "Unit Commitment", Power Apparatus and Systems, IEEE Transactions on, vol.PAS-85, no.5, pp.417–421, 1966.
9. G.B. Sheble, and G.N. Fahd, "Unit commitment literature synopsis," Power Systems, IEEE Transactions on, vol.9, no.1, pp.128–135, 1994.
10. N.P. Padhy, "Unit commitment-a bibliographical survey," Power Systems, IEEE Transactions on, vol.19, no.2, pp.1196- 1205, 2004.
11. C. Dang, and M. Li, "A floating-point genetic algorithm for solving the unit commitment problem", European Journal of Operational Research vol.181, pp.1370–1395, 2007.
12. S. Salam, "Unit Commitment Solution Methods", World Academy of Science, Engineering and Technology 35, pp.320–325, 2007.
13. Y. Ikeda, T.Ikegami, K.Kataoka, and K.Ogimoto, "A Unit Commitment Model with Demand Response for the Integration of Renewable Energies", Power and Energy Society General Meeting, 2012 IEEE, pp.1–7, July 2012
14. Y. Ikeda and K. Ogimoto, Unit Commitment Model with Active Demand and Energy Storage for Renewable Energy Grid Integration (In Japanese), IEEJ Transactions on Power and Energy, vol. 133, issue 7, pp. 598–605, (2013)
15. Y. Ikeda and K. Ogimoto, "Cross-correlation of output fluctuation and system-balancing cost in photovoltaic integration", The Journal of Engineering, Volume 2014, Issue 10, October 2014, p. 551–559

16. R. Barth, H. Brand, P. Meibom, and C. Weber, "A Stochastic Unit-commitment Model for the Evaluation of the Impacts of Integration of Large Amounts of Intermittent Wind Power", International Conference on Probabilistic Methods Applied to Power Systems 2006 (PMAPS 2006), pp.1–8, 11–15 June 2006.
17. J. Wang, A. Botterud, V. Miranda, C. Monteiro, and G. Sheble, "Impact of Wind Power Forecasting on Unit Commitment and Dispatch", 8th Int Workshop on LargeScale Integration of Wind Power into Power Systems Bremen Germany, 2009.
18. S. Bu, F. R. Yu, and P. X. Liu, "Stochastic unit commitment in smart grid communications", IEEE Conference on Computer Communications Workshops (INFOCOM WKSHPS) 2011, pp.307–312, 10–15 April 2011.
19. E.M. Constantinescu, V.M. Zavala, M. Rocklin, S. Lee, and M. Anitescu, "Unit Commitment with Wind Power Generation: Integrating Wind Forecast Uncertainty and Stochastic Programming", ANL/MCS-TM-309, September 2009.
20. C. Monteiro, R. Bessa, V. Miranda, A. Botterud, J. Wang, and G. Conzelmann, "Wind Power Forecasting: State-of-the-Art 2009", ANL/DIS-10-1, November 6, 2009.
21. L.B. Lave, and K. Spees, "Demand Response and Electricity Market Efficiency", The Electricity Journal, Vol.20, Issue 3 pp.69–85, 2007.
22. A. Faruqui, R. Hledik, S. Newell, and H. Pfeifenberger, "The Power of 5 Percent", The Electricity Journal, Vol. 20, Issue 8, pp.68–77, 2007.
23. A. Faruqui, R. Hledik, and J. Tsoukalis, "The Power of Dynamic Pricing", The Electricity Journal, Vol. 22, Issue 3, pp. 42–56, April 2009
24. J. Baia, H.B. Gooia and L.M. Xiaa, "Probabilistic Reserve Schedule with Demand-side Participation", Electric Power Components and Systems, Volume 36, Issue 2, pp.138–151, 2008.
25. M. Parvania, and M. Fotuhi-Firuzabad, "Demand Response Scheduling by Stochastic SCUC," Smart Grid, IEEE Transactions on, vol.1, no.1, pp.89–98, 2010.
26. S. Ramchurn, P. Vytelingum, A. Rogers, and N. Jennings, "Agent-Based Control for Decentralised Demand Side Management in the Smart Grid", Agent-Based Control for Decentralised Demand Side Management in the Smart Grid. In: The Tenth International Conference on Autonomous Agents and Multiagent Systems (AAMAS 2011), 2–6 May 2011, Taipei, Taiwan. pp. 5–12, 2011.
27. J.M. Arroyo, and A.J. Conejo, "Optimal Response of a Thermal Unit to an Electricity Spot Market", IEEE TRANSACTIONS ON POWER SYSTEMS, VOL.15, NO.3, pp.1098–1104, 2000.
28. S. de la Torre, J.M. Arroyo, A.J. Conejo, and J. Contreras, "Price Maker Self-Scheduling in a Pool-Based Electricity Market: A Mixed-Integer LP Approach", IEEE TRANSACTIONS ON POWER SYSTEMS, VOL.17, NO.4, pp.1037–1042, 2002.
29. A. Borghetti, A. Frangioni, F. Lacalandra, C.A. Nucci, and P. Pelacchi, "Using of a cost-based Unit Commitment algorithm to assist bidding strategy decisions", IEEE Power Tech Conference Proceedings, Bologna, 2003.
30. M. Ventosa, A. Baillo, A. Ramos, and M. Rivier, "Electricity market modeling trends", Energy Policy. Vol.33, Issue 7, pp.897–913, 2005.
31. A. Weidlich, and D. Veit, "A critical survey of agent-based wholesale electricity market models", Energy Economics 30, pp.1728–1759, 2008.
32. A. Frangioni, C. Gentile, and F. Lacalandra, "Solving unit commitment problems with general ramp constraints", Electrical Power and Energy Systems Vol.30, pp.316–326, 2008.
33. A. Frangioni, C. Gentile, and F. Lacalandra, "Tighter Approximated MILP Formulations for Unit Commitment Problems", IEEE TRANSACTIONS ON POWER SYSTEMS, Vol.24, No.1, pp.105–113, 2009.
34. D.S. Kirschen, "Demand-Side View of Electricity Markets", IEEE TRANSACTIONS ON POWER SYSTEMS, VOL.18, NO.2, pp.520–527, 2003.

35. M. Braun, Y.-M. Saint-Drenan, T. Glotzbach, T. Degner, and S. Bofinger, "VALUE OF PV ENERGY IN GERMANY -BENEFIT FROM THE SUBSTITUTION OF CONVENTIONAL POWER PLANTS AND LOCAL POWER GENERATION", 23rd European Photovoltaic Solar Energy Conference, Valencia, Spain, pp.3645–3652, 1–5 September 2008.
36. K. Ogimoto, K. Kataoka, T. Ikegami, S. Nonaka, H. Azuma, S. Fukutome, "Demand–Supply Balancing Capability Analysis for a Future Power System", Electrical Engineering in Japan, Volume186, Issue2,Pages 21–30, 2014

Chapter 7
Evolving Microgrid Network and Power Market

7.1 Complex Networks

Let us consider a system consisting of elemental constituents and relationships between constituents. We will ignore all the specific properties of the constituents and the interrelationships among them, and let us describe the components as points and the interrelationships as being connected. We call the point components nodes or vertices and the connections as links or edges. The system described using nodes and links is called a network or graph [1]. Suppose the network consists of N nodes and L links.

In nature and society, we have many different types of networks. For instance, a crystal lattice in which all atoms have the same number of neighboring atoms. This type of network is called a simple network.

Link is an inter-atomic interaction that is symmetrical in the exchange of neighboring atoms. The symmetrical link is called an undirected link; therefore, the network is called an undirected network. However, undirected networks are wider than simple networks. For example, co-authoring relationships for academic papers, relationships of actors performing together in movies, and interactions between amino acids in proteins do not have the same number of neighboring persons and amino acids. These are not categorized as simple networks.

On the other hand, links may have a direction. For instance, examples of directed links include URLs on the WWW where a website is cited from other websites, phone calls, and metabolic reactions, where one reaction occurs and is the cause of the subsequent reaction. The directed link is often expressed as arcs. Networks consisting of links with a direction are called directed networks.

Figure 7.1 shows examples of the simple network. The nodes are simple points. We often use node labels for better visibility. The links are drawn as lines for the undirected network and arcs for the directed network. For these networks, we have $N = 4$ and $L = 4$.

Y. Ikeda, *Data Science of Renewable Energy Integration*,
Evolutionary Economics and Social Complexity Science 30,
https://doi.org/10.1007/978-981-99-8779-5_7

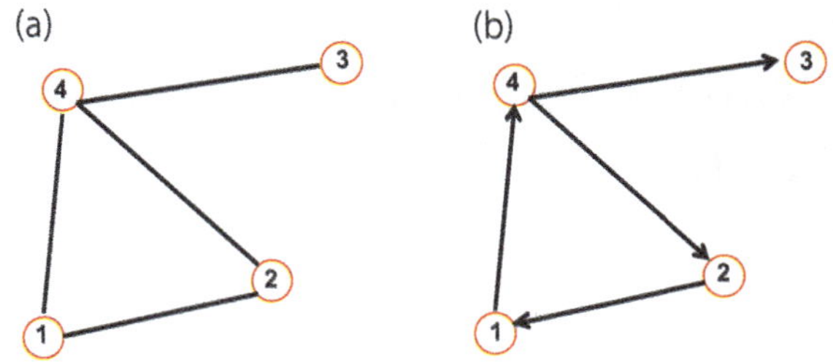

Fig. 7.1 Simple networks: (**a**) Undirected network, (**b**) Directed network

7.1.1 Adjacency Matrix

An adjacency matrix A, which is $N \times N$ square matrix, specifies the network structure. If the elements of the adjacency matrix A are 1 or 0, the network is called a binary network. For the undirected binary network, we have

$$\begin{cases} A_{ij} = 1 \text{ if there is a link between node } i \text{ and } j, \\ A_{ij} = 0 \quad \text{ if nodes } i \text{ and } j \text{ are not connected.} \end{cases} \tag{7.1}$$

For the directed binary network, we have

$$\begin{cases} A_{ij} = 1 \text{ if there is a link pointing from node } j \text{ and } i, \\ A_{ij} = 0 \quad \text{ if there is no link pointing from } j \text{ to } i. \end{cases} \tag{7.2}$$

7.1.2 Node Degree

Next, we define node degree k_i, which is the number of links connected to node i. For undirected networks,

$$\begin{gathered} k_i = \sum_{j=1}^{N} A_{ij}, \\ k_j = \sum_{i=1}^{N} A_{ij}, \\ k_i = k_j. \end{gathered} \tag{7.3}$$

For a directed network, we can define an in-degree

$$k_i^{in} = \sum_{j=1}^{N} A_{ij} \tag{7.4}$$

and out-degree

$$k_j^{out} = \sum_{i=1}^{N} A_{ij}. \tag{7.5}$$

The (total) degree k_i for a directed network is the sum of in-degree and out-degree,

$$k_i = k_i^{in} + k_i^{out}. \tag{7.6}$$

A node with $k_{in} = 0$ is called a source, and a node with $k_{out} = 0$ is called a sink.

The average degree of undirected network is obtained by

$$\langle k \rangle = \frac{1}{N} \sum_{i=1}^{N} k_i = \frac{2L}{N}. \tag{7.7}$$

About directed network, average degrees are calculated for in-degree and out-degree separately by

$$\begin{aligned} \langle k^{in} \rangle &= \tfrac{1}{N} \textstyle\sum_{i=1}^{N} k_i^{in}, \\ \langle k^{out} \rangle &= \tfrac{1}{N} \textstyle\sum_{i=1}^{N} k_i^{out}, \\ \langle k^{in} \rangle &= \langle k^{out} \rangle. \end{aligned} \tag{7.8}$$

Degree distribution $p(k)$ is defined as the probability that a randomly chosen node has degree k. In the discrete representation, p_k is the probability that a node has degree k, calculated by $p_k = N_k/N$, where N_k is the number of nodes with degree k. The normalization condition $\sum_0^{\infty} p_k = 1$ is applied. In the continuum representation, $p(k)$ is the probability density function of the degrees, where $\int_{k_1}^{k_2} p(k)dk$ represents the probability that a node's degree is between k_1 and k_2. The normalization condition $\int_{k_{min}}^{\infty} p(k)dk = 1$ is applied. Here K_{min} is the minimal degree in the network.

7.1.3 Path Length and Clustering Coefficient

We consider distance in a network. The distance (path length) d between two nodes is defined as the number of edges along the shortest path connecting them. Note that if the two nodes are disconnected, the distance is infinite.

In directed networks, each path needs to follow the direction of the arrows. Thus, in Fig. 7.1, the distance from node 1 to 2 (on a 1-4-2 path) is generally different from the distance from node 2 to 1 (on a 2-1 path). The path length is calculated using a search algorithm, such as the breadth-first search algorithm.

Diameter d_{max} is the maximum distance between any pair of nodes in the network. Average path length $\langle d \rangle$ is

$$\langle d \rangle = \frac{1}{N(N-1)} \sum_{i,j \neq i} d_{ij}, \tag{7.9}$$

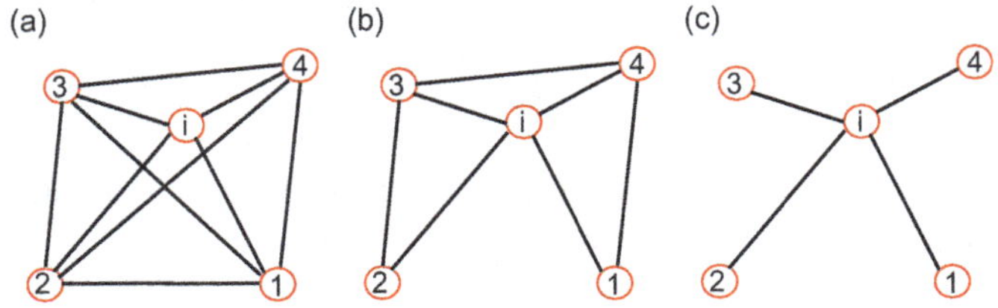

Fig. 7.2 **Clustering Coefficient:** (**a**) $C_i = 1.0$, (**b**) $C_i = 0.5$, (**c**) $C_i = 0.0$

where d_{ij} is the distance from node i to node j. In an undirected graph $d_{ij} = d_{ji}$, so we only need to count them once,

$$\langle d \rangle = \frac{1}{N(N-1)} \sum_{i,j \geq i} d_{ij}. \tag{7.10}$$

Clustering coefficient C_i means what fraction of your neighbors are connected, which is defined by

$$C_i = \frac{2e_i}{k_i(k_i - 1)}, \tag{7.11}$$

where e_i is the number of links among nodes neighboring node i, and k_i is the degree of node i. Range of C_i is [0, 1]. The average clustering coefficient C is an average value of Ci for all nodes. Figure 7.2 shows an example of the clustering coefficients.

7.1.4 Weight Matrix

The network is called a weighted network if the links between nodes are not restricted to binary values of 1 or 0 but have real-number weights. In a weighted network, the adjacency matrix A and the weight matrix W are required. However, when there is no link between node i and node j, both the adjacency matrix element a_{ij} and the weight matrix element w_{ij} are equal to zero. Hence, the adjacency matrix and the weight matrix have redundant parts. For the undirected weighted network, we have

$$\begin{cases} W_{ij} = w_{ij} & \text{if there is a link between node } i \text{ and } j, \\ W_{ij} = 0 & \text{if nodes } i \text{ and } j \text{ are not connected.} \end{cases} \tag{7.12}$$

For the directed weighted network, we have

$$\begin{cases} W_{ij} = w_{ij} \text{ if there is a link pointing from node } j \text{ and } i, \\ \quad W_{ij} = 0 \qquad \text{if there is no link pointing from } j \text{ to } i. \end{cases} \tag{7.13}$$

7.1.5 Node Strength

Weighted networks require not only degree k_i but also strength s_i. For undirected weighted networks,

$$\begin{aligned} s_i &= \textstyle\sum_{j=1}^{N} a_{ij} w_{ij} = \sum_{j=1}^{N} w_{ij}, \\ s_j &= \textstyle\sum_{i=1}^{N} a_{ij} w_{ij} = \sum_{i=1}^{N} w_{ij}, \\ & s_i = s_j. \end{aligned} \tag{7.14}$$

For a directed weighted network, we can define an in-strength

$$s_i^{in} = \sum_{j=1}^{N} w_{ij} \tag{7.15}$$

and out-strength

$$s_j^{out} = \sum_{i=1}^{N} w_{ij}. \tag{7.16}$$

The (total) strength s_i for a directed network is the sum of in-strength and out-strength.

Strength distribution $p(s)$ is a probability that a randomly chosen node has strength s. Weight distribution $p(w)$ probability that a randomly chosen link has weight w. In most real systems, $p(k)$, $p(s)$, and $p(w)$ are fat tailed.

7.1.6 Small-World Network

When all the nodes that constitute a network have the same degree k, the network is called a regular network. A set of nodes is randomly selected and linked together. This procedure is repeated until all nodes are linked, resulting in a random network or a random graph. The degree distribution of a random network is known to be a Poisson distribution. We can see that a regular network has a uniform degree distribution, whereas a random network has a Poisson degree distribution:

$$\begin{aligned} P(X=k) &= \tfrac{\lambda^k e^{-\lambda}}{k!}, \\ E[X] &= \lambda, \\ V[X] &= \lambda. \end{aligned} \tag{7.17}$$

Therefore, there is a network that falls into an intermediate category between these two types of networks. The small-world network is located between a regular network and a random network.

We refer to a social experiment carried out by Stanley Milgram in 1967. In this experiment, letters were passed from acquaintance to acquaintance. As a result, he showed that letters sent from the Midwestern region of the USA were delivered to intended acquaintances in the Eastern region of the USA through an average of six people. This connection between people is called a small-world network with six degrees of separation. Why are small-world networks surprising? We try to understand small-world networks by comparing them to regular networks and random networks.

We calculate distance (path length) in random networks. Random networks tend to have a tree-like topology with almost constant node degrees. Starting from a specified node, we have $\langle k \rangle$ nodes at distance one ($d = 1$). Similarly, $\langle k \rangle^2$ nodes at distance two ($d = 2$), $\langle k \rangle^3$ nodes at distance three ($d = 3$), and so on. In general, we have $\langle k \rangle^d$ nodes at a distance d. Therefore, we obtained the following relation:

$$N = \langle k \rangle + \langle k \rangle^2 + \cdots + \langle k \rangle^{d_{max}} = \frac{\langle k \rangle^{d_{max}+1} - 1}{\langle k \rangle - 1} \approx \langle k \rangle^{d_{max}}. \tag{7.18}$$

Maximum distance d_{max} is estimated by

$$d_{max} = \frac{\log N}{\log k}. \tag{7.19}$$

In most networks, this offers a better approximation to the average distance between two randomly chosen nodes $\langle d \rangle$, than to d_{max}:

$$\langle d \rangle = \frac{\log N}{\log k}. \tag{7.20}$$

In a regular network, the average path length depends linearly on the system size N. On the other hand, a small-world network is known to have the property that the average path length depends logarithmically on the system size N. This property is due to the fact that random processes generate links in small-world networks. The $1/\log\langle k \rangle$ term implies that the denser the network, the smaller will be the distance between the nodes.

Table 7.1 Summary of small-world property

Network type	Clustering coefficient	Path length
Regular network	$C(Reg) = \frac{3k-6}{4k-2}$	$d(Reg) \propto N$
Small-world network	$C(SW) = \frac{(1-p)^3(3k-6)}{4k-2}$	$d(SW) \propto \ln N$
Random network	$C(Ran) = \frac{<k>}{N}$	$d(Ran) \propto \ln N$

Next, we calculate the clustering coefficient in a random network. The definition of clustering coefficient C_i of node i is given by Eq. (7.11). Since nodes are independently linked with probability p so that they are proportional to degree in a random network, we have

$$e_i \simeq p\frac{k_i(k_i-1)}{2}, \tag{7.21}$$

where e_i is the number of links among nodes to neighboring node i, and k_i is the degree of node i. Therefore, we obtained the clustering coefficient for a random network.

$$C_i = \frac{2e_i}{k_i(k_i-1)} = p = \frac{<k>}{N}. \tag{7.22}$$

This means the average clustering coefficient C is small and decreases with the system size N for random networks. C_i is independent of a node's degree k_i for random networks . In most real-world networks, C has a larger value compared with the random networks, $C/<k>$ is independent with the system size N, and C_i decreases when the node's degree k_i increases, in contrast to the random network.

The small-world (SW) property is summarized in Table 7.1 by comparing with regular network (Reg) and random network (Ran). Table 7.1 means that for cluster coefficient $C(Reg) \simeq C(SW) > C(Ran)$, and characteristic path length $d(Reg) > d(SW) \simeq d(Ran)$.

7.1.7 Strogatz β Model

One of the network models to generate a small-world network is the Strogatz β model.

This model generates a small-world network starting from a regular network by rewiring links with rewiring probability β. If we use large β, the obtained network becomes a random network. The concept of the model is shown in Fig. 7.3. The algorithm is as follows:

1. Decide rewiring with probability β for four links of each node using a uniform random number $x[0-1]$. If $x < \beta$, rewire the link.

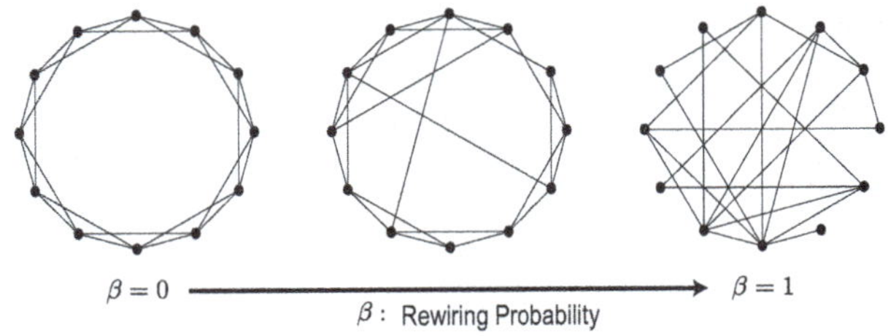

Fig. 7.3 Strogatz β Model: A regular network becomes a small-world network by rewiring links with rewiring probability β, if β is moderate

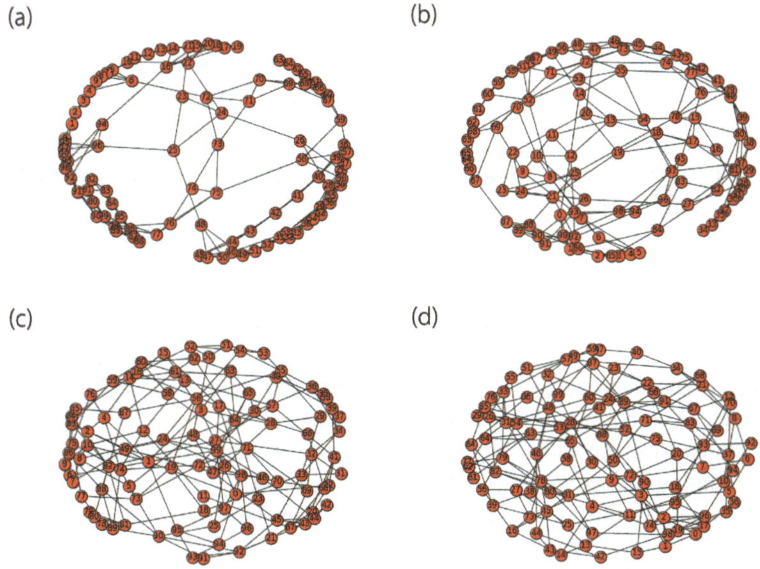

Fig. 7.4 The β **dependence of Strogatz** β **Model:** (**a**) $\beta = 0.05$, (**b**) $\beta = 0.10$, (**c**) $\beta = 0.30$, (**d**) $\beta = 0.50$

2. In the case of rewiring, decide destination node $int[x \cdot N] + 1$ using a uniform random number $x[0 - 1]$.
3. Repeat the above procedure until enough shortcut links are obtained.

Figure 7.4 shows a numerical simulation of the SW network for a several β. Figure 7.5 explains β dependence of the SW network. If the magnitude of the betas is appropriate, a small-world network with large cluster coefficients and small path lengths can be obtained.

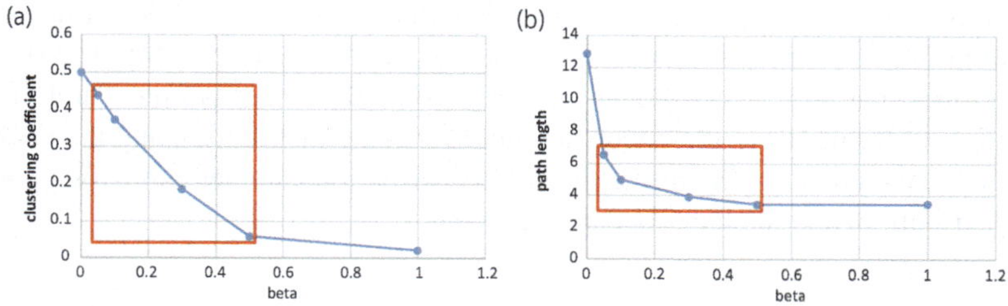

Fig. 7.5 The β **dependence of Strogatz** β **Model:** (**a**) Clustering coefficient, (**b**) Path length

7.1.8 Scale-Free Network

The degree distribution of a network that exhibits a power law is called a scale-free network:

$$p(k) = ak^{-\gamma}, \tag{7.23}$$

where γ is a power-law exponent. A small number of nodes have substantial degrees in the power-law distribution. Small-world networks and scale-free networks are collectively called complex networks.

If we draw a graph of a power-law distribution in a double-logarithmic plot,

$$\log p(k) = \log a - \gamma \log k, \tag{7.24}$$

the shape of the distribution looks like a straight line descending from left to right. The slope of the line corresponds to a power-law exponent γ. When we change the scale of the degree k by a factor c, we find that the shape of the distribution does not change. This property is called the scale-invariant:

$$p(ck) = a(ck)^{-\gamma} = c^{-\gamma} p(k) \propto p(k). \tag{7.25}$$

As node degrees are always positive integers, the discrete formalism captures the probability p_k that a node has exactly k links:

$$\begin{aligned} p_k &= Ck^{-\gamma}, \\ C &= 1/\textstyle\sum_{k=1}^{\infty} k^{-\gamma}. \end{aligned} \tag{7.26}$$

In the continuum formalism, it is often convenient to assume that the degree distribution $p(k)$ can take up any positive real value in the range $[k_{min} - \infty]$:

$$\begin{aligned} p(k) &= Ck^{-\gamma}, \\ C &= 1/\textstyle\int_{k_{min}}^{\infty} k^{-\gamma} dk. \end{aligned} \tag{7.27}$$

An essential characteristic of a power-law distribution is the 80/20 rule or the Pareto principle proposed by Vilfredo Federico Damaso Pareto (1848–1923). This rule implies that 80% of consequences come from 20% of the causes. Pareto made essential contributions to the understanding of income distribution and the analysis of individuals' choices. The former is Pareto distribution, another name for a power-law distribution, and the latter is Pareto efficiency.

7.1.9 Barabasi–Albert Model

One of the network growth models to generate a scale-free network is the Barabasi–Albert model. In this model, a scale-free network with the power law $p(k) \propto k^{-3}$ is obtained by applying the following two rules. One is "growth", which means to add new nodes and links to a network. Another is "preferential attachment", which means to choose existing nodes to be linked from the added nodes with probability proportional to the degree of the existing nodes in the network. The algorithm of the model is as follows:

1. Set an initial network (e.g., 3 nodes and 3 links).
2. Make adjacent matrix of the network $A[\text{node } i][\text{node } j] = 1(\text{linked}), 0(\text{nolink})$.
3. Calculate degree k_i for node i using the adjacent matrix A.
4. Calculate probability of the preferential attachment $p_i \propto k_i$ for node i.
5. Add m nodes and m links to a node chosen using random numbers $[0-1]$.
6. Return to 2. and repeat 2. to 5. until the number of nodes reaches a specified number of nodes.

An analytical derivation of the degree distribution in this model gives

$$p(k) = \frac{2m(m+1)}{k(k+1)(k+2)} \propto k^{-3} \tag{7.28}$$

for large k. The power-law exponent is $\gamma = 3$, which is independent of m.

As the power-law distribution describes systems of different ages and sizes, a correct model is expected to provide a time-independent degree distribution. Indeed, asymptotically, the degree distribution of the BA model is independent of time and the system size N. This implies that the network reaches a stationary scale-free state.

Figure 7.6 explains β shows numerical simulation of the BA model. The left panel is for $m = 3$, and the left $m = 10$. The power-law exponents for both distributions are $\gamma \approx 3$. This confirms that the power-law exponent $\gamma = 3$ is independent of m in the BA model.

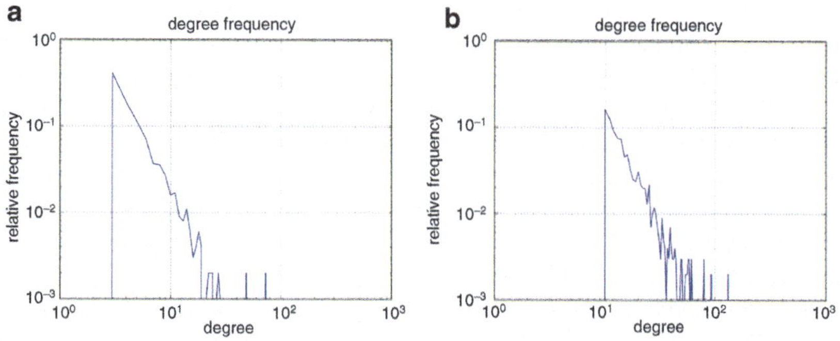

Fig. 7.6 Barabasi–Albert model: (**a**) $m = 3$, (**b**) $m = 10$. The power-law exponent $\gamma = 3$ is independent of m

7.2 Microgrids

The goal of the microgrid is to put into practice five functions: (i) sustainability, (ii) dependability, (iii) flexibility, (iv) affordability, and (v) scalability. "Sustainability" means avoiding climate change and limiting the use of fossil fuels and other natural resources. "Dependability" means to supply stable and quality power for use in technology-intensive industries, such as semiconductor device manufacturing and automotive industries. "Flexibility" is also related to the sustainability and stability of the power system. Suppose we integrate variable renewable energy, such as wind and solar power. In that case, we need the flexibility to establish a demand and supply balance using dispatchable power sources, such as thermal and hydropower plants. "Affordability" is obtained by avoiding costly technologies, such as nuclear fusion reactors, global super-grids, space solar PV, and artificial photosynthesis. "Scalability" is crucial for the development of a microgrid. Scalability means that this evolution can be accomplished at a reasonable cost proportional to the system size. This scalability is obtained if each microgrid has the four key technologies from the early stage of the evolution. In particular, the market mechanism is essential even if the power system size is very small.

7.2.1 Small Island Developing States

The population size of Small Island Developing States (SIDS) is limited. SIDS have a combined population of approximately 65 million [2], which is approximately 1% of the world's population. In SIDS, nearly 30% of the population lives at elevations less than 5 m above sea level. Therefore, SIDS are often said to be the most vulnerable to climate change due to the currently increasing CO_2 emissions.

The Intergovernmental Panel on Climate Change (IPCC) has reported that emissions from human activities substantially increase atmospheric concentrations of greenhouse gases, resulting, on average, in additional warming to the Earth's surface [3]. Based on IPCC reports, policymakers in various countries, including advanced countries, emerging economies, and SIDS, have proposed energy policies to introduce as much renewable energy as possible to curtail CO_2 emissions. Therefore, renewable energy is considered as a key to curtail CO_2 emission. However, skepticism persists due to the high investment cost in renewable energy and its integration into power grids.

Here, we briefly review the sustainable development goals (SDGs) [4] and discuss the synergy and trade-off between these goals. The SDGs consist of 17 goals and 169 targets. Some targets are common to different goals. Le Blanc expressed the SDGs as a bipartite network between the 17 goals and 169 targets [5]. Contracting the bipartite network, we obtain a network that only consists of targets. The links of the obtained network are weighted, even though the links of the original bipartite network are binary. Many high-degree nodes located at the center of the network, e.g., SDG 1 "poverty", SDG 10 "inequality", SDG 12 "sustainable consumption and production", and SDG 8 "growth and employment". The synergy between SDGs is identified as pairs of nodes connected by large weight links, e.g., SDG 1 "poverty" and SDG 10 "inequality", SDG 5 "gender" and SDG 4 "education", and SDG 16 "peaceful and inclusive", and SDG 10 "inequality". Conversely, a trade-off between SDGs is identified as pairs of nodes without links, e.g., SDG 7 "energy" and SDG 13 "climate change". Here, we point out the need for a concrete example studying the trade-off between energy goals and climate change.

In the following, we consider a power grid with highly renewable energy in the SIDS economy and discuss the nexus of energy, environment, and economic growth. This section aims to estimate the electricity price for a power grid with high renewable energy for a SIDS economy and to discuss how the estimated electricity price affects economic growth.

7.2.2 *Renewable Energy in Microgrid*

Our model of grid integration for renewable energy is described in this section [6–9]. We assume here that the thermal power plants in a SIDS economy consist primarily of diesel power, i.e., less coal and natural gas thermal power. We also assume that the renewable energy in the SIDS economy is mainly solar photovoltaic (PV), i.e., less wind power, because the low altitude area where most SIDS are located is rich in solar potential and poor in wind potential [10].

7.2.2.1 Evolving Microgrid

A typical conventional power grid today includes thermal power plants on the supply side and factories, office buildings, and individual residences on the demand side. The demand varies and fluctuates with time. This fluctuation needs to be balanced by changing the output of the thermal power plants.

The evolution of power grids is underway and moving toward the 1st phase of the smart grid. Solar PV panels are starting to be installed in various locations on power grids on both the supply and demand sides. The installation of renewable energy introduces additional fluctuation into a power grid. This requires additional balancing power on the supply side, even though the demand for electricity on the supply side decreases due to the additional renewable energy capacity. Therefore, we need a specific capacity for electricity storage to obtain sufficient balancing power for the stable operation of the power grid. Currently, the cost of electricity storage is high; therefore, the price of electricity must increase.

Shortly, demand side management systems, which consist of a smart meter, a home energy management server, and appliances applicable to the server, will become popular, and we will obtain a new source of balancing power on the demand side. This new balancing power will reduce the electricity storage capacity required to balance fluctuations in renewable energy. We call this the 2nd phase of the smart grid. However, it might be delayed to diffuse the technology of the 2nd phase of the smart grid due to the complicated relationships between power grid stakeholders.

Our grid model aims to describe the 1st phase of a smart grid to estimate the electricity price for a power system with a high level of renewable energy integration in a SIDS economy.

7.2.2.2 Microgrid Model

We formulate our grid model as an optimization model. The concept of our grid model is shown in Fig. 7.7. First, we define the objective function, which is equal to the fuel expenditure required to operate the thermal power plants:

$$\sum_{t=1}^{T}\sum_{i=1}^{N} b_i p_t^i, \tag{7.29}$$

where b_i is the fuel cost used to obtain a unit of electricity from a thermal power plant i and p_t^i is the output electricity at a time t from the plant i. T and N are the time period of the calculation and the number of thermal power plants, respectively. In this chapter, the period is one year, and the time step is one hour; therefore, $T = 24 \times 365 = 8760$.

The objective function in Eq. (7.29) is minimized under the global constraints of the demand–supply balance and some local constraints on the electricity storage.

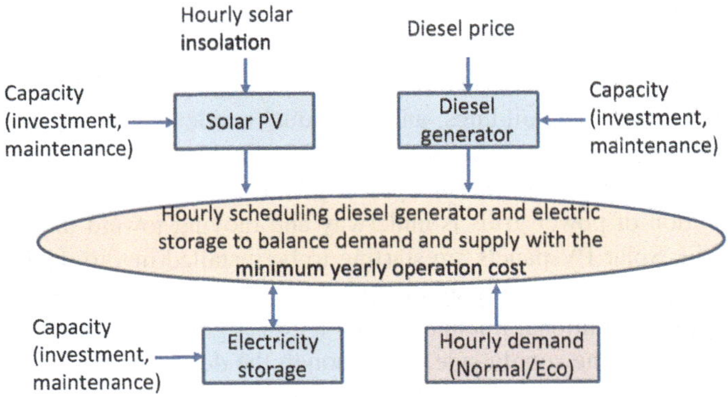

Fig. 7.7 Concept of microgrid model: The model is formulated using Eqs. (7.29)–(7.35)

The global constraints of the demand–supply balance are given as

$$\frac{\sum_{i=1}^{N} p_t^i + pv_t^{(f)} + g_t - h_t - d_t^{(f)}}{\sqrt{\sigma_d^2 + \sigma_p^2}} \geq \phi^{-1}(\alpha), \tag{7.30}$$

where $pv_t^{(f)}$, g_t, h_t, $d_t^{(f)}$, σ_d^2, and σ_p^2 are the solar PV output forecast at t, the discharged electricity from the electricity storage at t, the electricity charged to storage at t, the demand forecast at t, the variance of the demand fluctuation, and the variance of the fluctuation in the solar PV output, respectively. Note that the fluctuations of the demand and renewable energy are stochastically considered. α and $\phi(\alpha)$ are the probability of ensuring the supply–demand balance and the cumulative distribution function, respectively. We use $\alpha = 1.28$, which means that the demand–supply balance is satisfied with a probability of 90%. The cumulative distribution function is written using the error function $erf[\cdot]$ as

$$\phi(x) = \frac{1}{2}\left(1 + erf\left[\frac{x-\mu}{\sqrt{2\sigma^2}}\right]\right). \tag{7.31}$$

Here $\mu = 0$ and $\sigma = 1$ because the left-hand side of Eq. (7.30) is normalized. The local constraints on electricity storage are given as

$$v_t c_{min} \leq g_t \leq v_t c_{max}, \tag{7.32}$$

$$(1 - v_t)c_{min} \leq h_t \leq (1 - v_t)c_{max}, \tag{7.33}$$

$$R_{min} \leq \sum_{s=1}^{t}(h_s \eta - g_s)\Delta_t \leq R_{max}, \tag{7.34}$$

where v_t, c_{min}, c_{max}, R_{min}, R_{max}, η, and Δ_t are the state variable of the electricity storage ($v_t = 1$: discharge, $v_t = 0$: charge), minimum discharge power, maximum discharge power, minimum stored energy, maximum stored energy, efficiency, and time step, respectively.

Next, we describe the system-wise levelized cost of electricity (LCOE). The system-wise LCOE is estimated using thermal power plants, solar PV, and electricity storage. Therefore, the system-wise LCOE is interpreted as a measure of the electricity price, even though the cost of transmission and distribution and the profit of the utility company are not included. The system-wise LCOE is defined by the aggregated expenditure ($) to supply a unit amount of electricity (MWh) to the consumer:

$$\text{LCOE} = \frac{\sum_{y=1}^{Y} \frac{I_y+M_y+F_y}{(1+r)^y}}{\sum_{y=1}^{Y} \frac{E_y}{(1+r)^y}}, \tag{7.35}$$

$$E_y = \sum_{i=1}^{N} P_y^i + PV_y^{(f)} + G_y - H_y \tag{7.36}$$

$$P_y^i = \sum_{t=1}^{T} p_t^i, \tag{7.37}$$

$$PV_y^{(f)} = \sum_{t=1}^{T} pv_t^{(f)}, \tag{7.38}$$

$$G_y = \sum_{t=1}^{T} g_t, \tag{7.39}$$

$$H_y = \sum_{t=1}^{T} h_t, \tag{7.40}$$

where I_y, M_y, F_y, r, and T are the investment expenditures (including finance) in year y, the operation and maintenance expenditures in year y, the fuel expenditures in year y, the discount rate, and the life of the system, respectively. Note that I_y and M_y are aggregated for all the thermal power plants, renewable energy, and electricity storage on the supply side. Conversely, F_y is only aggregated for all thermal power plants. We assume that P_y^i, $PV_y^{(f)}$, G_y, and H_y are constant during the life of system Y.

7.2.3 Analysis Conditions

We assume a representative case to analyze a power grid with a high level of renewable energy in a SIDS economy. The details of the analysis conditions for a representative case are explained in this section.

7.2.3.1 Representative Case

We quantitatively estimated electricity prices for power grids in the Temburong area of Brunei with different shares of renewable energy using the grid model described in Sect. 7.2.2.2. Even though Brunei is not categorized as a SIDS area precisely, the Temburong area is an isolated enclave of Brunei, and its geographical and demographical conditions are similar to many SIDS areas. For this reason, we assume that the Temburong area is equivalent to SIDS in terms of power grid analysis.

The solar radiation time series in Brunei is shown in Fig. 7.8 for January 2016. Each line corresponds to the temporal change in solar radiation from 6:00 am to 20:00 pm daily. Figure 7.8 shows that the fluctuation can be considerable depending on the weather conditions, e.g., the coefficient of variation is 28% at noon.

The estimated electricity demand is shown in Fig. 7.9 for weekdays and weekends. The first 24 hours show the demand for a weekday, and the last 24 hours show that for a weekend day. The shape of the estimated electricity demand is relatively flat during the daytime and shows peaks in the early evening. The demand on a weekend day is considerably lower than on a weekday. The blue and red curves are the normal and eco-demand, respectively. Here, the eco-demand is estimated for energy-saving buildings. The peak of the eco-demand is approximately 78% that of the normal demand. Ref. [8] gives a detailed description of this demand estimation.

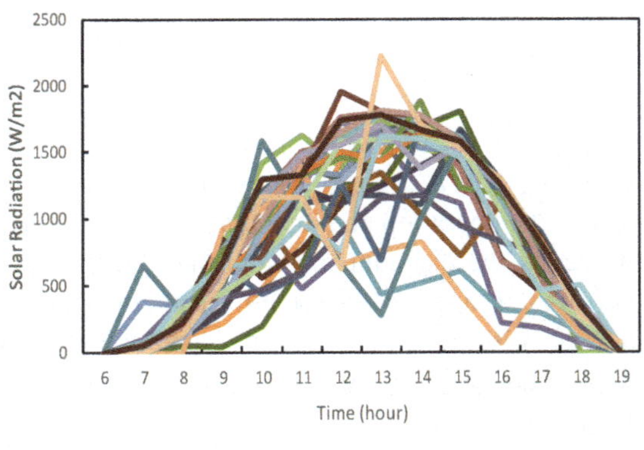

Fig. 7.8 Time series of the solar radiation in Brunei: Each line corresponds to the temporal change in the solar radiation from 6:00 am to 20:00 pm in a single day

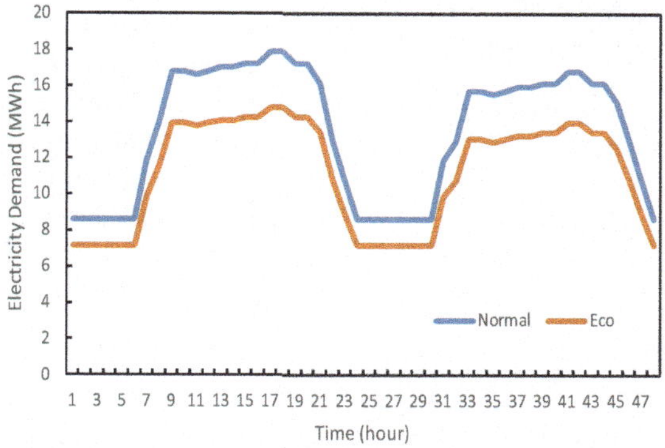

Fig. 7.9 Estimated electricity demand for a weekday and a weekend day: The eco-demand is estimated for energy-saving buildings. A detailed description of the demand estimation is given in Ref. [8]

Table 7.2 Parameters of diesel generator

Parameter	Value
Fuel price ($/litter) (Diesel price in Brunei, 1/3 of US)	0.32
Rated power (MW)	3.0
Road factor (Three-month maintenance after one-year operation)	0.8
Initial investment cost ($/kW)	650
Operation and maintenance cost ($/MW/year)	15 × 1000

Table 7.3 Parameters of solar PV

Parameter	Value
Installed capacity (MW)	0.0 to 192.0
Rated power of unit module (MW)	6.0
Road factor	0.18
Initial investment cost ($/kW)	1500 (Generation cost 5c/kWh)
Operation and maintenance cost ($/MW/year)	15 × 1000

7.2.3.2 Parameters on the Supply Side

The parameters for the diesel generators are shown in Table 7.2. The number of diesel generators N will change over 0–7. We assume that the diesel fuel price in Brunei is 1/3 the price in the USA. However, this assumption might only be valid for some SIDS economies. We use this assumption to provide a generic value in this representative case. In addition, when estimating the load factor of the diesel generators, we assume that the three-month maintenance is scheduled after one year of generator operation.

The solar PV parameters are shown in Table 7.3. The installed solar PV capacity varies in the 0.0–192.0 (MW) range. Note that the initial investment cost 1500 $/kW corresponds to the generation cost 5 c/kWh. This means that the generation cost of solar PV is reasonable today. The load factor 0.18 is relatively high for solar PV generation. We assume this value is plausible for most SIDS economies because most SIDS are located at low latitudes.

Table 7.4 Parameters of electricity storage

Parameter	Value
Minimum discharge power (MW)	0.0
Maximum discharge power (MW)	0.0 to 95.0
Minimum electricity storage capacity (MWh)	0.0
Maximum electricity storage capacity (MWh)	0.0 to 570.0
Efficiency	0.7
Initial investment cost ($/kWh)	200

The electricity storage parameters are shown in Table 7.4. Here, we consider battery-type electricity storage, such as Li-ion batteries, NaS batteries, lead acid batteries, and vanadium redox flow batteries. We assume a proportionality between the maximum discharge power x (MW) and the maximum stored electricity $6 \cdot x$ (MWh). Even though the efficiency 0.7 varies depending on the technology used for the storage, we use this as a generic value.

For the LCOE estimation in Eq. (7.35), we assume a life of the system of $Y = 20$ (years) and a discount rate of $r = 0.05$. Note that the fuel expenditures F_y and the electricity generation E_t calculated by aggregating the results of the optimization of our grid model described in Sect. 7.2.2.2 at an hourly time step over one year are used throughout the 20 years of the life of the system.

7.2.4 Estimation of System-Wise LCOEs

The estimation results for the grid operation and system-wise LCOE using the grid model described in Sect. 7.2.2.2 are shown and discussed regarding the nexus of energy, environment, and economic growth for a SIDS economy in this section.

The electricity supply required to satisfy a given demand was estimated throughout a year with a given output from the solar PV of 12 MW. Five diesel generators and a maximum electricity storage capacity of 30 MWh are required on the supply side to satisfy the normal demand throughout the year. The estimation results for the electricity supply on a weekday and a weekend in January are shown in Fig. 7.10. Here, the given electricity demand is indicated by a red curve for weekdays and weekends. The number of diesel generators is reduced from seven to five using a maximum electricity storage capacity of 30 MWh. In Fig. 7.10, p_t^1–p_t^5 show the outputs of the five diesel generators, h_t shows the excess output from solar PV charged into the electricity storage, and g_t shows the discharged electricity.

The estimated system-wise LCOEs are shown in Fig. 7.11 as a function of the eco-demand case's solar PV and electricity storage capacities. The white area in Fig. 7.11 indicates that the optimization described in Sect. 7.2.2.2 is not feasible for the specified solar PV and electricity storage capacities. For the systems in panel (a), the installed diesel generator and solar PV capacities are 12 MW and

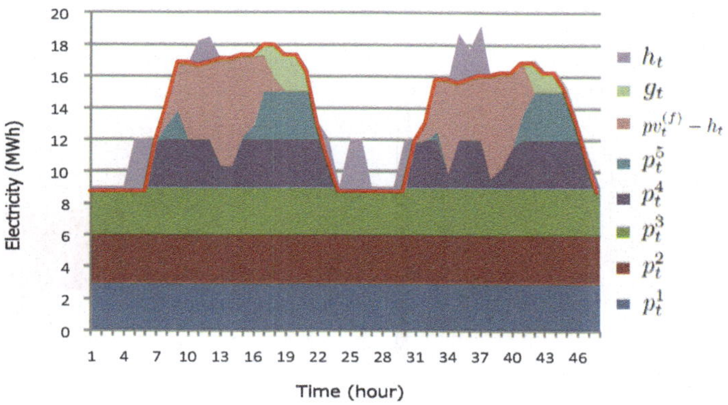

Fig. 7.10 Electricity supply on a weekday and a weekend in January: The electricity demand is indicated by a red curve for a weekday and a weekend day. p_t^1–p_t^5 show the outputs of the five diesel generators. h_t shows the excess output from solar PV charged into the electricity storage, and g_t shows the discharged electricity

6 MW, respectively. The maximum stored electricity is 48 MWh, and the lowest system-wise LCOE is estimated to be 123 $/MWh. For the systems in panel (b), the installed diesel generator and solar PV capacities are 6 MW and 72 MW, respectively, the maximum stored electricity is 180 MWh, and the lowest system-wise LCOE is estimated to be 245 $/MWh. For the systems in panel (c), the installed diesel generator and solar PV capacities are 0 MW and 120 MW, respectively, the maximum stored electricity is 450 MWh, and the lowest system-wise LCOE is estimated to be 397 $/MWh. The 12-MW and 0-MW diesel generator cases correspond to the business-as-usual and 100% While 100% renewable energy, i.e., no diesel generator, is technologically possible with electricity storage, the estimated LCOE is 397 $/MWh. Note that this estimated LCOE is substantially higher than electricity prices for residential consumers in the USA (125 $/MWh) and Japan (253 $/MWh) [11].

7.2.5 Nexus of Energy, Environment, and Economic Growth

The nexus of energy, environment, and economic growth is discussed here to provide invaluable insights toward realizing SDGs. The susceptibility analyses of the electricity price on economic growth can be broken down into two stages. The first stage analyzes the relationship between electricity demand and economic growth. The second stage is an analysis of the price elasticity of the electricity demand.

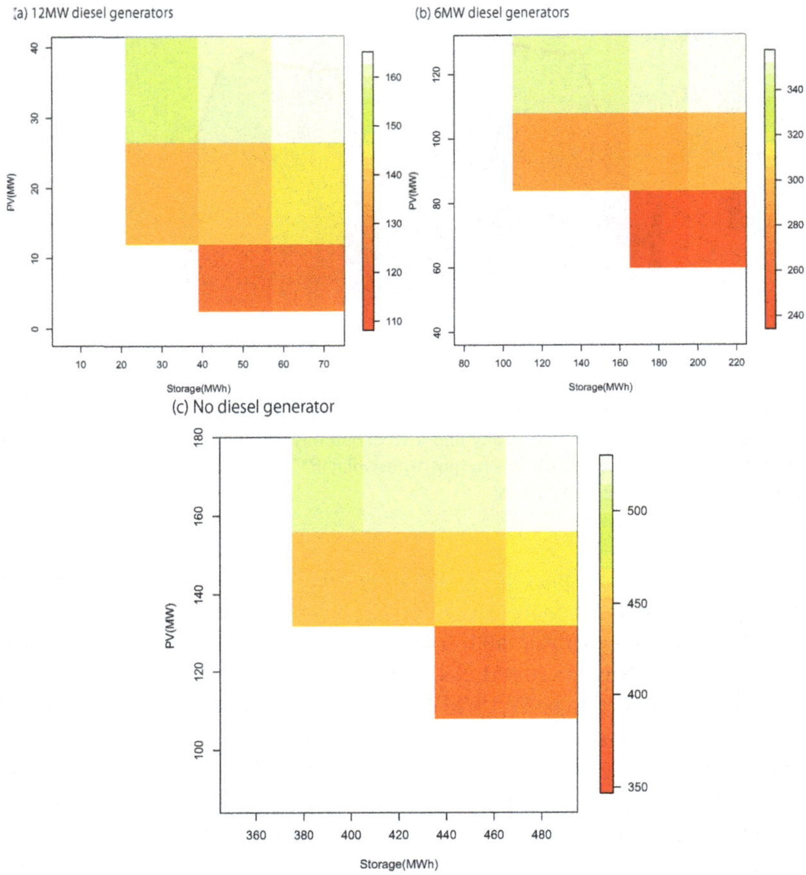

Fig. 7.11 System-wise LCOE as a function of the solar PV and electricity storage capacities for the eco-demand case: The white area indicates that the optimization is not feasible for the specified solar PV and the electricity storage capacities. The lowest system-wise LCOEs are 123 \$/MWh, 245 \$/MWh, and 397 \$/MWh in panel (**a**), (**b**), and (**c**), respectively

Studies in the first stage can be summarized as follows:

- In 17 Taiwanese industries, a study of electricity consumption and economic growth indicated that a 1% increase in electricity consumption boosts the real GDP by 1.72% [12].
- In China, from 1950 to 1980, electricity consumption's elasticity on economic growth increased from 2.33 to 2.72, which is significantly greater than 1. From 1990 to 2013, the growth rate of electricity consumption was highly correlated with change in the GDP growth [13].
- In the USA, the electricity consumption growth and GDP growth occurred at a similar rate from 1949 to 1996; however, from after 1996 to today, this correlation

gradually decreased [14]. This change might have been caused by structural changes in US industries toward the digital economy.

The studies in the second stage can be summarized as follows:

- In the USA, the estimated long-run price elasticity of the demand for commercial electricity varied from −3.11 to −0.497 depending on the states [15]. The average elasticity was −1.43. Conversely, the estimated long-run price elasticity of the demand for residential electricity varied from −0.618 to −0.058. The average elasticity was −0.282.
- In East Asia, an analysis of the long-run price elasticity of energy consumption, which is slightly different from electricity consumption, between 2000 and 2010 showed that the elasticity was +0.212, −0.125, +0.212, −0.0837, −0.343, −0.997, and +4.17 for Australia, Japan, China, India, Philippines, Thailand, and Singapore, respectively [16]. The average was −0.188: Here, Singapore was excluded from the average.

The first stage studies show that correlations between electricity demand and economic growth are very high. Therefore, we assume that the generic value of the correlation is approximately 1.0. The second stage studies show that the price elasticity of the electricity demand is a small negative value. Therefore, we assume that the generic value of the elasticity is approximately −0.05. The susceptibility analyses of the electricity price on the economic growth indicate that a 1% increase in electricity prices is associated with an approximately 0.05% decrease in economic growth. This assumption was confirmed by a study on the impact of electricity prices on economic growth in South Africa [17]. This study showed that a 1% We adopt the assumption that a 1% increase in electricity prices is associated with an approximately 0.05% decrease in economic growth for SIDS economies. These analyses imply that the estimated LCOE increases of 223% (=(397−123)/123) for the 100% renewable power grid correspond to an as high as 11% decrease in economic growth. This decrease would result in significant negative impacts on the SIDS economy.

The estimated high LCOE is primarily due to the high initial investment cost of the electricity storage shown in Table 7.4. Therefore, we need an innovation to permit the installation of electricity storage with a low initial investment cost. For example, grid-to-vehicle (G2V) and vehicle-to-grid (V2G) technologies might provide an additional capacity for electricity storage with a low cost [18], or excess solar PV supply could be used to produce hydrogen via the electrolysis of water [19, 20]. The produced hydrogen could then be stored in a hyperbaric chamber. Hydrogen combustion in gas turbines would reduce fuel consumption, and as a result, it would be possible to curtail CO_2 emissions. This would reduce the electricity storage capacity needed to balance the supply and demand. Consequently, a lower LCOE and, therefore, a higher economic growth would be attained with less CO_2 emissions.

7.2.6 Evolving Microgrid for SIDS Economies

We estimated the system-wise LCOE for a power grid with a high level of renewable energy to discuss how the estimated LCOE affects the economic growth in a SIDS economy. The estimation results of the system-wise LCOE using the grid model were discussed in terms of the nexus of energy, environment, and economic growth in a SIDS economy. While 100% renewable energy is technologically possible with the usage of electricity storage, the estimated LCOE is as high as 397 $/MWh. This estimated LCOE is substantially higher than electricity prices for residential consumers in the USA (125 $/MWh) and Japan (253 $/MWh).

Susceptibility analyses in South Africa, China, and the USA show that a 1% increase in electricity prices is associated with an approximately 0.05% decrease in economic growth. These analyses imply that the estimated cost increase of 223% with a 100% renewable power grid corresponds to an as high as 11% decrease in economic growth. This decrease in economic growth would result in a significant negative impact on SIDS economies.

The obtained high LCOE is primarily due to the high initial investment cost of the electricity storage. Therefore, we need an innovation to allow electricity storage with a low initial investment cost. For example, the excess solar PV supply could be used to produce hydrogen via water electrolysis. The produced hydrogen could be stored in a hyperbaric chamber. Hydrogen combustion in gas turbines would reduce the consumption of diesel fuel, and as a result, it would be possible to curtail CO_2 emissions. This would reduce the electricity storage capacity required to balance the supply and demand. Consequently, a lower LCOE and therefore higher economic growth would be attained with less CO_2 emissions.

Figure 7.12a shows a microgrid in its early stage. Here, most microgrids are independent, and transmission lines connect a few. Then, regional microgrids are formed by connecting adjacent microgrids, as depicted in Fig. 7.12b. Finally, many transmission lines are added between regional microgrids to form a national grid in Fig. 7.12c. Scalability means that this evolution can be accomplished at a reasonable cost proportional to the system size. This scalability is obtained if each microgrid

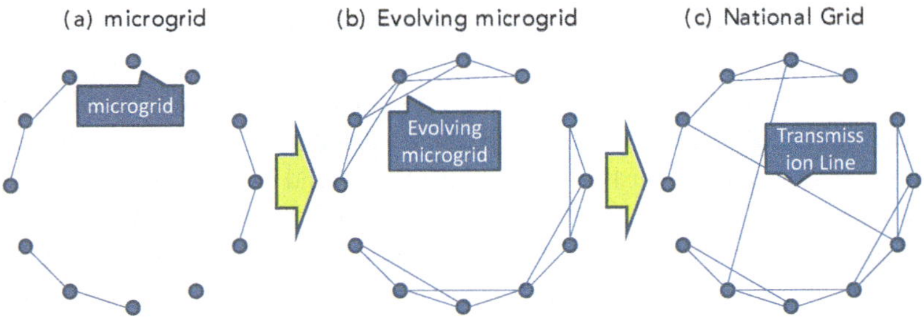

Fig. 7.12 Scalable Evolution of microgrid: (**a**) microgrid, (**b**) evolving microgrid, (**c**) national grid

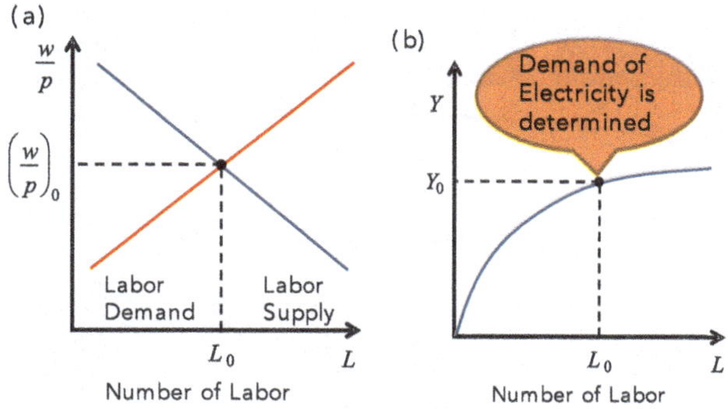

Fig. 7.13 Industrial Production and Demand of Electricity: (**a**) Labor Market, (**b**) Production Function

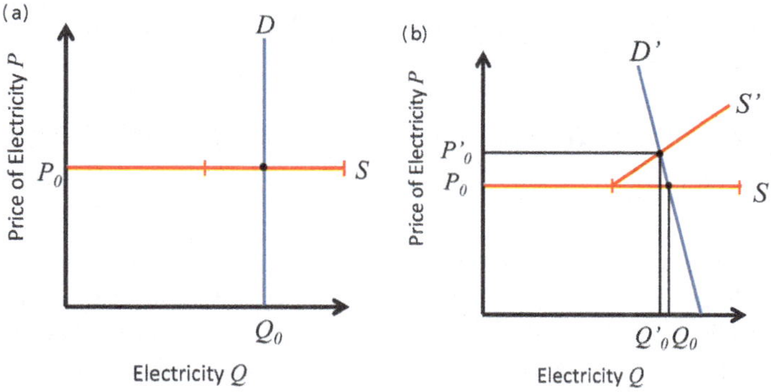

Fig. 7.14 Demand and Price of Electricity: (**a**) No Market Mechanism, (**b**) Supply and Demand Adjustment by Market Mechanism

has the four key technologies from the early stage of the evolution. In this evolution of the microgrid network, the transmission lines of the power grid are reconfigured according to Strogatz's β model, allowing for the flexibility of electricity between distant microgrids. In particular, the market mechanism is essential even if the power system size is very small.

The number of laborers and their wages is determined for an industry in the labor market as shown in Fig. 7.13a. Once the number of laborers is given, the supply of goods is determined by the production function in Fig. 7.13b. This production in the industry requires electricity Q_0. When we have no market mechanism, the electricity demand is presented by a vertical line D, which has no price elasticity of energy demand (see Fig. 7.14a). A utility company must supply Q_0 regardless of price P_0. On the other hand, when we have a market mechanism, the electricity demand is presented by an aslant line D', which has a price elasticity of energy demand (see Fig. 7.14b). If supply decreases due to output fluctuation of solar PV, a standby thermal power plant with higher cost supply electricity S'. As a result, price increases from P_0 to P_0', and demand decreases from Q_0 to Q_0'.

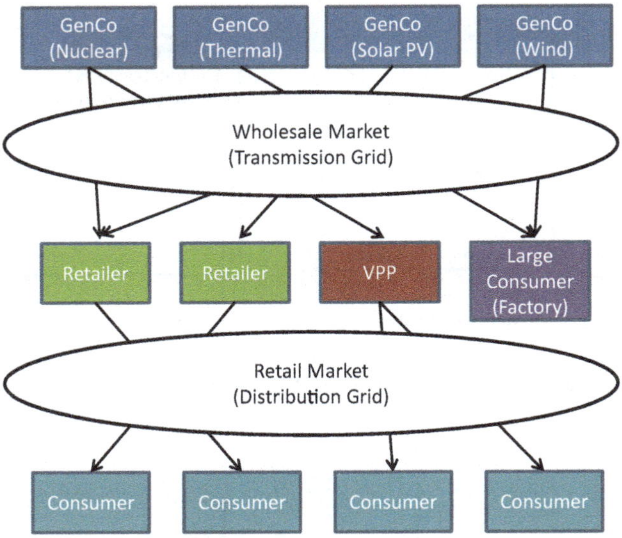

Fig. 7.15 Competitive Power Market: VPP stands for "Virtual Power Plant". VPP mainly works in balancing the market

The utility sector has been liberalized, and power markets have been introduced in Europe and the USA since the 1990s. Before the liberalization, regional monopoly was an ordinary business style and lacked the flexibility to integrate renewable energy. Fig. 7.15 shows a competitive power market after the liberalization was introduced [21]. Here, VPP stands for "Virtual Power Plant". VPP mainly works in balancing the market. They produce power by aggregating small-scale generating companies to supply balancing power. Similarly, they produce negative power (decrease demand power) by aggregate consumer demand response to supply balancing power. This flexibility accelerates the integration of PV and wind power by promoting the investment for balancing power into the system. However, we must pay attention to new issues arising in the power market. The liberalization is not always directly linked to the lower electricity price. In Europe, electricity prices increased because of the fuel price increase after the deregulation in the late 90s. Furthermore, it is pointed out in Europe that more than investment for transmission and balancing capabilities will be required in the near future.

Finally, we briefly review the history of the power market in the UK and Norway. The power market in the UK is summarized below: In 1990, the restructuring of the power industry was started. A mandatory pool was introduced to enhance competition. In 2002, the New Electricity Trading Arrangements, NETA, was established based on the over-the-counter trade. In 2005, NETA evolved into the British Electricity Trading Arrangements, BETTA. Power Market in Norway is summarized below: In 1991, restructuring and deregulation of the power industry were started. In 1993, a power market, NORD POOL, was established. In 2000, three Norwegian countries joined the NORD POOL.

7.3 Summary

A microgrid is an essential component of a fully functioning microgrid or a system of microgrids. A microgrid system allows optimal interaction of critical elements such as distributed generation, demand response, information technology, and market mechanism. It ensures that the overarching goal of establishing a sustainable society would be achieved in each microgrid.

The deployment of microgrid technologies, however, faces a significant challenge. These challenges include funding microgrid investments, interoperability of technologies, and consumer participation. The deployment, optimal operation, and utilization of smart technologies, therefore, require policy and regulatory interventions.

Electric utilities are mainly responsible for investments in key microgrid technologies. Demonstration projects could be funded by grants from either public or private entities, but replicating these projects in a system of microgrids requires a sustainable funding source. Policy and regulatory mechanisms, therefore, need to be established to ensure that utilities would recover their investments, whether on direct microgrid investments or to recover lost revenues due to energy efficiency improvements from microgrid deployment.

Microgrid technologies include communicating sensors and devices that exchange information and interoperate. Standards exist for some of these technologies but are still evolving for some. Regulatory agencies should ensure that microgrid technologies used in microgrids conform with existing international standards and provide clear guidance and directions regarding technologies with new or evolving standards.

Smart-grid technologies also elicit automated end-user demand and energy efficiency response. In order to promote efficient consumption, this must be accompanied by the introducing of consumer incentives through pricing schemes. Time-differentiated pricing schemes are found to stimulate behavioral changes and trigger demand response.

To further promote consumer participation in demand response, regulatory agencies must also assure consumer protection, especially related to privacy, ownership, and security issues related to access to detailed consumer data and other issues related to the social safety net associated with the introduction of new tariff rates, and protection associated with remote disconnections made by smart-grid technologies.

Decentralized renewable electricity in microgrids, on the other hand, could not only be supplied by utilities or independent power producers but also by electricity consumers who can generate their own supply or supply to the grid. Policy and regulatory interventions, such as feed-in tariffs and net metering schemes, would also be required to incentivize consumers to invest in RE technologies and be allowed to interconnect to the grid.

Considering the required level of investments and the evolution of technologies, the deployment of smart-grid technologies should progress incrementally. Developing a smart-grid road map for microgrids is therefore critical [22–25]. A road map could vary from one microgrid to another, which will be influenced mainly by the priority objectives. If the main objective is for higher deployment of renewable energy technologies, microgrid technologies that could be rolled out initially would facilitate higher renewables integration. On the other hand, if the objective is to improve energy efficiency, then smart metering and other technologies that facilitate demand response would be prioritized for implementation.

Overall, in order to achieve a fully functioning microgrid system in a microgrid that facilitates interaction between variable energy supply and flexible demand through a smart distribution network, a strong policy and regulatory intervention is required to incentivize (i) on the supply side consumers to become producers of variable renewable electricity supply, (ii) utilities to invest on standardized microgrid technologies, and (iii) on the consumption side, consumers to modify consumption patterns in response to time-differentiated pricing schemes.

References

1. A. Barabasi and M. Posfai, "Network Science", Cambridge University Press, July, 2016, ISBN 978-1107076266
2. UN-OHRLLS, "SIDS in Numbers Climate Change Edition 2015", (2015)
3. IPCC, "IPCC 1990 First Assessment Report Overview Chapter", (1990)
4. J. D. Sachs, "The Age of Sustainable Development", Columbia Univ Press, New York (2015)
5. D. Le Blanc, "Towards integration at last? The sustainable development goals as a network of targets", DESA Working Paper No. 141, March (2015)
6. Y. Ikeda and K. Ogimoto, "Cross-correlation of output fluctuation and system-balancing cost in photovoltaic integration", Journal of Engineering 2014, pp.1–9, https://doi.org/10.1049/joe.2014.0235 (2014)
7. S. Kimura, R. Pacudan, H. Phoumin, M. N. F. Matussin, M. H. Sapar, L. K. Seong, L. S. Meng, T. Kusajima, and Y. Ikeda, "Development of the Eco Town Model in the ASEAN Region through Adoption of Energy-Efficient Building Technologies, Sustainable Transport, and Smart Grids", ERIA Research Project Report 2015, No. 20, chapter 4, March (2015)
8. S. Kimura, M. N. F. Matussin, L. S. Meng, and Y. Ikeda, "Simulation Study on Energy Mix for Power Generation in Temburong Eco Town", ERIA Research Project Report 2017, No.02, chapter 3, October (2018)
9. Y. Ikeda, "Power grid with 100% renewable energy for small island developing states", Evolutionary and Institutional Economics Review, 17, pp.183–195 (2020).
10. S. Chatzivasileiadis, D. Ernst, G. Andersson, "The Global Grid", Renewable Energy 57, pp. 372–383 (2013)
11. International Energy Agency, "Energy prices and taxes quarterly", 4th Quarter (2015)
12. W. Lu, "Electricity Consumption and Economic Growth: Evidence from 17 Taiwanese Industries", Sustainability, 9, 50; https://doi.org/10.3390/su9010050 (2017)
13. C. Zhanga, K. Zhoua, S. Yanga, Z. Shaoa, "On electricity consumption and economic growth in China", Renewable and Sustainable Energy Reviews 76, pp.353–368 (2017)
14. R. F. Hirsh and J. G. Koomey, "Electricity Consumption and Economic Growth: A New Relationship with Significant Consequences?", The Electricity Journal, Vol. 28, Issue 9, pp.72–83, November (2015)
15. M. A. Bernstein and J. Griffin, "Regional Differences in the Price-Elasticity of Demand for Energy", NREL/SR-620-39512, February (2006)

16. H. Phoumin and S. Kimura, "Analysis on Price Elasticity of Energy Demand in East Asia: Empirical Evidence and Policy, Implications for ASEAN and East Asia", ERIA-DP-2014-05, April (2014)
17. H. B. Khobai, G. Mugano, and P. Le Roux, "The Impact of Electricity Price on Economic Growth in South Africa", International Journal of Energy Economics and Policy, 7(1), pp.108–116, (2017)
18. S. Heinen, D. Elzinga, S-K. Kim, Y. Ikeda, "Impact of smart grid technologies on peak load to 2050", International Energy Agency, Working Paper, OECD Publishing, August (2011)
19. S. E. Hosseini, M. A. Wahid, "Hydrogen production from renewable and sustainable energy resources: Promising green energy carrier for clean development", Renewable and Sustainable Energy Reviews 57, pp.850–866 (2016)
20. IRENA, "Hydrogen from renewable power: Technology outlook for the energy transition", International Renewable Energy Agency, Abu Dhabi (2018)
21. D. Kirschen and G. Strbac, "Fundamental of Power System Economics", John Wiley & Sons (20040.
22. IEA, Technology Road Maps: Smart Grids, International Energy Agency (2011).
23. IEA, How2Guide for Smart Grids in Distribution Networks, International Energy Agency (2015).
24. WEF, Accelerating Smart Grid Pilots, World Economic Forum (2010).
25. IEC, IEC Smart Grid Standardization Road Map. International Electrotechnical Commission (2010).

Chapter 8
Foreign Direct Investment in Renewable Energy

8.1 Causal Inference

Causal inference is the statistical estimation of the causal effects of events based on incomplete information obtained from observational data. In this chapter, we explain one method of causal inference, a methodology to identify factors using exploratory factor analysis (EFA) and to infer causal relationships using the structural equation model (SEM; also called the covariance structure analysis, CSA).

8.1.1 Exploratory Factor Analysis

Exploratory factor analysis (EFA) is used in many fields, including behavioral and social sciences, medicine, economics, and geography. The purpose of EFA is to interpret a new space by reducing the dimensions of the original space and reducing the number of new dimensions that may underlie the old dimensions [1]. The output of factor analysis gives a clearer view of the data and can be used for the subsequent analysis [2]. Before modeling by SEM, an EFA is often performed first.

The factor analysis aims to find a small number of "unobserved" latent common factors (latent variables) in the correlations among a large number of variables observed in an experiment (observed variables). It is classified into exploratory factor analysis and confirmatory factor analysis [3]. Confirmatory factor analysis is used when there is a prior hypothesis about the number of factors and which items are related to which factors, and it is used to verify the hypothesis. EFA is used when there is no clear hypothesis about the factors.

Y. Ikeda, *Data Science of Renewable Energy Integration*,
Evolutionary Economics and Social Complexity Science 30,
https://doi.org/10.1007/978-981-99-8779-5_8

The following are the basic procedures of EFA:

Confirmation of the Existence of a Factor
The test measures sampling adequacy for each variable in the model and for the complete model. Variables with a Measure of Sampling Adequacy (MSA) of 0.8 or greater are considered appropriate for inclusion in the factor analysis.

Determination of the Number of Factors
In order to determine the optimal number of factors, the minimum average partial correlation (MAP) criterion is used [4]. Partial correlation coefficients between observed variables with the principal components as control variables are obtained. The number of principal components with the minimum average partial correlation is the number of factors.

Estimation of Factor Loadings
The magnitude of the influence of a latent variable on multiple observed variables is called "factor loadings." The factor loadings are the regression coefficients between the observed variables and the latent variables, and their estimation can be done using either maximum likelihood or principal axis factorization (PAF), depending on the degree to which the data are normally distributed. In PAF, the first factor explains as much variance as possible; then the second factor explains the following most variance. Selecting an appropriate factor axis rotation method to facilitate factor interpretation is also necessary. For factor interpretation, the factors should have a simple structure. Simple structure means that the factor loadings for each item are large for only one particular factor and close to zero for the others. The rotation methods are roughly classified into orthogonal rotation and oblique rotation. Linear rotation assumes no correlation between factors, while oblique rotation assumes that there is a correlation. In practice, since there is a correlation between the factors, oblique rotation is often a better fit for the data.

Interpretation of Factors
After selecting an appropriate rotation method and estimating the factor loadings using it, the factors are interpreted. We identify the items strongly related to each factor (with large factor loadings) and interpret what the factor implies.

8.1.2 Structural Equation Modeling

Structural equation modeling (SEM) is a comprehensive approach to testing hypotheses about the relationship between observed and latent variables [5–8]. Its purpose is to understand patterns of correlation/covariance among a set of variables and to explain as much of that variance as possible in a given model. In the multi-regression analysis, the analysis is based on a fixed model. However, in structural equation modeling, it is possible to flexibly construct a model that matches the causal relationship of the data. It is characterized by the fact that it quantitatively estimates the influence of endogenous variables while examining the relationships among variables, i.e., it simultaneously performs factor analysis and multiple-regression analysis to express the causal relationship of the data.

The following are the basic procedures of SEM:

Variables
There are three types of variables: observed variables, latent variables, and error variables. Observed variables and latent variables are sometimes collectively referred to as structural variables. Factors obtained by exploratory factor analysis (EFA) are used as latent variables.

Measurement Equation
Based on the results obtained in the exploratory factor analysis (EFA), the influence from the observed variable to the latent variable is expressed.

Structural Equation
Based on the analyst's mind model, the structural equation expresses the causal relationships among latent variables. The direction from cause to effect is considered.

Covariance Relation
The covariance relation expresses the correlation between observed variables. It considers that there are correlations between observed variables at different points in time.

Parameter Estimation
The maximum likelihood estimation or the Bayesian estimation is used for parameter estimation. Thus, in addition to verifying the causal model that explains the entire data, it is possible to evaluate which observed variables significantly influence the latent variables. The fit of the model to the data can be considered good if some of the following conditions are met: SRMR (Standardized Root Mean Square Residual) < 0.05, AGFI (Adjusted Goodness of Fit Index) > 0.95, RMSEA (Root Mean Square Error of Approximation) < 0.05, CFI (Comparative Fit of Index) > 0.95, and TLI (Tucker–Lewis Index) > 0.95.

Path Diagram
Path diagrams are often used to represent SEM models. A path diagram usually consists of nodes that serve as variables and arrows that indicate the relationship between the variables. In the path diagram, latent variables (factors) are represented by "ellipses," observed variables by "rectangles," and parameters indicating the strength of the relationship between variables by "arrows." Error variables affect the observed variables but not the latent variables.

8.1.3 Illustrative Example

We perform the factor analysis using "PoliticalDemocracy" data in R library "semTools." The data consist of 75 observations of 11 variables, summarized in Table 8.1.

Table 8.1 Variables of "PoliticalDemocracy" data

Variable	Description
y1	Expert ratings of the freedom of the press in 1960
y2	The freedom of political opposition in 1960
y3	The fairness of elections in 1960
y4	The effectiveness of the elected legislature in 1960
y5	Expert ratings of the freedom of the press in 1965
y6	The freedom of political opposition in 1965
y7	The fairness of elections in 1965
y8	The effectiveness of the elected legislature in 1965
x1	The gross national product (GNP) per capita in 1960
x2	The inanimate energy consumption per capita in 1960
x3	The percentage of the labor force in the industry in 1960

Table 8.2 Factor loadings

	MR2	MR1	MR3
y1			0.71
y2		0.76	
y3			0.87
y4		0.50	0.36
y5			0.59
y6		0.92	
y7			0.57
y8		0.60	
x1	0.91		
x2	0.97		
x3	0.89		

In the factor analysis, we used R libraries "psych" and "GPArotation." The Kaiser–Meyer–Olkin (KMO) factor adequacy is a sampling adequacy criterion. The KMO factor was more significant than 0.8 for each variable, and the overall MSA was 0.88. Therefore, it is considered possible to find the factors by the factor analysis.

The Velicer MAP was 0.05 from the factor analysis, indicating two factors. The BIC was −89.72, a factor of two, while the sample-size-adjusted BIC was −2.58, a factor of five. Therefore, the number of factors may be larger than two, and we assumed that the number of factors was three in this analysis.

Assuming that the number of factors is three, we estimated the factor loading. The root mean square of the residuals (RMSR) was 0.022, the Tucker–Lewis Index of factoring reliability was 0.9833, and the RMSEA index was 0.0476. These figures indicate that the estimated results of the factor loading are statistically reliable. Table 8.2 shows the estimated factor loadings and Fig. 8.1 visualizes the results of these estimations. The obtained factors MR1, MR2, and MR3 are interpreted as "politics," "economy," and "democracy," respectively.

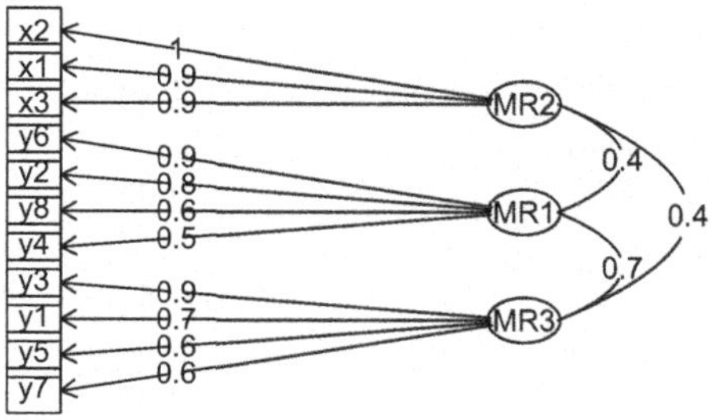

Fig. 8.1 Factor Analysis. The obtained factors MR1, MR2, and MR3 are interpreted as "politics," "economy," and "democracy," respectively

In the structural equation modeling, we used R libraries "semTools" and "semPlot." Based on the results of the factor analysis, the following measurement equations are obtained:

$$\begin{aligned} \text{economy} &\rightarrow \text{x1} + \text{x2} + \text{x3} \\ \text{democracy} &\rightarrow \text{y1} + \text{y3} + \text{y5} + \text{y7} \\ \text{politics} &\rightarrow \text{y2} + \text{y4} + \text{y6} + \text{y8}. \end{aligned} \tag{8.1}$$

How to give a causal relationship between factors is very important. In this analysis, we assume that the economy influences democracy and that the economy and democracy influence politics, as in the following structural equations:

$$\begin{aligned} \text{democracy} &\leftarrow \text{economy} \\ \text{politics} &\leftarrow \text{economy} + \text{democracy}. \end{aligned} \tag{8.2}$$

Furthermore, for the covariance relation between variables, we assumed that there are correlations between the same type of variables at different time points, as follows:

$$\begin{aligned} \text{y1} &\leftrightarrow \text{y5} \\ \text{y3} &\leftrightarrow \text{y7} \\ \text{y2} &\leftrightarrow \text{y6} \\ \text{y4} &\leftrightarrow \text{y8}. \end{aligned} \tag{8.3}$$

The parameter estimation of the structuring equation modeling was performed. The Comparative Fit Index (CFI) was 0.975, the Tucker–Lewis Index (TLI) was 0.963, the Root Mean Square Error of Approximation (RMSEA) was 0.078, and the Standardized Root Mean Square Residual (SRMR) was 0.052. These figures indicate that the estimated parameters are statistically reliable. The results of the parameter estimation are shown in Tables 8.3, 8.4, 8.5, and 8.6.

Table 8.3 Parameter estimation of the measurement equations

Objective var	Explanatory var	Estimate	Std.Err	z-value	P(>|z|)	Std.lv	Std.all
Economy	→						
	x1	1.000				0.670	0.921
	x2	2.177	0.139	15.714	0.000	1.459	0.972
	x3	1.817	0.152	11.979	0.000	1.218	0.872
Democracy	→						
	y1	1.000				2.193	0.842
	y3	1.085	0.157	6.926	0.000	2.380	0.730
	y5	0.955	0.104	9.185	0.000	2.095	0.807
	y7	1.245	0.148	8.439	0.000	2.731	0.837
Politics	→						
	y2	1.000				2.890	0.737
	y4	1.038	0.138	7.506	0.000	3.000	0.902
	y6	0.909	0.110	8.294	0.000	2.628	0.785
	y8	1.000	0.134	7.462	0.000	2.891	0.897

Table 8.4 Parameter estimation of the structural equation

Objective var	Explanatory var	Estimate	Std.Err	z-value	P(>|z|)	Std.lv	Std.all
Democracy	←						
	Economy	1.640	0.389	4.216	0.000	0.501	0.501
Politics	←						
	Economy	0.315	0.370	0.852	0.394	0.073	0.073
	Democracy	1.140	0.194	5.867	0.000	0.865	0.865

Table 8.5 Parameter estimation of the covariance relation

	Estimate	Std.Err	z-value	P(>|z|)	Std.lv	Std.all
y1 ↔ y5	0.403	0.373	1.082	0.279	0.403	0.187
y3 ↔ y7	0.413	0.613	0.674	0.500	0.413	0.104
y2 ↔ y6	1.664	0.792	2.102	0.036	1.664	0.302
y4 ↔ y8	−0.768	0.552	−1.393	0.164	−0.768	−0.375

Finally, the SEM results are summarized in a path diagram, as shown in Fig. 8.2. From this path diagram, we can read the causal relationships among the latent variables (factors), the relationships between the latent and observed variables, the variance of each variable, and the correlations among the variables. Although various assumptions are necessary, as explained above, it is possible to statistically infer causal relationships from the observed data using EFA and SEM.

Table 8.6 Variances

	Estimate	Std.Err	z-value	P(>|z|)	Std.lv	Std.all
x1	0.081	0.020	4.134	0.000	0.081	0.152
x2	0.122	0.071	1.732	0.083	0.122	0.054
x3	0.467	0.090	5.167	0.000	0.467	0.240
y1	1.979	0.459	4.306	0.000	1.979	0.292
y3	4.956	0.951	5.211	0.000	4.956	0.467
y5	2.347	0.505	4.647	0.000	2.347	0.349
y7	3.197	0.718	4.452	0.000	3.197	0.300
y2	7.018	1.274	5.509	0.000	7.018	0.457
y4	2.070	0.717	2.887	0.004	2.070	0.187
y6	4.316	0.821	5.256	0.000	4.316	0.385
y8	2.033	0.682	2.980	0.003	2.033	0.196
Economy	0.449	0.087	5.178	0.000	1.000	1.000
Democracy	3.600	0.865	4.162	0.000	0.749	0.749
Politics	1.531	0.646	2.371	0.018	0.183	0.183

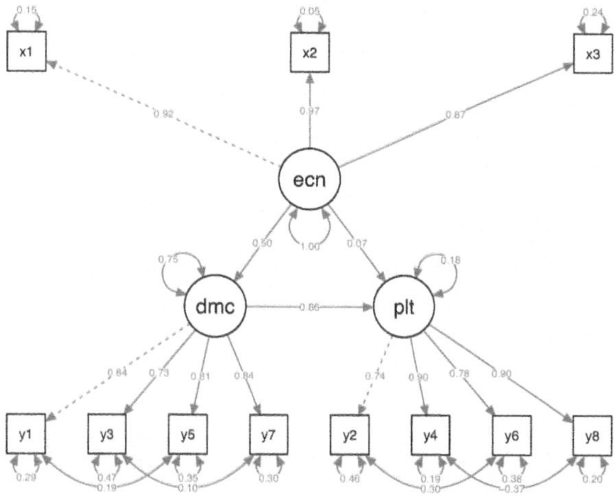

Fig. 8.2 Path diagram. Latent variables (factors) are represented by "ellipses," observed variables by "rectangles," and parameters indicating the strength of the relationship between variables by "arrows"

8.2 Carbon Pricing and Investment to Developing Countries

8.2.1 Development Assistance by Governments

Development cooperation refers to international cooperative activities by governments and government-related organizations with the primary objective of developing regions. Public funds for development cooperation are called Official Development Assistance (ODA). Governments or their implementing agencies provide ODA to developing countries or international organizations to develop developing countries, including peacebuilding, governance, promotion of fundamental human rights, and humanitarian assistance.

8.2.2 Foreign Direct Investment by the Private Sector

In recent years, development assistance has reached a turning point in the international debate, shifting from a project-centered to a poverty reduction-centered approach and from a loan-centered to a grant-centered approach. The question is whether development assistance promotes economic growth. Some empirical studies have concluded that aid does stimulate growth if the recipient country's policies are favorable, but the estimates could be more robust. The United States, for example, emphasizes political aspects of aid, while Nordic countries, for example, focus on poverty reduction. While receiving aid itself, China is actively providing aid to Africa and other countries.

In the world of development assistance, it is increasingly recognized that aid alone is not sufficient for economic growth and that private capital flows are also significant. Concerning foreign direct investment (FDI), which is the flow of funds from the private sector to developing countries, it has been shown that if the level of education in the recipient country is high, FDI can stimulate growth or have a positive impact through technological spillovers, even if there is no direct effect. Therefore, development assistance is not a direct response to the growth of the recipient country.

The question then arises whether development assistance induces FDI in recipient countries. The effects of development assistance on FDI include "infrastructure effects" and "rent-seeking effects." The infrastructure effect is the positive effect of improving developing countries' economic and social infrastructure. The rent-seeking effect is an adverse effect caused by the creation of regulations by lobbying politics and bureaucracy, which stimulates economic agents to act to change the environment in which they operate to their advantage. Previous empirical studies have found no apparent FDI-inducing effect on development assistance (Table 8.7).

8.2.3 Types and Objectives of Foreign Direct Investment

Russia's invasion of Ukraine has triggered a global review of supply chains, and there is growing concern about the hollowing out of domestic industry. Why do companies establish subsidiaries in foreign countries and transfer industries to

Table 8.7 Types and objectives of FDI

Type	Purpose	Representative study
Horizontal FDI	Savings in transportation costs	[9]
Vertical FDI	Reduction of production costs	[10]
Export-based FDI	Savings in production costs	[11]
Combined FDI	Savings in production and transportation costs	[12]

foreign countries through FDI? By organizing the types of FDI by firms, we will sort out the objectives of firms' overseas relocation (FDI strategies).

Horizontal FDI
The typical objective is to save transportation costs by moving the production site closer to the market. In order to sell cars to the US consumers, it is better to make and sell cars at a local subsidiary established through FDI than to make cars in Japan and export them to the USA, thereby saving transportation costs. FDI aimed at reducing transportation costs in this way is called horizontal FDI.

Vertical FDI
Another typical objective is to save on production costs. Wages in Japan and Thailand are different. Therefore, production costs can be reduced by producing in Thailand, where wages are lower, and importing to Japan. FDI that aims to reduce production costs in this way is called vertical FDI.

Export-Based FDI
In reality, FDI may not be categorized as either vertical or horizontal FDI. To save transportation and production costs, some Japanese firms relocated their production bases to China and supplied products to the Japanese and Chinese markets. One way for Japanese firms to supply the US market with automobiles is to export automobiles to the USA that were manufactured by a subsidiary in Mexico, where wages are lower. This strategy is called export-based FDI.

Complex FDI
Another possible strategy is for a Japanese company to manufacture auto parts at a subsidiary in Mexico, where wages are lower, and to do final assembly at a subsidiary in the USA. This is called complex FDI. This is called a "complex integration strategy." This strategy reduces production costs by transferring the intermediate goods production process to a low-wage country. At the same time, transportation costs are also reduced by conducting the final goods production process in the country that serves as the market. Both production and transportation costs can be reduced.

8.2.4 Foreign Direct Investment and New Trade Theories

In order to elucidate the international development of the firms, a new trade theory (firm heterogeneity model) is being studied.

In the traditional trade theory, countries concentrate on producing goods with a comparative advantage and import comparably inferior goods through trade from foreign countries [13]. Ricard's comparative advantage theory focuses on labor productivity due to differences in technological levels and considers only labor as a production factor. Heckscher and Olean's theory of comparative advantage considers both labor and capital. Capital-rich countries produce and export capital-intensive goods, while labor-rich countries produce and export labor-intensive goods, and the profits from the import and export of these goods are why international trade occurs. As a result, international trade from comparative advantage is inter-industry trade between different industries.

However, in recent years, trade has been centered on intra-industry trade; Krugman postulated a monopolistic competitive market to reflect the characteristics of differentiated goods and explained why intra-industry trade in terms of diversity goods [14, 15]. Furthermore, Melitz focused on firms' productivity and showed that firms with low productivity exit the international market because they cannot bear the fixed costs of exporting. In contrast, firms with high productivity earn higher profits from the international market by bearing the fixed costs [16]. Following these trends, research was conducted on the relationship between firm productivity and overseas business strategies [17, 18]. Here, differences in firm productivity are referred to as firm heterogeneity. The overseas business strategy of a firm also refers to export or foreign direct investment (FDI).

FDI has traditionally been classified into two major categories, horizontal FDI and vertical FDI, based on the two-country model of the home country and host country [19], and other variants, as explained above. Horizontal FDI is a direct investment in developed countries to establish a base abroad that replicates the domestic market's production and sales processes. Horizontal FDI is selected when it is more advantageous than exporting for saving trade costs (trade barriers) and transportation costs. On the other hand, vertical FDI is a direct investment in developing countries, which divides the production process into different production stages and implements an international division of labor to control production costs by taking advantage of disparities in the cost of labor capital. The objective is to transfer labor-intensive production processes to developing countries with low wages.

In recent years, there has been an increase in the type of trade in which intermediate goods are procured from the host country, home country, or third country, and the final goods are sold to the host country, home country, or third country. FDI that establishes an overseas base for this type of trade is networked FDI [20]. As a result of widespread networked FDI, each production process is now connected through trade across borders. This production process form is called a global production network [21, 22].

Traditionally, imports and exports between developing and developed countries have been considered inter-industry trade in which primary commodities and industrial products are imported and exported. In contrast, imports and exports between developed countries are considered intra-industry trade. In recent years, economic integration in Europe has led to the development of a production pattern characterized by industrial integration. In East Asia, on the other hand, intra-industry trade, which corresponds to the cross-border division of labor between developing countries and developed countries or between developing countries, has been increasing.

Also, there has been an increase in research on the robustness and resilience of the global production networks for exogenous shocks such as the Asian financial crisis (1997–1998), the United States subprime mortgage crisis (2008–2009), the severe floods in Thailand (2011), and the Great East Japan Earthquake (2011) [23–27].

8.2.5 Carbon Pricing Mechanism

The potential of carbon pricing mechanism (emissions trading system) as a driver for new economic growth is attracting attention. Large amounts of CO_2 are being emitted, accelerating climate change and causing damage due to various abnormal weather conditions. Carbon pricing is positioned at the center of CO_2 emission mitigation measures. Carbon pricing will change corporate behavior to mitigate CO_2 emissions. The question is how to create innovation to put existing technological seeds into practical use.

An emission trading system is one in which the government sets a cap on emissions, and emitters subject to the system trade allowances in the market as needed. The carbon price is determined as a result of trading allowances. Each emitter has three options depending on its emission reduction costs.

Option 1 Reduce emissions by improving production efficiency or converting to energy with lower carbon emissions to reduce their emissions.

Option 2 Reduce emissions by purchasing allowances from other businesses and from businesses with surplus allowances (businesses with low emission reduction costs).

Option 3 Depending on the offset credit utilization system, some emission reductions can be covered by credits created by emission reduction projects.

For Option 3, the concept of carbon offsetting is essential. Citizens and companies, after identifying the part of their emissions that their CO_2 emission reductions cannot be sufficient, purchase credits for CO_2 emission reductions achieved in other places. This is called carbon offsetting.

8.2.6 Carbon Pricing and FDI

Some developing countries argue that developed countries should take responsibility for global warming because it is caused by CO_2 emitted by developed countries in the past. Comparing the historical cumulative CO_2 emissions of a group of more than 147 developing countries and a group of more than 20 developed countries around the world, the historical cumulative CO_2 emissions were 41% and 39%, respectively, in 1990, about the same level. On a per-country basis, this can be interpreted as a result of the economic development of the Western industrialized countries since the Industrial Revolution of the eighteenth century, which the consumption of fossil fuels has supported.

Developing countries are urging developed countries to take more initiative in responsibility for global warming and reduce their CO_2 emissions. This historical emissions-based approach is expressed in the term "common but differentiated responsibility" (CBDR) in the United Nations Convention on Climate Change. CBDR means that all countries are responsible for halting global warming, but the weight of that responsibility varies from country to country.

In recent years, however, there has been a change in the composition of such a conflict between developed and developing countries. The historical cumulative CO_2 emissions of the developing country group more than doubled to 61% in 2016, compared to 28% for the developed country group. Rapid economic development and population growth in many countries of the developing country group are the leading causes of this change. In response to these changes, developed countries, which bear a heavy responsibility, have begun to voice their discontent. In contrast, some developing countries argue that "we also have the right to emit CO_2 for poverty reduction and economic development." The situation on the part of developing countries is not uniform. Small island nations in the Pacific Ocean, which face an imminent threat of national extinction due to rising sea levels caused by global warming, are calling for immediate global warming countermeasures. On the other hand, the "emerging economies" such as the BRICS (Brazil, Russia, India, China, and South Africa), whose population, economic scale, and CO_2 emissions are all off the charts, continue to prioritize economic growth, believing that they will not suffer much damage from global warming.

In recent years, global-scale disasters, including extreme weather events caused by global warming, have been causing severe damage across national borders, regardless of whether they occur in developed or developing countries. It is a pressing issue for not only developed countries but also developing countries, whose infrastructures and coping capacities (financial, technological, and human) are poorer than those of the developed countries, to cope with disasters caused by global warming.

Carbon pricing and FDI are considered to be effective in resolving these issues. In particular, carbon offsetting is expected to play a significant role. Suppose that carbon taxes, the price of emission credits, are higher in developed countries than in developing countries. A company in an industrialized country may make an FDI to implement an emission reduction project in a developing country. Companies in developed countries can purchase emission reduction credits from this project to compensate for some of their emissions through carbon offsets. In other words, using FDI CO_2 emission reduction projects and carbon offsets is expected to help resolve the conflict between developed and developing countries represented by CBDR.

In the following sections of this chapter, FDI to implement CO_2 emission reduction projects using renewable energy will be discussed in detail.

8.3 Investment and Development in Renewable Energy

The future development of the global economy is closely linked to the sustainable, efficient, and safe use of raw materials and energy based on sustainable production concepts that are ecologically and economically appropriate for the long-term future of society [28]. The transition to an energy system based on sustainable production technologies is critical in these challenges. The twentieth century was characterized mainly by the rise of carbon-based economies that relied heavily on fossil fuel consumption; in the twenty-first century, many countries have recognized the importance of increasing the deployment of renewable energy to address climate change and air pollution and to create new economic opportunities and have yet to provide access to electricity increasingly provide sustainable energy for those who do not have it. Developing countries, in particular, face significant challenges in meeting growing electricity demand and providing electricity to all as their populations grow and their economies expand rapidly [29]. Considering the expected increase in electricity demand and CO_2 emissions in developing countries, the energy systems of developing and developed countries must be transformed.

The transformation will only work if technology transfer is adequately implemented and financial resources are appropriately allocated. A closer look at the flow of foreign direct investment (FDI) in renewable energy in developing countries is needed to examine the factors affecting the allocation of this critical type of financial resource to the renewable energy sector and the extent to which renewable energy support policies affect the allocation of FDI financial resources. Financing terms and conditions heavily influence the cost of renewable energy projects. Therefore, policymakers need to consider the impact of renewable energy support policies on the allocation of FDI [30].

Concerning FDI, developing countries' allocations to renewable energy are approaching those of developed countries. However, FDI allocations vary widely among developing countries. Understanding what factors determine the allocation of FDI to renewable energy in developing countries is necessary. Given the importance of support measures specific to the renewable energy sector, such as political, economic, and regulatory support, in promoting renewable energy, these sector-specific factors influence the allocation of FDI and the traditionally accepted determinants of FDI. Most empirical studies on the determinants of FDI examine this issue from the perspective of FDI as a whole, but very few studies focus on specific industries or sectors [31]. Since different industries and sectors have other factors in attracting funds, there are no precise results regarding FDI. In 2014, FDI in the renewable energy sector was about 7% of the total FDI (45,000 billion yen) [32], and among all, it is in the top five. Therefore, investigating the determinants of FDI in renewable energy in developing countries not only reveals the effectiveness of renewable energy support measures but also demonstrates how important sector-specific factors are compared to traditionally accepted determinants and provides an opportunity to focus on specific industries and sectors. We want to clarify the need to investigate the determinants of this international financial resource that is focused on a particular industry or sector.

We use causality modeling to analyze the impact of renewable energy support measures on the allocation of FDI compared to conventionally accepted determinants. The results show that renewable energy support policies are equally or more effective than traditional determinants such as level of governance, price stability, access to finance, corruption control, and GDP growth rate. This result demonstrates the effectiveness of renewable energy support policies and the importance of analyzing the determinants of FDI, focusing on specific sectors rather than looking at FDI as a whole.

The following section provides an overview of empirical studies on the determinants of FDI. It serves as a basis for variable selection for the quantitative analysis of the determinants of FDI in wind power in developing countries.

8.4 Precedent Studies on the Determinants of FDI

There are numerous precedent studies on the determinants of FDI. Regardless of the hypothesis, prior studies have examined various combinations of variables and yielded mixed results regarding statistical significance and causality direction. Due to the mixed results of prior studies, it is difficult to reach a consensus on which explanatory variables can be perceived as the "true" determinants of FDI. The following subsections provide widely accepted determinants explained in light of prior research and renewable energy sector-specific determinants that may be incorporated into subsequent analyses of the determinants of FDI in wind energy in developing countries [52].

8.4.1 Traditional FDI Determinants

Market Potential

Some researchers argue that market size/potential as indicated by GDP per capita or GDP can be considered the most robust determinant of FDI [33]. Some studies also show that the GDP growth rate is a statistically significant explanatory variable, whereas GDP is not [34].

Price Stability

Price stability usually refers to maintaining a stable and low inflation rate over a long period. Stable and low inflation, or price stability, is considered to have a statistically positive and significant relationship with the allocation of FDI [35]. Given that many investments in renewable energy, including wind power investments, require a long payback period, price stability may strongly influence investors' decisions.

Trade Protection
Kojima argues that the impact of trade protection on FDI depends on the specific nature of the investment [36]. Suppose the objective is to supply the domestic market and overcome trade barriers. In that case, tighter trade protection increases the likelihood that firms will replace affiliated production with exports to avoid trade protection costs, commonly referred to as tariff-jumping FDI. On the other hand, if investment viability is highly dependent on imports, FDI would be encouraged if trade protection were less stringent. However, trade policies can limit the ability of investors to import, increase transaction costs, and adversely affect production efficiency, thus discouraging the implementation of FDI [37]. Previous empirical studies have shown evidence supporting both assumptions, and the expected results are not precise [38].

Investment Restrictions on Foreign Investors
Possible investment restrictions on foreign investors include (1) foreign business ownership, (2) industries and firms open to foreign investors, and (3) performance requirements for foreign firms. The correlation between investment restrictions and FDI would be intuitive. Indeed, without barriers, capital will flow to countries with higher return on investment and productivity rates. Firms tend to invest in countries with less restrictive regulations on capital flows.

Access to Finance
Access to finance is another important determinant of FDI empirically examined in the previous studies [39]. Some researchers have empirically shown that the relative deterioration of Japanese firms' access to finance in the 1990s has strongly influenced the decline in Japanese FDI [40].

Labor Cost/Labor Quality
Labor cost and labor quality are important determinants of FDI that influence potential investors' decisions [41]. Minimum wages and regulatory aspects such as restrictions on hiring and firing are also considered essential determinants.

Corruption
The level of corruption in the host country is recognized as one of the critical determinants of FDI allocation. In theory, corruption is considered an additional tax on profits [42] and harms the profitability of the investment. Thus, corruption can have a significant impact on FDI allocation decisions.

Government Effectiveness
Government effectiveness can positively impact investors' business by reducing heavy bureaucracy and the overall time and procedures it takes to finish it [43, 44]. Therefore, government effectiveness, as measured by the quality of policy formulation and implementation, government commitment to policy, and the quality of public services, is one of the critical determinants of FDI.

Regulatory Quality
Regulatory quality is another important determinant for foreign investors; Fazio and Talamo find that quality regulation promotes FDI by reducing the negative effects

of unfriendly market policies such as restrictions on capital movement, government intervention, and price controls [45].

Rule of Law
The rule of law has a substantial impact on long-term asset values because the rule of law protects future returns [46]. In particular, since renewable energy investments, including wind power, require a long payback period, the existence of the rule of law may have a strong influence on investors' decision-making.

Political Stability
Political stability ensures continuity, especially for projects heavily influenced by existing policies [47]. Also, since most FDI projects are long-term investments, foreign investors will avoid investing if a threat could adversely affect future returns [45]. Meier has also empirically shown that when political risks are high, many MNCs avoid FDI in a country [48]. In some cases, political instability has led to changes in regulations and economic support, posing a significant threat to FDI projects in the renewable energy sector.

8.4.2 Determinants of FDI in the Renewable Energy Sector

Government involvement is an essential determinant of FDI in the renewable energy sector. Government involvement is essential during the emergence phase when renewable energy technologies must be protected from direct competition from existing technologies to promote their diffusion. During the emergence phase, market forces alone will not be sufficient to promote the widespread diffusion of renewable energy technologies: As of early 2015, more than 164 countries had renewable energy targets, and about 145 countries have introduced some form of support policies for renewable energy [49]. Government support policies are sometimes perceived as a means to correct the negative externalities of fossil fuel use and promote technology shifts [49]. These policies can significantly impact the development of renewable energy technologies. For wind power, some studies argue that "it is the nature of the policy instrument, rather than the resource base for wind power, that has the greatest impact on wind power deployment" [49].

Renewable energy support can be divided into three types: policy, regulatory, and economic. Despite their theoretical and empirical significance for the diffusion of renewable energy, no study has examined the impact of these types of renewable energy support on the distribution of FDI. Policy support includes institution building, strategy development, and goal setting. Regulatory support includes the presence of audits, priority grid access, and norms and standards. Financial support includes the presence of feed-in tariffs (FiTs), renewable portfolio standards (RPSs), tradable renewable energy certificates (RECs), and tax incentives. Although focused on something other than FDI, some studies have examined the impact of these various renewable energy supports on renewable energy deployment at the national and state levels. That study examined the economic efficiency of various renewable energy economic support policies for renewable energy technology development, comparing quantity-based and price-based approaches and concluded that FiT is the best economic support policy in terms of economic efficiency [50]. There is also a study of the diffusion of renewable energy technologies (excluding hydro energy) in 108 developing countries between 1980 and 2010 [51]. They found that economic and regulatory support policies strongly impact the diffusion of renewable energy technologies. In contrast, policy support harms the diffusion of renewable energy technologies.

8.5 Analysis of FDI in Wind Power Plants

In this section, we review the results of a study of FDI in wind power plants in developing countries using EFA and SEM [52].

8.5.1 Data of Wind Power Plants

Data on existing wind power plant installations in developing countries were obtained from GlobalData (2015). The data include the power plant's location, ownership, installation year, and capacity. Based on ownership data, plants owned by companies headquartered in a different country (home country) than the country where the plant is located (host country) are counted as foreign direct investments in wind power plants. If a plant is owned by more than one company, the capacity of the plant is divided by the number of owners for convenience. Power plant capacity and the number of direct investment projects are treated as proxies for the amount of direct investment.

Data on wind power plants installed between 2008 and 2014 were used in the analysis: 190 FDI projects were implemented between 2008 and 2014, representing 7,950 MW of wind power plants in developing countries. Wind power plants owned by domestic firms were also analyzed for comparison with the results of FDI: as with FDI, plants owned by domestic firms were counted as domestic investments based on ownership data. Some plant data do not have ownership information, but these are also counted as domestic investments in the analysis: between 2008 and 2014, 2,626 projects were implemented, occupying 52,790 MW of wind power plants.

Data from GlobalData on the ownership structure of wind power plants in developing countries show that the share of direct investment in total investment varies widely from country to country. China and India have larger installed capacities than other countries, but these two countries need to stand out concerning the amount of FDI.

8.5.2 Analysis Results

The renewable energy industry is one of the fastest-growing industries, attracting significant amounts of FDI, attracting more than 7% (450 billion yen) of the total FDI in 2014, making it one of the top five industries in terms of FDI allocation. Given the importance of this industry and the large differences in FDI allocations in renewable energy in developing countries observed in data obtained from Global-Data, this analysis focuses on wind energy in developing countries. It examines how the determinants of this industry differ from traditional determinants. Determinants include theoretically rejected and empirically validated variables and determinants specific to the renewable energy industry. After conducting an exploratory factor analysis (EFA), variables constructed two latent factors, Governance and Business Environment. Then, variables specific to renewable energy were added as exogenous variables, and structural equation modeling (SEM) was conducted.

First, an exploratory factor analysis (EFA) was performed using data from the years in which the firms received FDI and the years in which they did not receive FDI. The two latent factors obtained, F1 and F2, refer to Governance and Business Environment, respectively; the Economy and Regulation factors are structural as exogenous variables, given the theoretical and empirical importance of these factors in both the FDI and domestic investment cases incorporated into the structural equation modeling (SEM) analysis. Similarly, GDP growth, a measure of market potential, was also included in the analysis as an exogenous variable, given its importance as a determinant of FDI. Renewable energy policy did not yield load factors above the cutoff line in the analyses.

Next, based on the EFA results, a structural equation model (SEM) analysis was performed, including the latent variables governance and business environment and the variables RE Economic, RE Regulatory, GDP growth rate, and capacity. Capacity represents the amount of investment in wind farms and was used as a proxy variable for investors' investment decisions. The structural equation model was constructed based on the following hypotheses:

Hypothesis 1 Countries with better governance will receive more investment.

Hypothesis 2 Countries with a good business environment will receive more investment.

Hypothesis 3 Countries with regulations and financial support for renewable energy will attract more investment.

Hypothesis 4 Larger potential markets will attract more investment.

Hypothesis 5 Better governance will promote a better business environment.

Hypothesis 6 Countries with better governance and a better business environment will adopt regulatory and economic support for renewable energy.

Hypotheses 1–4 are based on the previous studies. Hypothesis 5 is based on the argument that a country's investment climate is largely determined by its level of governance. Similarly, we examine the impact of better governance and a better

Table 8.8 Statistical significance of each path in FDI

Variable	Factor	p-Value
Economic	Institutional	0.021
RE Regulatory	Economic	0.034
RE Regulatory	Institutional	0.843
RE Economic	Economic	0.047
RE Economic	Institutional	0.480
TF (Trade Freedom)	Economic	0.016
IF (Investment Freedom)	Economic	0.014
FF (Financial Freedom)	Economic	0.017
CC (Control of Corruption)	Institutional	0.000
RoL (Rule of Law)	Institutional	0.000
GE (Government Effectiveness)	Institutional	0.000
Capacity	Economic	0.000
Capacity	Institutional	0.044
Capacity	RE Regulatory	0.000
Capacity	RE Economic	0.062
Capacity	GDP growth	0.462

business environment on the development of regulation and financial support for renewable energy.

The results of the FDI analysis will be described. The structural model fit measures were at acceptable levels for all three indicators. A high correlation of 0.508 is found between RE Economic and RE Regulatory. This is a reasonable result given that the correlation between the two can be interpreted as "government willingness to support renewable energy." Table 8.8 shows the statistical significance of each path depicted by the model. The paths from governance to electricity economics and electricity regulation are not statistically significant, with p-values of 0.48 and 0.84, respectively. Furthermore, the path from GDP growth to capacity is also insignificant (p-value: 0.46). Based on this model, market potential is not a strong determinant of direct investment in wind energy in developing countries.

Governance is shown to influence the business environment strongly, and the business environment is shown to influence economic and regulatory support. After accounting for these indirect effects, the effect of each variable on investment capacity is 0.29 for energy regulation and 0.12 for the renewable energy economy. The direct effect of governance on capacity shows a very small but statistically significant negative effect (value: −0.14; $p < 0.05$). The implication of this result is that the lower the level of governance, the more deals can be done. This result should be interpreted as governance having little influence on investment decisions since the effect is very small. This is true even when indirect effects are taken into account, with a total effect of governance of 0.04.

This section describes the analysis results for the comparator, domestic investment. The goodness-of-fit measures of the model are at acceptable levels. As for the statistical significance of each path in Table 8.9, as in the case of FDI, the paths from Governance to Wind Power Economy and Wind Power Regulation are statistically

Table 8.9 Statistical significance of each path in domestic investment

Variable	Factor	p-Value
Economic	Institutional	0.004
RE Regulatory	Economic	0.012
RE Regulatory	Institutional	0.819
RE Economic	Economic	0.022
RE Economic	Institutional	0.459
TF (Trade Freedom)	Economic	0.003
IF (Investment Freedom)	Economic	0.002
FF (Financial Freedom)	Economic	0.002
CC (Control of Corruption)	Institutional	0
RoL (Rule of Law)	Institutional	0
GE (Government Effectiveness)	Institutional	0
Capacity	Economic	0
Capacity	Institutional	0.247
Capacity	RE Regulatory	0.738
Capacity	RE Economic	0
Capacity	GDP growth	0

insignificant with p-values of 0.46 and 0.82, respectively, unlike in the case of FDI, the path from "Regulator" to "Production Capacity" is also statistically insignificant (p-value: 0.738).

It is worth noting that GDP growth has a significant ($p < 0.01$) and a strong (0.24) effect on productive capacity in the case of domestic investment but is not statistically significant for productive capacity in the case of FDI. This emphasizes that market potential is not an important variable for FDI in wind power in developing countries. Regarding domestic investment, the RE economy had the strongest impact on the investment decision (0.28), followed by GDP growth (0.24). For FDI, regulatory support for renewable energy has the strongest direct effect on investment decisions. In contrast, regulatory support is statistically insignificant for domestic investment, and economic support has the strongest effect. Finally, the business environment has a statistically significant and negative effect. A country may adopt economic support for renewable energy as an industrial strategy despite the low quality of the business environment.

The results of the analysis are summarized as follows:

1. Economic support for renewable energy substantially impacts investment decisions for both direct and domestic investment.
2. Regulatory support for renewable energy is particularly important for foreign investors, but for domestic investors, economic support has a more significant effect than regulatory support.
3. Traditional FDI determinants such as financial access, trade openness, and general investment regulations still strongly influence foreign investors' investment decisions.
4. Governance aspects weakly influence foreign investors' investment decisions.

5. The more a country's market grows, the more active domestic investment is. However, market potential has no statistically significant effect on foreign investors' investment decisions.

These results suggest that more than the mere presence of economic support policies, including price signals such as FiTs, is needed to create an attractive environment for foreign investors. Since foreign investors perceive the lack of adequate regulatory support as a major risk, well-structured and credible regulatory support policies are needed in addition to economic support policies to attract foreign investment in the renewable energy sector.

8.6 Summary

Causal inference is the statistical estimation of the causal effects of events based on incomplete information obtained from observational data. In this chapter, we explain one method of causal inference, a methodology to identify factors using exploratory factor analysis (EFA) and to infer causal relationships using the structural equation model (SEM) (also called the covariance structure analysis, CSA).

We discussed that carbon offsets and FDI are considered to be effective in resolving these issues. The use of FDI CO_2 emission reduction projects and carbon offsets is expected to help resolve the conflict between developed and developing countries represented by CBDR.

We analyzed FDI on wind energy in developing countries and examined how this industry's determinants differ from traditional ones. After conducting an exploratory factor analysis (EFA), variables constructed two latent factors: Governance and Business Environment. Then, variables specific to renewable energy were added as exogenous variables, and structural equation modeling (SEM) was conducted.

References

1. Rietveld, T., Van Hout, R., "Statistical techniques for the study of language and language behaviour", Walter de Gruyter,. (1993).
2. Kokubo, Y., Suzuki, M., Nagata, M., Sato, K., Kawatsuki, N., Uchida, H., "Lecture Evaluation Questionnaire Items Desired by the Alumni of the Faculty", Journal of Japanese Society for Engineering Education, 54. (2006).
3. Leandre R. Fabrigar, Duane T. Wegener, "Exploratory Factor Analysis", Oxford University Press, December 22, 2011, ISBN 978-0199734177
4. Velicer, W. F, "Determining the number of components from the matrix of partial correlations", Psychometrika,41,321–327, (1976)
5. Morgan, S., Winship, C., "Counterfactuals and Causal Inference: Methods and Principles for Social Research", New York: Cambridge University Press. (2007) ISBN 978-0-521-67193-4
6. Demirhan, E., Masca, M., "Determinants of foreign direct investment flows to developing countries: a cross-sectional analysis", Prague economic papers, 4, 356–369, (2008).

7. Reinartz, Werner, Michael Haenlein, and Jörg Henseler. "An empirical comparison of the efficacy of covariance-based and variance-based SEM", International Journal of Research in Marketing 26.4 (2009): 332–344.
8. Kline, R.B., "Principles and practice of structural equation modeling", Guilford publications. (2015).
9. Markusen, James R. (1984) "Multinationals, Multi-plant Economies, and the Gains from Trade", Journal of International Economics, 16(3-4):205–226.
10. Helpman, Elhanan. (1984) "A Simple Theory of International Trade with Multinational Corporations", Journal of Political Economy, 92(3):451–471.
11. Ekholm, Karolina, Rikard Forslid and James R. Markusen. (2007) "Export-Platform Foreign Direct Investment", Journal of the European Economic Association, 5(4):776–795.
12. Yeaple, Stephen Ross. (2003) "The Complex Integration Strategies of Multinationals and Cross Country Dependencies in the Structure of Foreign Direct Investment", Journal of International Economics, 60:293–314.
13. Krugman P. Obstfeld, M., Melitz, M.: International Trade. Theory and Policy, 10th Edition. Pearson (2015)
14. Krugman, P.: Scale economies, product differentiation, and the pattern of trade. The American Economic Review 70(5), 950–959 (1980)
15. Helpman, E., Krugman, P.: Market Structure and Foreign Trade. Increasing Returns, Imperfect Competition, and the International Economy. Cambridge, MA: MIT Press (1985)
16. Melitz, M.J.: The impact of trade on intra-industry reallocations and aggregate industry productivity. Econometrica 71(6), 1695–1725 (2003)
17. Helpman, E., Melitz, M.J., R., Y.S.: Exports versus FDI with heterogenous firms. The American Economic Review 94(5), 300–316 (2004)
18. Antras, P., Helpman, E.: Global sourcing. Journal of Political Economy 112(3), 552–580 (2004)
19. Helpman, E.: A simple theory of international trade with multinational corporations. Journal of Political Economy 92(3), 451–471 (1984)
20. Baldwin, R., Okubo, T.: Networked FDI: Sales and sourcing patterns of Japanese foreign affiliates. The World Economy 37(8), 1051–1080 (2014)
21. Ernst, D., Kim, L.: Global productions networks, knowledge diffusion, and local capability formation. Research Policy 31(8), 1417–1429 (2002)
22. Kimura, F., Ando, M.: Two-dimensional fragmentation in east Asia: Conceptual framework and empirics. International Review of Economics and Finance 14(3), 317–348 (2005)
23. Obashi, A.: Stability of production networks in East Asia: Duration and survival of trade. Japan and the World Economy 22(1), 21–30 (2010)
24. Obashi, A.: Resiliency of production networks in Asia: Evidence from the Asian crisis. In: S. Evenett, M. Mikic, R. Ratnayake (eds.) Trade-led Growth: A Sound Strategy for Asia, chap. 3, pp. 29–52. United Nations Economic and Social Commission for Asia and the Pacific (ESCAP) (2011)
25. Ando, M., Kimura, F.: How did the Japanese exports respond to two crises in the international production networks? the global financial crisis and the great east Japan earthquake. Asian Economic Journal 26(3), 261–287 (2012)
26. Okubo, T., Kimura, F., Teshima, N.: Asian fragmentation in the global financial crisis. International Review of Economics and Finance 31, 114–127 (2014)
27. Todo, Y., Nakajima, K., Matous, P.: How do supply chain networks affect the resilience of firms to natural disasters? evidence from the great east Japan earthquake. Journal of Regional Science 55(2), 209–229 (2015)
28. Dovi, V. G., Friedler, F., Huisingh, D., Klemes, J. J., "Cleaner energy for sustainable future", Journal of Cleaner Production, 17.10: 889-895. (2009).
29. Dincer, I., "Renewable energy and sustainable development: a crucial review", Renewable and Sustainable Energy Reviews, 4, 157–175, (2000).
30. Wiser, R.H., Pickle, S.J., "Financing investments in renewable energy: the impacts of policy design", Renewable and Sustainable Energy Reviews, 2, 361–386. (1998).

31. Mullen, J.K., Williams, M., "Foreign direct investment and regional economic performance", Kyklos, 58, 265–282. (2005).
32. FDiintelligence, 2015. The fDireporT 2015 Global greenfield investment trends.
33. Artige, L., Nicolini, R., "Evidence on the determinants of foreign direct investment: the case of three European regions" (2005).
34. Khan, R.E.A., Nawaz, M.A., "Economic Determinants of Foreign Direct Investment in Pakistan", Journal of Economics, 1, 99–104. (2010).
35. Benacek, V., Lenihan, H., Andreosso-O'Callaghan, B., Michalíkova, E., Kan, D., "Political risk, institutions and foreign direct investment: how do they relate in various European countries?", The World Economy, 37, 625–653, (2014).
36. Kojima, K., "International trade and foreign investment: substitutes or complements", Hitotsubashi Journal of Economics, 16, 1–12. (1975).
37. Drabek, Z., Payne, W., "The impact of transparency on foreign direct investment", Journal of Economic Integration, 777–810. (2002).
38. Caetano, J., Galego, A., "FDI in the European Union and Mena Countries: Institutional and Economic Determinants", CEFAGE-UE Working Papers, 9, (2009).
39. Garcia-Herrero, A., Navia, D., "Determinants and impact of financial sector FDI to emerging economies: a home country's perspective", (2003).
40. Klein, M., Peek, J., Rosengren, E., "Troubled banks, impaired foreign direct investment: the role of relative access to credit", (2000).
41. Kinoshita, Y., Campos, N.F., "Why does FDI go where it goes? New evidence from the transition economies", (2003).
42. Al-Sadig, A., "Effects of Corruption on FDI Inflows", The. Cato J., 29, 267, (2009).
43. Amal, M., Tomio, B.T., Raboch, H., "Determinants of Foreign Direct Investment in Latin America" */Determinantes de la inversión directa externa en Latinoamérica/Determinantes do investimento estrangeiro directo na América Latina. Revista de Globalización, Competitividad y Gobernabilidad, 4, 116, (2010).
44. Sedik, W.M., Seoudy, H., "The Impact of Country Risk and New Institutional Economics on Foreign Direct Investment" (2012).
45. Fazio, G., Talamo, C., "How "Attractive" is good governance for FDI", International Finance Review, 9, 33–54. (2008).
46. Hoff, K., Stiglitz, J.E., "The creation of the rule of law and the legitimacy of property rights: the political and economic consequences of a corrupt privatization", (2005).
47. Asiedu, E., "On the determinants of foreign direct investment to developing countries: is Africa different?", World Development, 30, 107–119, (2002).
48. Meier, L., "Factors driving US foreign direct investment", (2009).
49. REN21, "Renewables 2015 Global Status Report" (2015).
50. Menanteau, P., Finon, D., Lamy, M., "Prices versus quantities: choosing policies for promoting the development of renewable energy", Energy Policy, 31, 799–812. (2003).
51. Pfeiffer, B., Mulder, P., "Explaining the diffusion of renewable energy technology in developing countries", Energy Economics, 40, 285–296. (2013).
52. Keeley, A. R, Ikeda, Y. Determinants of Foreign Direct Investment in Wind Energy in Developing Countries. Journal of Cleaner Production, Vol.161, pp.1451–1458, (2017).

Part III
Our Future World

Chapter 9
Hydrogen Energy Storage and Nuclear Energy

9.1 Hydrogen as Energy Storage

The International Energy Agency (IEA) has proposed the NZE 2050 scenario, which calls for net-zero greenhouse gas emissions by 2050 [1]. Achieving this will require major changes in the energy sector. Renewable energy sources such as solar PV and wind power, which have become increasingly popular in recent years, effectively reduce CO_2 emissions. However, other sources must compensate for their output fluctuations. The fluctuation and uncertainty of output and the need for flexible resources to balance supply and demand are not a problem unique to renewable energy [2]. However, as the deployment of renewable energy increases, the need to address it in all time domains, from real-time operation to long-term system planning, will increase [3]. The need for flexibility in the power system is becoming increasingly important, requiring the efficient use of a combination of different power sources [2, 4, 5].

Various electricity storage technologies have been developed to increase the power system flexibility with the increase of renewable energy sources [6]. The massive installation of solar PV power generation to reduce CO_2 emissions and the installation of energy storage devices to address the associated grid problems were considered. In addition, the use of storage batteries and hydrogen co-firing in thermal power generation and two technologies were proposed to reduce CO_2 emissions and the total costs. Pumped hydropower generation is already widespread, and hydrogen-based power storage technologies are promising for promoting large-scale grid integration of renewable energy sources in the future [6]. Using these power and energy storage devices, 100% renewable energy is at least technically feasible [7–9]. However, the use of storage batteries has been studied [7, 10], but there are currently issues regarding price and lifetime.

Hydrogen is attracting attention as a next-generation clean energy storage method for various applications ranging from energy transportation and storage to industrial

Y. Ikeda, *Data Science of Renewable Energy Integration*,
Evolutionary Economics and Social Complexity Science 30,
https://doi.org/10.1007/978-981-99-8779-5_9

use [11]. However, since hydrogen does not exist in nature as a single substance, a hydrogen production technology that does not generate CO_2 is needed. Such production technologies include electrolysis (alkaline water electrolysis, AWE), thermochemical methods (Sulfur–Iodine Cycle, SI cycle), photolysis, and biolysis [12]. For large-scale hydrogen production, it is desirable to use renewable energy and energy sources that do not emit CO_2, such as nuclear power (Light Water Reactor, LWR). The High-Temperature Gas Reactor (HTGR) [14], a next-generation nuclear reactor, is suitable for thermochemical hydrogen production due to its high coolant temperature of 750–1000 °C at the core outlet. This chapter will investigate hydrogen production by a combination of renewable energy and AWE [13] and hydrogen production by a combination of HTGR and SI cycle. This type of power generation and hydrogen production in HTGR is called hydrogen cogeneration.

In this chapter, a feasible and ideal future power supply configuration was analyzed for the TEPCO service area based on an economic evaluation by minimizing the total cost with CO_2 reduction as a constraint, considering the use of hydrogen [15]. The analysis was intended to be realistic, limiting the analysis to technologies that are expected to be put into practical use shortly and technologies with relatively low costs. In addition, since this study focuses on the economic evaluation of capital investment, the analysis of grid integration of renewable energy with output fluctuation using the unit commitment model [4, 8, 9] was not performed.

9.2 Analysis Conditions

9.2.1 Load Duration Curve

A load duration curve is a time-independent order of magnitude of the load within a certain period of "time stamp." In this study, the order of magnitude of the load duration curve, which is independent of time, is called "time" and is distinguished from "time stamp," which represents the actual passage of time. This study covers the TEPCO service area from 2030 to 2050, assuming that electricity demand in 2030 is the same as in 2019 and using 1 year of actual generation data from April 1, 2019 to March 31, 2020. The future solar PV and wind power generation were calculated by dividing the 2019 generation by the actual installed capacity in 2019 [16] and multiplying by the estimated installed capacity in 2030 [5]. The same assumptions were used for the years 2040 and 2050.

9.2.2 Time and Generation Technologies

The variables are distinguished by two subscripts t and j. First, t denotes time. Time t is a variable of the load duration curve and is distinguished from the time stamp.

Table 9.1 Index number and type of generation technology

j	Type of generation technology
1	Solar PV
2	Wind
3	Pumped hydro
4	Hydropower (general hydropower)
5	Coal-fired power generation
6	Liquefied Natural Gas (LNG) fired power generation
7	Hydrogen blending with liquefied natural gas
8	Nuclear (Light Water Reactor, LWR) power generation
9	Nuclear (High-Temperature Gas-cooled Reactor, HTGR) power generation

This study divides 1 year (8784 hours) into 2^7. The time step is $h_t = 8784/2^7$ and time t is the value of

$$t = h_t, 2h_t, 3h_t, \cdots, 2^7 h_t. \tag{9.1}$$

Also, j is the type of generation technology used ($j = 1, \cdots, N$). Table 9.1 shows the index number of j and the type of generation technology.

9.2.3 *Model Variables*

This section describes the variables used in the linear programming model.

9.2.3.1 Operating Output

The operating output $y_{tj}(t = 1, \cdots, T, j = 1, \cdots, N)$ (MW) is the output of power source j at time t. For example, $y_{6000,coal}$ is the output of coal-fired power generation at 6000 hours in the load duration curve.

9.2.3.2 New Installed Capacity

New installed capacity $x_j(j = 1, \cdots, N)$ (MW) is the capacity of newly constructed facilities in addition to the existing installed capacity of each source j. Existing facilities can be used without construction costs, only annual expenses and fuel costs, while new facilities require construction costs as well.

9.2.3.3 Fuel Cost

Fuel cost $FC_j(j = 1, \cdots, N)$, (¥) is the cost of fuel used in each source j in 1 year.

9.2.3.4 Pumped Storage Demand

Pumped storage demand $N_t(t = 1, \cdots, T)$ (MW) is the power required to pump water in pumped storage at time t and store it in the reservoir. In other words, it is the power used to "charge" the pumped storage power plant.

9.2.3.5 Hydrogen Production

The hydrogen production $H_{in,t}(t = 1, \cdots, T)(Nm^3)$ is the amount of hydrogen produced at time t. Here 1 Nm^3 is the volume of 1 m^3 of gas converted to standard conditions (0 deg C, 1 atm).

9.2.3.6 Hydrogen Consumption

Hydrogen consumption $H_{out,t}(t = 1, \cdots, T)(Nm^3)$ is the amount of hydrogen consumed at time t.

9.2.3.7 LNG Consumption

LNG (liquified natural gas) consumption $LNG_{out,t}(t = 1, \cdots, T)(ton)$ is the amount of LNG consumed at time t. It is used when determining CO_2 emissions for each year.

9.2.3.8 Coal Consumption

Coal consumption $Coal_{out,t}(t = 1, \cdots, T)(ton)$ is the amount of coal consumed at time t. This is also used when determining CO_2 emissions for each year.

9.2.4 *Model Parameters*

We describe the parameters used in the linear programming model. The parameters' values are given and treated as constants in the linear programming model. Table 9.2 summarizes the values of the parameters. Many values in Table 9.2 are based on the

Table 9.2 List of parameters and input values

Item name	Symbol	Hydrogen plant	Pumped hydro	LNG	Coal fired	General hydropower	Nuclear power (LWR)	Nuclear power (HTGR)	Wind power	Solar PV
Existing installed capacity in FY2030 (MW)	z_j	0	6142.4	21,480.8	2560	1756.8	10,089.6	0	0	0
Upper limit of installed capacity (MW)	zu_j	$1.00 \times 10^{1}9$	4936.7	1.00×10^8	1.00×10^8	4936.7	1.00×10^8	1.00×10^8	1.00×10^8	
Peak Availability Ratio	a_j	0.95	1	0.95	0.95	0.95	0.95	0.95	–	–
Annual expense ratio	b_j	0.0366/0.0225 (SI/AWE)	0.00996	0.0377401	0.0365768	0.00996	0.05214	0.03511	0.0368	0.02538
Construction unit price (10^3 yen/kW)	c_j	207/454(SI/AWE)	482	120	230	850	350	251	300	400
Upper limit of facility utilization rate	l_j	0.8	0.2	0.8	0.8	0.95	0.9	0.9	–	–
Fuel cost (yen/kWh)	p_j	0.160/0.0766	0.5067	8.6250	4.6933	0.506735	1.99	1.99	0	0
Years of operation	$year_j$	10	60	40	40	80	60	40	20	20
Decommissioning cost (construction cost rate)	$waste_j$	0.05	0.05	0.05	0.05	0.05	0.161904	0.161904	0.05	0.05

2011 Cost and Other Verification Committee estimates. The 2015 Power Generation Cost Verification Working Group estimates have extremely high coal and LNG prices. Given the current prices, the Working Group on Electricity Generation Cost Verification estimates are somewhat uncertain. Therefore, the Cost and Other Verification Committee estimates were used in this study.

9.2.4.1 Existing Installed Capacity

Installed capacity $z_j (j = 1, \cdots, N)$ corresponds to the initial value of the facility. It is assumed that as the facility is used, the installed capacity in 2030 will decrease to 80% of the 2015 actual value, the installed capacity in 2040 will decrease to 80% of the 2030 calculated value, and the installed capacity in 2050 will decrease to 80% of the 2040 calculated value.

9.2.4.2 Installed Capacity Limit

In this analysis, no installed capacity limit $zu_j (j = 1, \cdots, N)$ was set except for hydropower and pumped storage, which were assumed to be unlimited. In actual calculations, for example, a very large value such as 10^8 MW should be given as the upper limit. The total amount of hydroelectric and pumped hydropower generation was assumed to be 6142.4 MW and 1756.8 MW in TEPCO's service area in April 2020.

9.2.4.3 Facility Utilization Rate Limit

The real-world facility utilization rate limit $l_j (j = 1, \cdots, N)$ varies from power source to power source. Facility utilization rates were set for nuclear, general hydropower, and thermal power. For nuclear power, the maximum facility utilization rate was set at 90%. General hydropower is operated at a maximum of about 90–95%. Since the average of the facility utilization rate for general hydropower is 60% for the flow-in type and 45% for the storage type, the upper limit (0.95) for general hydropower in Table 9.2 is considered to be somewhat large. However, the calculation results in Fig. 9.2 reproduce the actual values for the composition ratio of hydropower, and the calculations with various conditions, as shown in Figs. 9.3, 9.5, 9.7, and 9.8, show no change in the composition ratio of hydropower, which is approximately 6%. These results indicate that there is no problem with this upper limit. By its principle, pumped hydro cannot have a high utilization rate. For peak use, the upper limit of the utilization rate would be about 20%. Five hours of operation in 24 hours would result in 20%. Five hours of operation would require $5/0.65 = 7.7$ hours for pumped water. Given the demand for electricity generated by pumping (peak load), it is unlikely that pumped storage capacity utilization would be more than 30–40%. There is no need to set the capacity utilization for solar and wind power because the electricity generated was given in the manner described in Sect. 9.2.1.

9.2.4.4 Facility Availability at Peak Demand

How to set the facility utilization rate $a_j(j = 1, \cdots, N)$ during peak hours is a matter of facility operation and cannot necessarily be a technical decision. Suppose electricity supply and demand are tight, and facilities are expected to be in short supply during peak hours. In that case, it is recommended that the repair and maintenance of facilities be moved to off-peak hours. It is also necessary to consider equipment breakdowns. For nuclear power, 95% was assumed by considering actual repair maintenance and breakdowns based on the flat operation. For hydropower, a similar assumption of 95% was made.

9.2.4.5 Annual Expense Ratio

The annual expense is the sum of operation and maintenance costs and overhead. The annual expense ratio $b_j(j = 1, \cdots, N)$ is used to convert the construction cost (yen/kW) into the cost of generating electricity per year. The construction cost was set to be multiplied by the annual expense ratio to arrive at the 1-year operating cost. The calculations for coal-fired, LNG-fired, general hydropower, light water reactor, wind power, and solar power were based on the figures from the Japan Cabinet Secretariat's Cost and Other Verification Committee. Pumped storage was set to the same value as general hydropower; HTGR was calculated based on calculations performed by the Japan Atomic Energy Agency. For the hydrogen plant, the SI cycle was calculated based on the JAEA report [17]. For alkaline electrolysis, the value of 2.25% was used from the literature [18].

9.2.4.6 Unit Construction Cost

The unit construction cost $c_j(j = 1, \cdots, N)$ is $yen/(Nm^3/h)$ for hydrogen plant only and yen/kW for other power sources. For coal-fired, LNG-fired, general hydropower, LWR, wind, and solar PV power, the figures from the Japanese government's Cost Estimation and Review Committee were used. For HTGR, an estimated value of 198 thousand yen/kW was published by JAEA in 2006 and revised in 2018 [19]. Including unit construction costs, operating costs, and feedstock costs, the estimated unit power generation cost per unit of electricity generated in 2006 was 4.14 yen/kW, and the estimated unit power generation cost in 2018 was 5.3 yen/kW, which is 1.28 times higher than the previous year. However, this unit power generation cost is based on the unit power generation cost (yen/kWh) = (capital cost + operation and maintenance cost + fuel cost)/(amount of electricity generated (transmission end)). Therefore, the unit cost of construction was also multiplied by 1.28 to 251,000 yen/kW. We used 482,000 yen/kW from reference [18] for the pumped hydropower generation equipment cost. The SI cycle was calculated from the JAEA report [17] for hydrogen power generation. The construction cost of an 85400 Nm^3/h hydrogen plant was assumed to be 20.7×10^4 yen/(Nm^3/h) based on the construction cost of 17.7 billion yen. Alkaline electrolysis is 454^3 yen/(Nm^3/h) using the construction cost of 810 Euro/kW_e in reference [18], the energy input per Nm^3 of hydrogen production of 4.3 kW_e/Nm^3, and an average exchange rate of 130.3 yen/Euro in 2018 when the paper was published.

9.2.4.7 Supply Reserves

The supply reserve d was assumed to be 8%. In recent years, the supply reserve ratio has been maintained at around 10%, although some utility companies have reduced it to around 3% in some cases. In this study, we used the literature value of 8%.

9.2.4.8 Fuel Price by Power Source

Fuel price by power source $fc_j(j = 1, \cdots, N)$ is expressed in yen/Nm^3 for hydrogen plants and yen/kW for all others. For coal-fired, LNG-fired, general hydropower, LWR, wind power, and solar PV power, the figures from the Japanese government's Cost Estimation Committee22 were used, as were the construction unit costs; for HTGR, the same values as for LWR were used. For all of them, the reprocessing model value of 1.99 yen/kWh was used based on the estimates in reference. The same value as for general hydropower was used for pumped storage. For the hydrogen plant, the feedstock cost of 7.66×10^{-2} yen/Nm^3 was calculated from the water rate of 95 yen/m^3 (ref24). The SI cycle also requires cooling water 9.37×10^{-3} yen/Nm^3, and catalyst 7.43×10^{-2} yen/Nm^3 will be required [17].

9.2.4.9 Operating Years

Operating years are $year_j(j = 1, \cdots, N)$. We used the figures from the Japanese government's Cost and Other Verification Committee for coal-fired, LNG-fired, general hydropower, LWR, wind power, and solar PV power. For hydrogen plants, the SI cycle was assumed to have a 10-year amortization period, and alkaline electrolysis was assumed to have a 10-year amortization period as in reference [18].

9.2.4.10 Decommissioning Cost Rate

The decommissioning cost rate $waste_j(j = 1, \cdots, N)$ is uniformly assumed to be 5% of the construction cost based on the Cost and Cost Estimation Committee, except for nuclear. For LWRs, the decommissioning cost rate is 16.19%, based on the construction and decommissioning cost of a single nuclear power plant in the same document [19]. Since HTGRs are simpler to manage and are expected to have lower decommissioning costs, the decommissioning cost rate is also set at 16.19%.

9.3 Analysis Model

9.3.1 Setting of Analysis Conditions

The calculations are performed to minimize the cost of operating the power supply for 1 year in TEPCO's service area for the fiscal years 2030, 2040, and 2050. A linear programming model was formulated for this operation plan and analyzed for each year.

9.3.1.1 Renewable Energy

Future solar PV and wind generation were calculated by dividing the 2019 generation by the actual installed capacity in 2019 [14] and multiplying by the expected installed capacity in 2030 [5] after giving the installed capacity for the 2030–2050 period. After subtracting that expected generation amount from the total demand, a load duration curve was created. From this load duration curve, calculations were made to assign the optimal operation of each power source. Figure 9.1 shows the annual demand (April 1–March 31) (in panel a) and its load duration curve (in panel b) for the year 2050. In this study, considering the time-dependent nature of PV and wind power output, the load $M_s (= D_s - PV_s - W_s)$ is the hourly demand D_s subtracted by the PV_s and W_s of PV and wind power output at the same time stamp, respectively, and the load duration curve was created. This load duration curve is accompanied by the fluctuation (variance) $\sigma^2_{M,s} (= \sigma^2_{D,s} + \sigma^2_{PV,s} + \sigma^2_{W,s})$

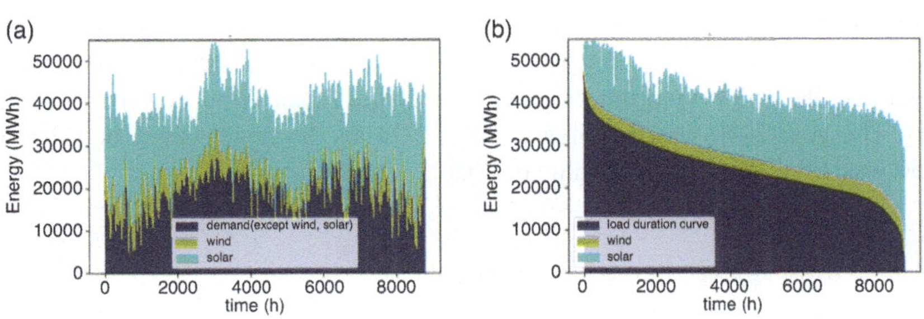

Fig. 9.1 Load duration curve: (**a**) annual demand (April 1–March 31) in 2050 and (**b**) load duration curve (April 1–March 31) in 2050

due to PV and wind generation. Here s is the time in which the demand is ordered in order of magnitude, and $\sigma_{D,s}$, $\sigma_{PV,s}$, and $\sigma_{W,s}$ are the fluctuation in demand, PV generation, and wind generation, respectively, over the same period. The actual calculations used the expected average demand M_t obtained for each time step h_t in Eq. (9.1). The fluctuation $\sigma_{M,t}$ per time step h_t is not considered. Figure 9.1a shows the demand for 1 year with time on the horizontal axis, and Fig. 9.1b shows the load duration curve. Wind and solar PV power generation were sorted by tying them to the time of day when sorting the demand in order of magnitude. It should be noted here that in Fig. 9.1a there is no PV output at night, and in Fig. 9.1b the rightmost portion with the lowest demand falls during the daytime of the May holiday weekend.

9.3.1.2 Hydrogen Production

We consider two methods of hydrogen production: HTGR and SI cycles or solar PV and alkaline electrolysis. Hydrogen is consumed as fuel mixed with LNG. If excess hydrogen is produced, it is assumed to be stored. Since a significant cost reduction is essential for the large-scale use of hydrogen in transportation, this study does not consider the use of hydrogen in transportation, as it only uses technologies that are expected to be commercialized shortly. This study does not consider large-scale use of storage batteries in the power system due to price issues [8–10].

9.3.1.3 Installed Capacity

We assumed that 80% of the previous base year's installed capacity, except for solar PV and wind, was available as existing capacity. Table 9.3 shows each year's established installed capacity. The installed capacity of solar PV and wind power in each year was set to increase as shown in Table 9.4; the installed capacity of solar PV and wind power at December 2019 was 13847 MW and 222 MW, respectively [16]. The installed capacity of PV power generation in 2030 was set at 26,000 MW by multiplying the installed capacity of the 2010 Basic Energy Plan by 1.5 and prorating it by the electricity demand of each electric utility company [5]. The installed capacity for wind power generation was set at 7,700 MW, based on the target installation of wind power generation by the Japan Wind Power Association [16]. The year 2040 was assumed to be 1.25 times the target value, and the year 2050 was 1.5 times the target value of 2030.

Table 9.3 Existing installed capacity in each fiscal year

	2030	2040	2050
Installed capacity	80% available in FY 2019	80% available in FY 2030	80% available in FY 2040

Table 9.4 Installed capacity of solar PV and wind power in each fiscal year

	2030	2040	2050
Solar PV(MW)	26,000	32,500	39,000
Wind (MW)	7700	9625	11,550

Table 9.5 CO_2 emission limits

	2030	2040	2050
No CO_2 emission limit	None	None	None
With CO_2 emission limit	Max. 73.3 $MtCO_2$	Max. 26.1 $MtCO_2$	Max. 0.05 $MtCO_2$

Table 9.6 Four future scenarios

	Hydrogen production method	CO_2 emission limitation
Scenario 1	HTGR-SI cycle	None
Scenario 2	HTGR-SI cycle	Available
Scenario 3	Solar-AWE	None
Scenario 4	Solar-AWE	Available

9.3.1.4 Next Generation Nuclear Reactors

The HTGR, a next-generation nuclear reactor, will be available from FY2040. According to the Japan Atomic Energy Agency, the HTGR is targeted to start construction in 2030 [19]. In this study, the amount of HTGR introduced in FY2030 is 0 and available from FY2040.

9.3.1.5 CO_2 Emission

The total electricity generated in 1 year in TEPCO's service area was 2.84×10^8 MWh, coal-fired power generation was 5.61×10^7 MWh, LNG-fired power generation was 1.63×10^8 MWh, and oil-fired power generation was 5.50×10^5 MWh. According to the World Bank report [20], emissions for each source are 879 gCO_2/kWh for coal-fired, 713 gCO_2/kWh for oil-fired, and 391 gCO_2e/kWh for LNG-fired power. From these conditions, we obtain CO_2 emissions of 1.136×10^8 tCO_2 and CO_2 emission factor of 0.4003 $kgCO_2$/kWh in the TEPCO service area for FY2019. This calculated value of CO_2 emission factor is consistent with the actual value of CO_2 emission factor of TEPCO in 2019 (see Table 9.5).

9.3.1.6 Four Future Scenarios

As shown in Table 9.6, four future scenarios were set up. First, two scenarios with and without limiting CO_2 emissions in Table 9.5 and two hydrogen production methods, HTGR-SI cycle and Solar-AWE, were considered. However, for the hydrogen/LNG mixed generation, the ratio of LNG to hydrogen in its vaporized state must be at least 70%. Therefore, CO_2 emissions of about 1/5000th of FY 2019 were considered acceptable.

9.3.2 *Linear Programming Model*

9.3.2.1 Objective Function

The objective function is the total cost for the year

$$\sum_{j=1}^{N} \frac{CaC_j + DisC_j}{year_j} + \sum_{j=1}^{N} (OpexC_j + FuC_j) + \sum_{j=LWR,HTGR} (RepC_j + SaC_j + PC_i). \quad (9.2)$$

The first term in Eq. (9.2) is the 1-year cost of new construction and its decommissioning divided by the number of available years (life). Since a reasonable discount rate estimation is difficult, a discount rate of 0 % was assumed in this study. A higher (lower) discount rate results in a larger (smaller) contribution from new capital investment costs and a smaller (larger) contribution from future decommissioning costs. The second term in Eq. (9.2) is the 1-year operating and fuel costs, and the third term is the 1-year reprocessing and safety costs and policy costs for nuclear LWRs and HTGRs. Here CaC_j, $DisC_j$, $OpexC_j$, FuC_j, $RepC_j$, SaC_j, and PC_j are the construction, decommissioning, operation and maintenance, fuel, reprocessing, safety, and policy costs, respectively. The CaC_j, $DisC_j$, and $OpexC_j$ are given as follows:

$$CaC_j = c_j \cdot x_j \ (j = 1, \cdots, N) \quad (9.3)$$

$$DisC_j = c_j \cdot x_j \cdot waste_j \ (j = 1, \cdots, N) \quad (9.4)$$

$$OpexC_j = c_j \cdot x_j \cdot b_j \ (j = 1, \cdots, N). \quad (9.5)$$

The $RepC_j$, SaC_j, and PC_j will be explained in Eqs. (9.27)–(9.29).

9.3.2.2 Constraints

Constraint 1: Power Sufficiency Equation
At each time t, demand M_t (MW) and pumping demand n_t (MW) are equal to the sum (MW)of power generated by all sources j:

$$M_t + n_t = \sum_{j=1}^{N} y_{t,j} \ (t = 1, \cdots, T). \tag{9.6}$$

Constraint 2: Facility Availability During Peak Hour
During peak hours (time $t = 1$), the power generated by each source j (MW) is the existing installed capacity z_j (MW) + the new installed capacity x_j (MW) multiplied by the peak hour availability ratio:

$$y_{t=1,j} \leq a_j \left(x_j + z_j\right) \ (j = 1, \cdots, N). \tag{9.7}$$

Constraint 3: Reserves During Peak Hours
During peak hours, the sum of existing installed capacity z_j (MW)+new installed capacity x_j (MW) of all power sources is greater than the sum of peak demand (MW) and reserve supply capacity (= total demand (MW) × supply reserve capacity %):

$$\sum_{j=1}^{N} a_j \left(x_j + z_j\right) \geq (1 + d)\, M_1. \tag{9.8}$$

Constraint 4: Operating Output
Since the load duration curve sorts the electricity demand in the order of magnitude, the larger the time t, the smaller the operating output y_{tj}:

$$y_{t,j} \geq y_{t+1,j} \ (t = 1, \cdots, T; \, j = 1, \cdots, N). \tag{9.9}$$

This constraint does not apply to solar PV and wind power because of their fluctuating output.
Constraint 5: Hydrogen Production
Hydrogen is produced using surplus electricity. Since demand decreases along the time t of the load duration curve, the system is set to produce more hydrogen at low demand:

$$H_{in,t} \leq H_{in,t+1} \ (t = 1, \cdots, T - 1). \tag{9.10}$$

Constraint 6: Power for Pumping
To account for the efficiency of pumped hydropower, we assumed that the total power output for pumped hydro for all hours equals m_0 times the total demand

for pumped storage for all hours. The value of m_0 was assumed to be 0.65 from reference:

$$m_0\left(\sum_{t=T/2+1}^{T}(n_t+n_{t+1})h_t - n_{T/2+1}\cdot h_{T/2+1}\right) = \sum_{t=1}^{T/2-1}\left(y_{t,PHydro}+y_{t+1,PHydro}\right)\cdot h_t - y_{\frac{T}{2},PHydro}\cdot h_{\frac{T}{2}-1}. \tag{9.11}$$

Constraint 7: Constructed Capacity Limit

For each source j, the sum of the new installed capacity x_j (MW) and the existing installed capacity z_j (MW) must be less than the preset installed capacity limit zu_j (MW):

$$x_j + z_j \leq zu_j \ (j = 1, \cdots, N). \tag{9.12}$$

Constraint 8: Upper Limit on Facility Utilization

For each source j, the total output for all hours (MWh) is less than the sum of the new installed capacity x_j (MW) and the existing installed capacity z_j (MW) times the total hours (h) and the facility utilization limit:

$$\sum_{t=1}^{T}\frac{1}{2}(y_{t-1,j}+y_{t+1,j})\cdot h_t \leq (x_j+z_j)\cdot l_j\cdot h_{max} \ (j=1,\cdots,N). \tag{9.13}$$

Constraint 9: Fuel Consumption

Fuel consumption for all hours is equal to the fuel cost p (yen/kWh) times the output y_{tj} (kW) at each hour times aggregated for each power source j:

$$FuC_j = \sum_{t=1}^{T}\frac{1}{2}(y_{t-1,j}+y_{t+1,j})\cdot h_t \ (j=1,\cdots,N). \tag{9.14}$$

Constraint 10: Hydrogen Sufficiency Equation

The total hydrogen consumption for all hours (m^3) equals the total hydrogen production for all hours (m^3)

$$\sum_{t=1}^{T}\frac{1}{2}(H_{in,t}+H_{in,t+1})\cdot h_t = \sum_{t=1}^{T}\frac{1}{2}(H_{out,t}+H_{out,t+1})\cdot h_t. \tag{9.15}$$

Constraint 11: Hydrogen Output

For hydrogen/LNG mixed combustion, vaporized hydrogen can be mixed up to 30 vol% of the total [21]. Hence, the condition that hydrogen can be added up to LNG (m^3): hydrogen (m^3) = 7:3 during liquefaction is necessary. Assuming that the volume multiplication factor during vaporization is $m_1 = 600$ times and the LNG volume-to-weight relationship $m_2 = 1360\,\text{m}^3/\text{ton}$, we have

$$LNG_{out,t} \cdot m_1 \cdot m_2 \cdot \frac{3}{7} \geq H_{out,t}. \tag{9.16}$$

Constraint 12: Hydrogen, LNG, and Coal Power Generation vs. Feedstock
Calculate the amount of electricity generated by hydrogen, LNG, and coal from $H_{out,t}(m^3)$, $LNG_{out,t}(ton)$, and $coal_{out,t}(ton)$ consumed. For LNG and coal, the heat per kg was multiplied by the thermal efficiency in power generation and converted to kWh after subtracting the onsite consumption from the generated electricity. The onsite rate is the percentage of the power output that is consumed by the plant itself (in percent):

$$y_{t,H_2} = H_{out,t} \cdot k_{H_2}, \tag{9.17}$$

$$y_{t,LNG} = LNG_{out,t} \cdot k_{LNG}, \tag{9.18}$$

$$y_{t,Coal} = Coal_{out,t} \cdot k_{Coal}, \tag{9.19}$$

where k is a coefficient calculated by the following equation. The amount of heat are $m_3 = 54.60\,\text{MJ/kg}, 25.70\,\text{MJ/kg}, 142.0/22.4\,\text{kJ/l}$ for LNG, Coal, and H_2 respectively, thermal efficiency are $m_4 = 0.51, 0.42, 0.51$ for LNG, Coal, and H_2 respectively, the in situ rate are $m_5 = 0.02, 0.062, 0.02$ for LNG, Coal, and H_2 respectively, and heat conversion rate $m_6 = 0.278\,\text{kWh/MJ}$:

$$k_{LNG\ or\ Coal} = m_3 \cdot m_4 \cdot (1 - m_5) \cdot m_6, \tag{9.20}$$

$$k_{H_2} = m_3 \cdot m_4 \cdot (1 - m_5) \cdot m_6 \cdot 10^{-3}. \tag{9.21}$$

Constraint 13: CO_2 Emissions Limit
The emissions limitation scenario sets a limit that linearly reduces CO_2 emissions by year; the allowed emissions for years 2030, 2040, and 2050 are 73.3, 26.1, and 0.05 Mt CO_2. In the no emission limit scenario, the upper limit is set to a sufficiently large value (e.g., 108 Mt CO_2). The emissions for each source are $m_7 = 879\,\text{g}\,CO_2/\text{kWh}$ for coal-fired, $713\,\text{g}\,CO_2/\text{kWh}$ for oil-fired, and $m_8 = 391\,\text{g}\,CO_2\,\text{e/kWh}$ for LNG-fired [20]:

$$CO_{2,\text{Coal}} + CO_{2,\text{LNG}} \leq \text{Emissions Limitation}, \tag{9.22}$$

$$CO_{2,\text{Coal}} = \sum_{t=1}^{T} m_7 \cdot \frac{1}{2}(y_{t-1,Coal} + y_{t+1,Coal}) \cdot h_t, \tag{9.23}$$

$$CO_{2,\text{LNG}} = \sum_{t=1}^{T} m_8 \cdot \frac{1}{2}(y_{t-1,LNG} + y_{t+1,LNG}) \cdot h_t. \tag{9.24}$$

Constraint 14: Constant Operation of LWRs
LWRs are more economical with higher facility utilization because of their higher construction costs and lower fuel costs compared to other power sources. Therefore, LWRs are widely used as a baseload power source. The operation at constant power output is technically accessible. In this study, the power output of an LWR at $t = 8760$ is set to be no less than 90% of the power output at $t = 1$:

$$y_{t,LWR} \cdot rate_{LWR} \geq y_{t+1,LWR} \ (t = 1, \cdots, T - 1), \tag{9.25}$$

$$rate_{LWR} = 0.9^{1/(T-1)}. \tag{9.26}$$

Constraint 15: Reprocessing, Safety Measures, and Policy Costs of Nuclear Power
Unlike other power sources, nuclear power (LWRs and HGTRs) requires additional costs for reprocessing, safety measures, and policy purposes. These costs increase in proportion to the amount of nuclear power used. Of these costs, the reprocessing cost of 0.95 yen/kWh and the policy cost of 1.3 yen/kWh are for both LWRs and HTGRs. On the other hand, the safety cost consists of an additional post-accident safety cost of 0.6 yen/kWh and an accident risk response cost of 0.3 yen/kWh. The accident risk response cost is required for both LWRs and HTGRs, while the additional safety cost is required only for LWRs:

$$RepC_j = \sum_{t=1}^{T} rep \cdot \frac{1}{2}(y_{tj} + y_{t+1,j}) \cdot h_t \ (j = LWR, HTGR), \tag{9.27}$$

$$SaC_j = \sum_{t=1}^{T} sac \cdot \frac{1}{2}(y_{tj} + y_{t+1,j}) \cdot h_t \ (j = LWR, HTGR), \tag{9.28}$$

$$PC_j = \sum_{t=1}^{T} pc \cdot \frac{1}{2}(y_{tj} + y_{t+1,j}) \cdot h_t \ (j = LWR, HTGR), \tag{9.29}$$

where $rep = 0.95 \times 10^3$, $sac_{LWR} = 0.9 \times 10^3$, $sac_{HTGR} = 0.9 \times 10^3$, and $pc = 0.3 \times 10^3$ yen/MWh.
Constraint 16: Hydrogen Production Method 1: HTGR–Sulfur–Iodine Cycle
The first hydrogen production method combines HTGR and a thermochemical method: the sulfur–iodine cycle (sulfur–iodine cycle, SI cycle). This hydrogen production method does not emit CO_2 from electricity and heat production to hydrogen production. The amount of hydrogen produced is given by

$$H_{in,t} \leq (x_{HTGR} + z_{HTGR} - y_{t,HTGR}) \cdot m_9 \ (t = 1, \cdots, T). \tag{9.30}$$

Equation (9.30) represents hydrogen cogeneration in HTGR, where at a given time t in the load duration curve, the total installed capacity subtracted by the HTGR generation for that time multiplied by a conversion factor equals the amount of hydrogen produced. The coefficient represents 85400 Nm^3 hydrogen per hour from HTGR 275 MW_e(600 MWth). This coefficient is $m_9 = 85400/275(Nm^3/h/MW_e)$. This corresponds to 3.22 kWh required per Nm^3 of hydrogen.

Constraint 17: Hydrogen Production Method 2, Solar–Alkaline Electrolysis

We consider a combination of solar PV and alkaline electrolysis as a second hydrogen production method. This method is another hydrogen production method that does not emit CO_2. For the coefficient, we used the value 4.3 kW_e/Nm^3 from reference [11], which calculates the cost of alkaline electrolysis. As $m_1 0 = 1/(4.310^{-3})(Nm^3/h/MW_e)$, the hydrogen production (sunlight + alkaline electrolysis) at time t is given by

$$H_{in,t} \leq y_{t,solar} \cdot m_{10} \; (t = 1, \cdots, T). \tag{9.31}$$

It should be pointed out that low-temperature alkaline electrolysis requires about 1.1 kW_e more power per 1 Nm^3 than the SI cycle. Therefore, low-temperature alkaline electrolysis is said to be suitable for regions with low electricity prices [10].

9.4 Analysis Results

9.4.1 Combination of Power Generation Sources in 2009

We reproduced the combination of power generation sources in 2009 to validate the model. Many nuclear power plants were temporarily shut down after the Fukushima Daiichi Nuclear Power Plant accident caused by the Great East Japan Earthquake. As of 2020, they have only partially resumed operations. In order to match the situation in 2009, the planning conditions, objective function, and assumptions were changed as follows.

9.4.1.1 Change of Settings

First, the HTGR and additional safety costs were excluded from the objective function presented in Eq. (9.32) of Sect. 9.3.2.1:

$$\sum_{j=1}^{N} \frac{CaC_j + DisC_j}{year_j} + \sum_{j=1}^{N} \left(OpexC_j + FuC_j\right) + \sum_{j=LWR} \left(RepC_j + PC_i\right). \tag{9.32}$$

In addition, all existing installed capacity was set to zero. This allows us to determine the economically optimal power supply configuration when all power sources are newly constructed. The newly installed capacity of HTGR was also set to zero.

9.4.1.2 Result

Figure 9.2 shows the combination of power generation sources (the amount of electricity generated) in 2009. Figure 9.2a shows the actual values for 2009, and (b) shows the value of this calculation. It can be seen that this calculation reproduces the overall characteristics of the actual supply and demand in 2009.

9.4.2 Scenario 1: HTGR-SI Cycle, Without Emission Limit

Figure 9.3 shows the combination of power generation sources for the HTGR-SI cycle without emission limits, and Fig. 9.4 shows the analysis of power operations for HTGR-SI cycle without emission limits. Power generation from fossil fuels decreased from FY2030 to FY2050 by 59%, 30%, and 12%, respectively (Fig. 9.3). This is due to the set increase in renewable energy and the increased introduction of HTGR, which will be introduced in FY2040 due to its cost advantage. Due to these factors, CO_2 emissions will decrease by about 30% from 2030 to 2050 even under the unlimited scenario.

In Fig. 9.4, it is shown that from FY2030 to FY2050, LNG-fired power generation, coal-fired power generation, and pumped hydropower generation play a role in adjusting demand fluctuations. In particular, pumped storage and LNG-fired power generation are consumed during high demand. On the other hand, coal-fired power generation is consumed not only during peak demand, as shown on panel (a) of the figure, but also during times when the slope of demand is moderate. On the other hand, coal-fired power generation is used not only during peak demand, as shown on the left side of the figure, but also during periods when demand is slower. Light water reactors and hydroelectric power generation operate at a constant output. Hydrogen production has been underway since FY2040 (Fig. 9.4b, c). Here,

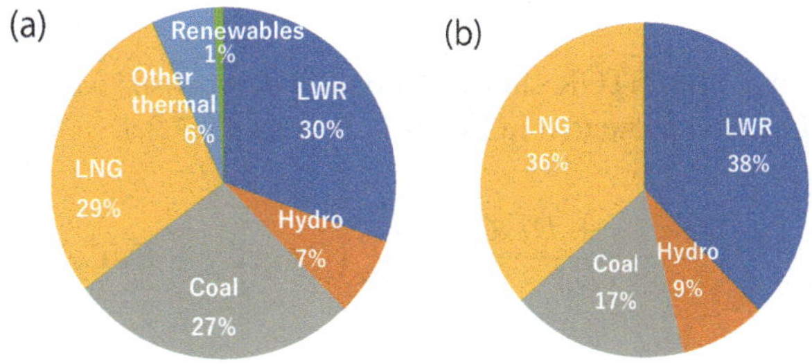

Fig. 9.2 Power source combination in 2009 (electricity generated): (**a**) actual values and (**b**) calculated values

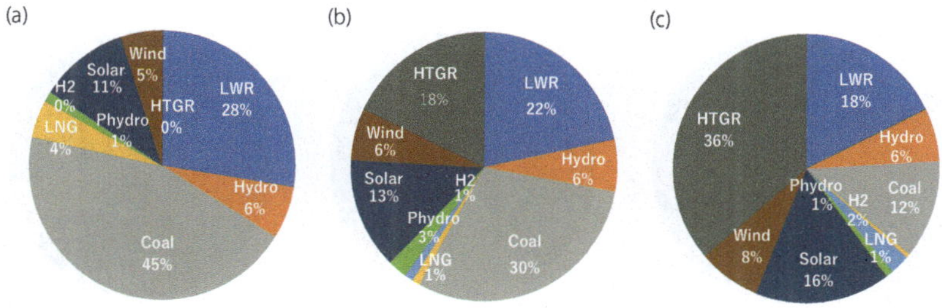

Fig. 9.3 Scenario 1: HTGR-SI cycle CO_2 power source combination without emission limitation (electricity generated): (**a**) FY2030, (**b**) FY2040, and (**c**) FY2050

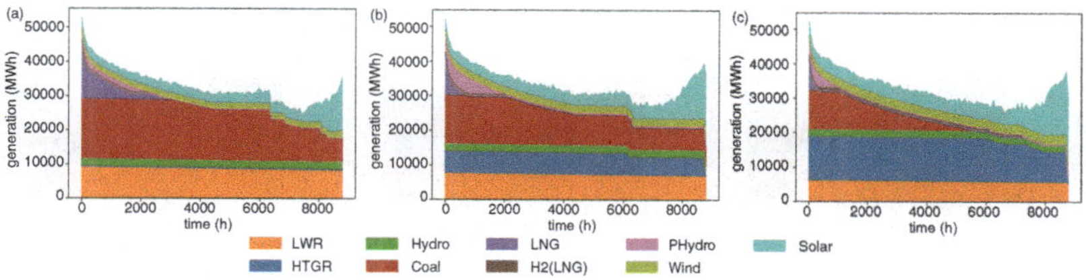

Fig. 9.4 Scenario 1: HTGR-SI cycle CO_2 operation without emission limitation: (**a**) FY2030, (**b**) FY2040, and (**c**) FY2050

hydrogen consumption by hydrogen and LNG mixed power generation, indicated by "H2 (LNG)," is characterized by its almost constant operation. This is because the only materials for hydrogen production are water and catalyst costs, which are inexpensive relative to the plant's construction cost, making it more cost-effective to continue operation. This characteristic is also valid for light water reactors and hydroelectric power generation.

9.4.3 Scenario 2: HTGR-SI Cycle, with Emission Limit

Figure 9.5 shows the combination of power generation sources for the HTGR-SI cycle with emission limitation, and Fig. 9.6 shows the analysis of power operation for HTGR-SI cycle with emission limitation. From Fig. 9.5, it is possible to read what is needed to achieve a steady reduction in CO_2 emissions. Thermal power must be reduced to 34% in FY2030. Under the no emissions limit scenario, thermal power in FY2030 is about 49%. In other words, thermal power generation must be significantly reduced in the case of restrictions. As for nuclear power, in FY2030, LWRs account for 43%; in 2040 and 2050, HTGRs and LWRs together account for 56% and 69% of the total, respectively. These results suggest that reducing CO_2 emissions to zero is very difficult.

In Fig. 9.5a, b, LNG-fired and coal-fired power generation, as well as pumped hydropower generation, play a role in adjusting demand fluctuations in FY2030 and

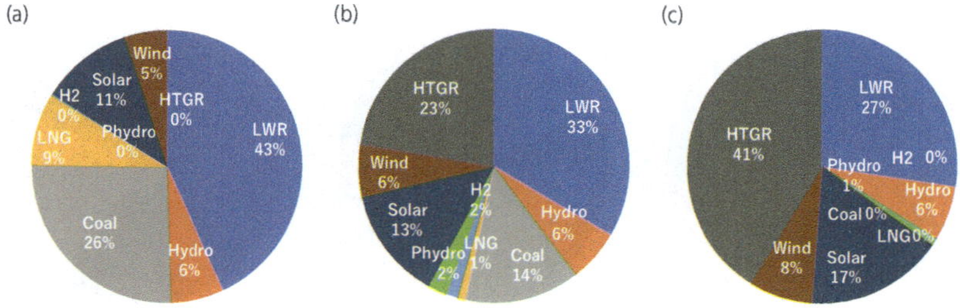

Fig. 9.5 Scenario 2: HTGR-SI cycle CO_2 power source combination with emission limitation (electricity generated): (**a**) FY2030, (**b**) FY2040, and (**c**) FY2050

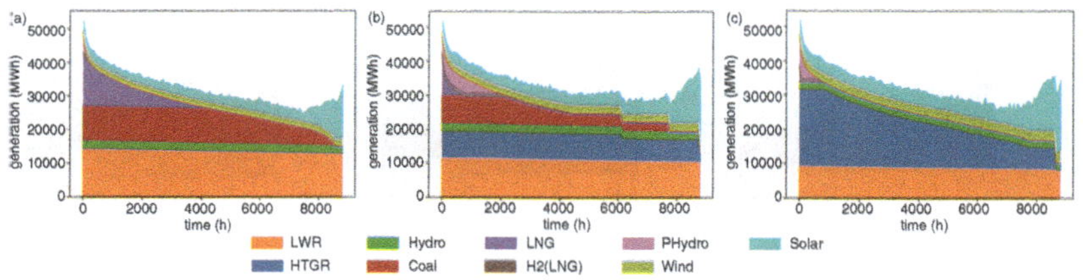

Fig. 9.6 Scenario 2: HTGR-SI Cycle CO_2 operation with emission limitation: (**a**) FY2030, (**b**) FY2040, and (**c**) FY2050

FY2040. This is the same as in Scenario 1 in Sect. 9.4.2: HTGR-SI Cycle without emissions limit. However, the picture is different in FY2050. Since CO_2 emissions are almost entirely prohibited, thermal power is reduced to almost invisible in Fig. 9.6c. Instead, hydrogen and pumped storage are replacing power generation in the high-demand peak areas and in the high slope areas that were once served by LNG pumped storage. Although hydrogen generation is not necessarily significant, covering the sharp peak portions shows that hydrogen can be an alternative regulating source to existing thermal power. In addition, HTGR is responsible for the slower slope hours. It is important to note that solar PV and wind power cannot replace flexible fossil fuels. This is because peak demand is usually in the evening, when sunlight shades and darkens, and the wind stops at calm. Thus, solar PV and wind power can generate only small outputs when demand is the highest. When CO_2 emissions are extremely limited, hydrogen replaces fossil fuels in the role of output regulation. The consumption of hydrogen as a fuel would be expected to reduce CO_2 emissions and accelerate the adoption of renewable energy.

9.4.4 Scenario 3: Solar-AWE, Without Emission Limit

The results of hydrogen production from solar PV and alkaline water electrolysis (AWE) are shown in Fig. 9.7. No hydrogen production took place throughout the entire process. It may be because it is cheaper to generate electricity using HTGR

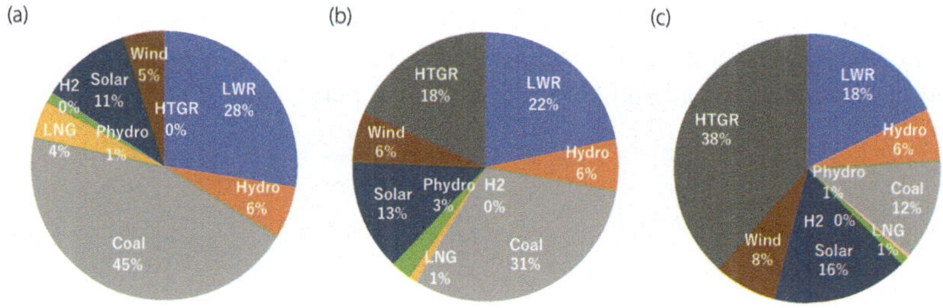

Fig. 9.7 Scenario 3: Solar-AWE CO_2 combination of power generation sources without emission limitation (electricity generated)

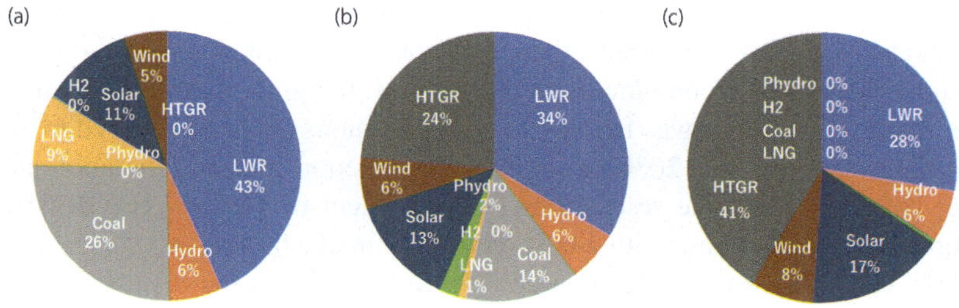

Fig. 9.8 Scenario 4: Solar-AWE CO_2 combination of power generation sources with emission limitation (electricity generated)

than hydrogen during high-demand hours. The construction cost per year is higher for hydrogen plants because the number of years they can be used is set as 10 years. However, the operation cost is cheaper because the raw material is water. For the SI cycle, we used the literature depreciation years of 10 years [11], but the construction cost is 207,000 yen/(Nm^3), which is less than half of the AWE cost of 454,000 yen/(Nm^3), so we believe that power was generated using hydrogen.

In FY2030, thermal power will account for nearly half of the output, at about 48%. LWRs account for about a quarter of the output at about 28% (Fig. 9.7a). In FY2040, HTGR will be introduced, and coal and LWRs will decrease by a certain percentage to compensate for this.

9.4.5 Scenario 4: Solar-AWE, with Emission Limit

The results for combining power generation sources with Solar-AWE emission limits are shown in Fig. 9.8. In this case, as in Sect. 9.4.4, Scenario 3: Without Solar-AWE Emissions Limits, no hydrogen production occurs. In the year 2030, Light Water Reactors account for 43% and Thermal Power for 34% (Fig. 9.8a). In the no emission limitation scenario, LWRs and thermal power were 28% and 48%, respectively. With the CO_2 emission limitation, 14% of the thermal power was changed to mainly LWRs. In FY2040, the difference from the no-emission limitation scenario is not significant. This is also due to the introduction of HTGRs, which in turn reduces coal and LWRs by a certain percentage (Fig. 9.8b).

9.5 Discussion

For each of the four future scenarios, the total costs and CO_2 emissions are compared. Figure 9.9 shows the cost comparison results for the four future scenarios. In FY2040, there is no significant difference between the two scenarios. In FY2050, HTGR-SI and Solar-AWE are about 200 billion yen more expensive with emission limitations. With CO_2 constraint, there is a cost difference of 200 billion yen per year in 2050, and the economic burden to achieve the goal of zero CO_2 emissions is not small. However, it was found that CO_2 emissions could be reduced to zero at about the same cost as in 2030 through the introduction of next-generation nuclear power called HTGR. The research and development of the new HTGR-SI cycle technology may encourage future cost reductions in electricity.

Next, CO_2 emissions are compared. Figure 9.10 shows the results of the CO_2 emissions comparison for the four scenarios, with HTGR-SI emissions limits and Solar-AWE emissions limits overlapping. Emissions also decrease in the case of no restrictions. This is due to the introduction of renewable energy and HTGR. However, the no emission limitation scenario emits 550 times more CO_2 in FY2050 than the emission limitation scenario.

The future role of hydrogen will be discussed. Methods of generating hydrogen without emitting CO_2 include a combination of HTGR and SI cycle and renewable energy and electrolysis. Renewable energies being introduced have several problems, such as fluctuating output and low output during peak demand hours. Hydrogen is expected to play an essential role in mitigating the various problems of renewable energy by producing hydrogen during low-demand hours and using it to generate electricity during high-demand hours (peak hours).

Furthermore, we must reexamine our future combination of power generation sources. Table 9.7 shows the actual values for FY 2019, the target of the power source combination in 2030 by the Agency for Natural Resources and Energy, and

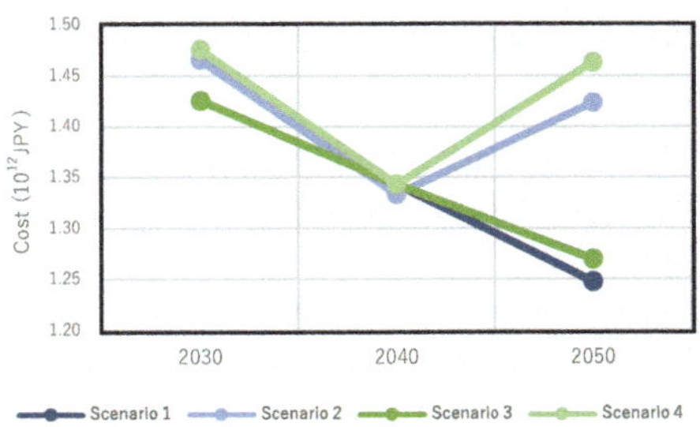

Fig. 9.9 For each of the four future scenarios, the total costs and CO_2 emissions are compared. In FY2050, HTGR-SI and Solar-AWE are about 200 billion yen more expensive with emission limitations

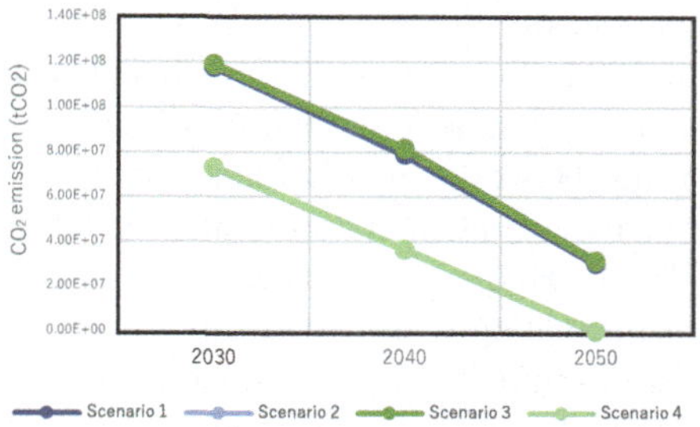

Fig. 9.10 CO_2 emissions are compared for the four scenarios, with overlapping HTGR-SI emissions limits and Solar-AWE emissions limits

Table 9.7 Actual combination of power generation sources and targets for FY2030

	FY 2019 (Japan)	FY 2030 (Japan) Agency for Natural Resources and Energy Targets	FY 2030 (TEPCO area) This study with emission limits scenario
Thermal power	78%	56%	34%
Nuclear power	6%	22–20%	43%
Renewable energy	16%	22–24%	22%

the combination of power generation sources in 2030 when CO_2 emissions are targeted to be 0 in 2050. The Agency for Natural Resources and Energy's target is 56% for thermal power and 22–20% for nuclear power, but if CO_2 emissions are to be reduced according to the CO_2 emission limits shown in Table 9.5, it is necessary to reduce fossil fuel to 33% and increase LWRs more than 40%. This analysis suggests that the target of the combination of power generation sources set by the Agency for Natural Resources and Energy is not sufficient to achieve CO_2 emissions zero in 2050. LWRs will be installed by 30–40% between 2030 and 2050. The prospects for nuclear power require further careful consideration. In addition, the introduction of HTGRs, which can produce hydrogen more safely and inexpensively, should be thoroughly discussed among various stakeholders.

9.6 Summary

Considering the future introduction of renewable energy and effective use of hydrogen, we analyzed the least-cost power system in TEPCO's service area regarding CO_2 reduction and economic efficiency. Four future scenarios were set up according to differences in hydrogen production methods and whether or not CO_2 emission limits are imposed. Based on the results of the analysis comparing the total cost and CO_2 emissions for each of the four future scenarios, the investment cost of reducing CO_2 emissions to zero in FY2050 is approximately 1.42 trillion yen. For the hydrogen production method, installing the HTGR-SI cycle was estimated to be 40 billion yen lower than that of Solar-AWE. Two hundred-billion-yen cost higher for the scenario with CO_2 emission limitation in FY2050 than the scenario without the limitation for HTGR-SI and Solar-AWE. The annual cost difference of 200 billion yen in 2050 with and without CO_2 emission limitation is not small, and the economic burden to achieve the goal of zero CO_2 emissions is not small at all. However, it was found that CO_2 emissions 0 can be achieved at the same overall cost as in 2030 by introducing the next generation of nuclear power called HTGR. Hydrogen is expected to play an essential role in mitigating the problem of fluctuating renewable energy output by using it as a power source during peak periods. Introducing new technologies such as HTGR may also reduce the economic burden to achieve CO_2 emissions of 0. The introduction of HTGR will need to be fully discussed among the various stakeholders.

References

1. IEA, "World Energy Outlook 2020," IEA (2020)
2. IEA, "Harnessing Variable Renewables," OECD/IEA, (2011)
3. IEA, "System integration of renewables Decarbonising while meeting growing demand", https://www.iea.org/topics/system-integration-of-renewables. (Cited 24 Feb 2020)
4. Y. Ikeda, K. Ogimoto, "Cross-correlation of output fluctuation and system-balancing cost in photovoltaic integration," The Journal of Engineering, https://doi.org/10.1049/joe.2014.0235, (2014) pp.1–9.
5. Y. Ikeda, K. Ogimoto, "A Unit Commitment Model with Demand Response and Electric Storage Device for the Integration of Variable Renewable Energies," IEEJ Transactions on Power and Energy Vol. 133 No. 7 (2013) pp. 598–605.
6. M. Beaudin, H. Zareipour, A. Schellenberglabe, W. Rosehart, "Energy storage for mitigating the variability of renewable electricity sources: An updated review," Energy for Sustainable Development 14 (2010) pp. 302–314.
7. Y. Ikeda, "Power grid with 100% renewable energy for small island developing states," Evolutionary and Institutional Economics Review (2020) 17:183–195, 2020.
8. Y. Ikeda, "Simulation Study on Energy Mix for Power Generation in Temburong Eco Town," ERIA Research Project Report 2017, No. 02 (2018).
9. Y. Ikeda, "Development of the Eco Town Model in the ASEAN Region through Adoption of Energy-Efficient Building Technologies, Sustainable Transport, and Smart Grids," ERIA Research Project Report 2015, No. 20 (2015)
10. C. Bordin, H. O. Anuta, A. Crossland, I. L. Gutierrez, C. J. Dent, D. Vigo, "A linear programming approach for battery degradation analysis and optimization in offgrid power systems with solar energy integration," Renewable Energy 101 (2017) pp. 417–430.
11. Hydrogen Council, Hydrogen scaling up A sustainable pathway for the global energy transition (2017).
12. IAEA, "Hydrogen Production," IAEA Nuclear Energy Series No. NP-T-4.2 (2013).
13. G. Locatelli, ''Cogeneration: An option to facilitate load following in Small Modular Reactors," Journal of Nuclear Science and Technology, 44:2(2007) pp. 147–154.
14. B. Yildiz, "Efficiency of hydrogen production systems using alternative nuclear energy technologies," International Journal of Hydrogen Energy 31 (2006) pp. 77–92.
15. H. Tanaka and Y. Ikeda, "Economical Evaluation of Hydrogen Cogeneration by Renewable Energies and Nuclear Energies (In Japanese)", Journal of Japan Society of Energy and Resources, Vol. 43, No. 2, pp. 33–44 (2022)
16. Agency for Natural Resources and Energy, Ministry of Economy, Trade and Industry https://www.fit-portal.go.jp/PublicInfoSummary. (Cited 12 Dec 2020) (in Japanese)

17. J. Iwatsuki, et al., "Economic Evaluation of HTGR IS Process Hydrogen Production System," JAEA-Review 2014-037(2014).
18. G. Locatelli, "Load following of Small Modular Reactors (SMR) by cogeneration of hydrogen: A techno-economic analysis," Energy (2018).
19. T. Nishihara, et al., "Excellent Feature of Japanese HTGR Technologies," Japan Atomic Energy Agency (2018).
20. The World Bank, "Understanding CO2 Emissions from the Global Energy Sector," The World Bank (2014).
21. K. Inoue, et.al, "Development of Hydrogen and Natural Gas Co-firing Gas Turbine," Mitsubishi Heavy Industries Technical Review Vol. 55 No. 2 (2018).

Chapter 10
Blockchain Energy Trade System

10.1 Crypto Asset

In recent years, decentralized and open information systems based on blockchain technology have attracted much attention. Blockchain technology can effectively solve various global issues known as sustainable development goals, SDGs, and energy issues.

10.1.1 Blockchain Technology and the Crypto Asset Bitcoin

10.1.1.1 From Web1 to Web2

The Internet was initially a physical infrastructure of wires and servers that allowed computers and the people in front of them to talk to each other. In 1969, the first messages were sent by the US government's ARPANET. Later, in 1991, the first Web, Web1, emerged with the ability (read) for users to navigate between pages via HTML and URLs.

In the early 2000s, the Internet became more interactive and the era of user-generated content (read/write). Social media was an essential feature of Web 2, with the rise of Facebook and Twitter. YouTube, Wikipedia, and Google expanded the ability to watch, learn, search, and communicate, along with the ability to comment on content.

The Web2 era was also an era of centralization. The companies accumulated tremendous wealth by collecting user data and using that data to target advertising. Services came to be offered "for free," but users did not initially understand what that meant. As the Web grew, centralized, and corporatized, many began to doubt the arrival of a better future.

Y. Ikeda, *Data Science of Renewable Energy Integration*,
Evolutionary Economics and Social Complexity Science 30,
https://doi.org/10.1007/978-981-99-8779-5_10

10.1.1.2 From Web2 to Web3

Web3 (Internet of value) was proposed as a new technology to solve the problems of Web2. "Read, write, and own" are the features of Web3. Are you worried about privacy? Web3 provides you with an encrypted wallet to protect your identity. Are you worried about censorship? Web3 provides you with decentralized databases to store everything in a highly transparent manner. Are you worried about centralization? Web3 provides you with the opportunity to vote and establish value-added decision-making.

The seeds of Web3 were planted in 1991 when W. Scott Stornetta and Stuart Haber launched the first blockchain, a project to time-stamp digital documents. However, the idea took hold in 2009, when Satoshi Nakamoto launched the crypto asset Bitcoin in the wake of the financial crisis. Ethereum, which appeared in 2015, is both a crypto asset and a platform for building systems. More than a decade later, proponents of the blockchain-based Web proclaimed the dawn of a new era, Web3.

10.1.1.3 Centralized and Distributed Systems

A centralized system is one in which a central computer and other computers are connected. If the central computer stops, the entire system stops working (single point of failure), easier to maintain system consistency and flexibility. However, data integrity (no data missing or wrong) can be guaranteed.

In the distributed system, multiple computers are interconnected with each other. No single point of failure reduces the possibility of system downtime. Computing power can be increased by adding more computers. The system cannot function without a network, requiring communication costs. It is not easy to achieve data integrity. Blockchain technology solves the data integrity problem.

10.1.1.4 Blockchain and P2P

Transaction data on the blockchain is transmitted from node to node like a bucket relay to the entire network. Nodes confirm the connection and the contents of the data to be transmitted to each other. A blockchain system consists of two different types of nodes. A "full node" is a node that holds all the data in the blockchain. This node alone can verify the integrity of the entire blockchain. A storage capacity of several hundred GB is required. Simplified Payment Verification (SPV) node is a node that holds only block header data. It queries the full node as necessary. SPV requires only 1/1000 of the storage capacity of the full node.

10.1.1.5 Cryptographic Hash Function

The cryptographic hash function is a function that converts input numerical values into numerical values according to specific rules and outputs them. Calculating hash values with hash functions is called hashing. The cryptographic hash function possesses the following properties: Irreversibility assures that can only compute in one direction from input to output and cannot compute the input from the output. Confidentiality assures that if the input data changes even slightly, the output data changes significantly. The fixed-length output assures the output data is the same length regardless of the length of the input data. The Fast Processing Speed assures that output data can be easily calculated from input data.

10.1.1.6 Common Key Cryptography

Plain text "Hello" is encrypted to cipher text "aA@43k," and cipher text "aA@43k" is decrypted to plain text "Hello," as shown in Fig. 10.1.

In public common key cryptography, encryption and decryption are performed using the same key. A different key is required for each party. The more the number of counterparties increases, the more complicated it becomes to manage the security of the key (Fig. 10.2).

10.1.1.7 Public Key Cryptography

Bitcoin uses public-key cryptography, which is used for address generation and transaction data. The public key encryption method uses different keys for encryp-

Fig. 10.1 Encryption and decryption. "Hello" ⇔ "aA@43k"

Fig. 10.2 Common key cryptography. A different key has to be used for each party

Fig. 10.3 Public key cryptography. The private key is held only by you and not shared with others

tion and decryption, as shown in Fig. 10.3. Create a public key from a private key. The private key cannot be reversed from the public key. The private key is held only by the user (you) and not shared with others. The public key is widely available, and everyone knows it. This solves the "key delivery problem." The sender encrypts the data with the public key, and the receiver (you) decrypts it with the private key.

10.1.1.8 Elliptic Curve Cryptography

Elliptic curve cryptography is used to generate public keys from private keys in blockchains. The elliptic curve is defined by $y^2 = x^3 + ax + b$. We use the elliptic curve with $a = 0$ and $b = 7$: $y^2 = x^3 + 7$. Elliptic curve cryptography is given by $y^2 = (x^3 + 7)$modp. Here, Modulo $x$modp is a remainder of x divided by p. Elliptic curve cryptography ensures that it is easy to compute y from x, and it is challenging to reversely compute x from y.

10.1.1.9 Electronic Signature

Electronic signature is a technique for verifying that data was generated by a particular sender, as shown in Fig. 10.4.

10.1.1.10 Blockchain Concept

Blockchain is a distributed bookkeeping system in which a certain number of transaction records are managed as a single block and then as a chain created from the blocks. Figure 10.5 shows a conceptual diagram of a blockchain [1]. Bitcoin, BTC, goes through a mining process before adding a new transaction record to the blockchain [2]. Mining is the process of approving a crypto asset transaction through the efforts of a third party to ensure the transaction's fairness. Proof of Work (PoW) is the approval process, and the person or organization performing the mining is known as a miner. The following steps are taken to record the transaction history. When a new transaction occurs, a hash value (fingerprint) is generated using

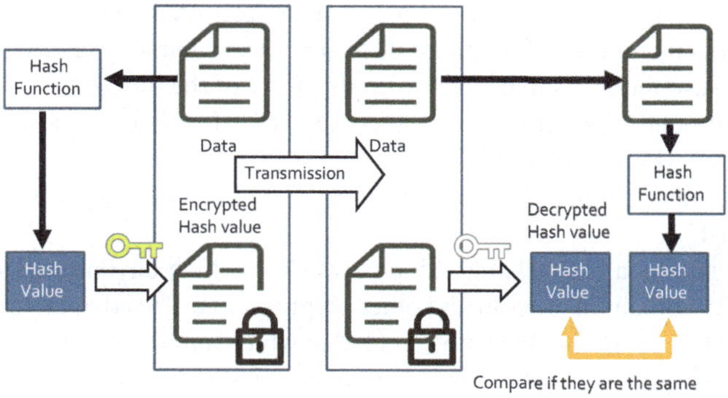

Fig. 10.4 Electronic signature. The signer creates the signature with his/her private key, and the verifier verifies it with the public key

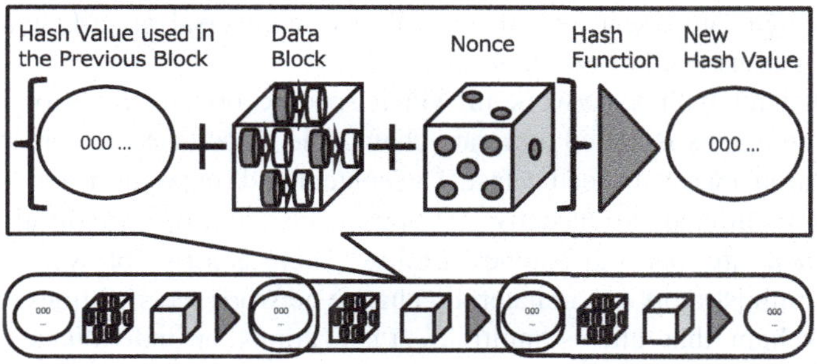

Fig. 10.5 Conceptual diagram of blockchain a certain number of transaction records are managed as a single block and then as a chain created from the blocks

a Nance (random number). Through trial and error, minors attempt to find the Nance that generates the hash value smaller than the target value. The miner who discovers the Nance that meets the criteria adds the transaction data block to the blockchain and notifies the other miners. The miner will be rewarded if the other miners confirm the data block and the Nance. Following this procedure, all transaction histories are recorded in the blockchain and approved by an unknown number of people to ensure that transactions are completed correctly. The use of alternatives in crypto asset BTC is supported by free-participation mining.

10.1.1.11 Blockchain Structure and Features

Multiple transaction data are grouped together to form a block. Each block contains the transaction data, a time stamp, and other information in the block header.

Blocks are chained together in the order of the recorded time to form a chain. Data integrity is maintained throughout the chain. Data cannot be rewritten (tamper-resistant) because there is a dependency on the order of the blocks.

Blockchain data is not stored on a specific server but shared and managed by many computers (P2P method).

There is no need to give "trust" to users from a third party. This is often called "Trustless trust."

A public chain is a blockchain with a network in which anyone can participate, such as the blockchain for Bitcoin and other crypto assets. The advantage of a public chain is that a fair environment can be created in which authority is not concentrated in the hands of specific users. Transparency and publicity can be maintained at a high level because it is under public view. The approval of specification changes and transactions can be done reasonably. The disadvantage of a public chain is that approval is time-consuming and resource-intensive (electricity and computing power). Settlement is never final and cannot be overturned (finality). If blocks of sufficient length are connected, the possibility of reversal is sufficiently low. In Bitcoin, it is considered to be six blocks.

A blockchain with a network in which specific organizations or individuals approve participants is called a private chain. The advantage of a private chain is that transaction approval speed is fast. The content and scope (users) of information sharing can be limited. No incentive for approval is required. The disadvantages of a private chain are that transparency are low. There is a risk of system use being stopped. Administrators can unilaterally change network specifications.

A blockchain that allows multiple organizations or individuals to approve participants in the network or to change specifications by consensus among a limited number of entities is called a consortium chain. It has a structure and functionality between a public chain and a private chain.

10.1.1.12 Transaction Approval: Proof of Work

Bitcoin (BTC) undergoes a mining process before new transaction records are added to the blockchain. Mining is the process of approving transactions of a crypto asset through a third party's efforts to ensure the transaction's fairness. Proof of work (PoW) is a prime example of such an approval process. The individual or organization that performs the mining is called a miner.

The transaction history is recorded in the following steps. When a new transaction occurs, a hash value (fingerprint) is generated using nonce (random number). Miners search for a nonce that generates a hash value smaller than the target value through trial and error. When a miner finds a nonce that satisfies the conditions, he/she adds the transaction data block to the blockchain and notifies other miners. The miner is rewarded if the other miners can verify the data block and nonce.

In this procedure, all transaction histories are recorded in the blockchain, approvals are recorded in the blockchain, and an unspecified number of people approve the transaction to ensure it is completed correctly.

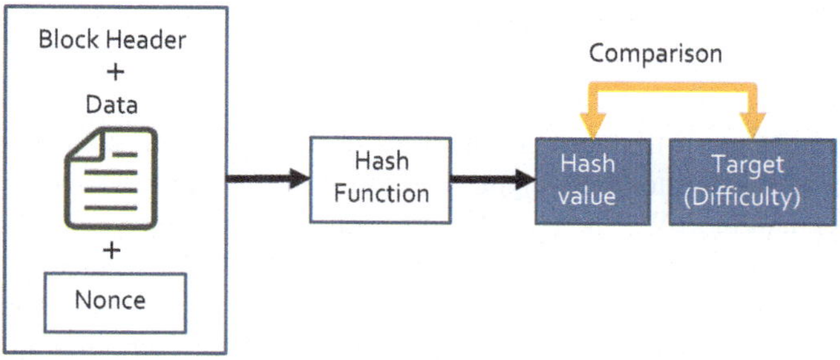

Fig. 10.6 Proof of work. It is necessary to perform a huge volume of calculations, and the electricity consumption is also enormous

10.1.1.13 Proof of Work

Figure 10.6 shows the concept of Proof of Work. The miner repeats the calculation by changing the nonce until it becomes smaller than the target value. The miner that finds a nonce that meets the conditions completes the block and reports it to the other nodes. The other nodes verify that the contents of the block and the nonce are correct. If correct, the miner is declared a "mining success." The miner receives a reward (12.5 BTC) for "mining success."

Due to the "confidentiality" of the hash function, a small change in the input value causes a large change in the output. Therefore, it is difficult to "guess" the nonce. Mining requires a lot of computing power (hash rate) and electricity.

10.1.1.14 Problems with PoW

In crypto asset transactions incorporating the PoW mechanism, anyone can freely participate in mining and receive crypto assets as a reward for mining. In this case, the crypto assets paid as rewards are newly issued.

In the case of BTC, miners with higher hash rates are rewarded more and thus have more influence on the network. Increasing the hash rate requires a large investment, such as dedicated computers and electricity. Such miners have great power because they contribute greatly to the activation of BTC. However, if this trend goes too far, the centralized nature of BTC becomes apparent. This is known as "51% attack": a malicious individual or organization with a majority of the computational capacity of the entire network justifies a fraudulent transaction or is denied a legitimate transaction.

For blockchain and crypto assets to be used to solve global problems, the price of crypto assets must be stable, the amount of energy required for transactions must be appropriate, the cost of transactions must be appropriate, and the transaction speed must be fast.

10.1.1.15 Other Consensus Algorithms

In chains that use "proof-of-stake" such as Ethereum, approval of transactions is done by consensus of the verifiers involved in the chain (the greater the number of assets held, the greater the weight), and power consumption is more efficient in this process than in PoW. The term "smart contract" was proposed by Nick Szabo, an American computer scientist and cryptographer, in 1997. A smart contract is a mechanism by which an agreement between parties to a contract can be executed automatically without the parties being present if the conditions are met. Ethereum enables the functionality of programming on the blockchain. The risk of tampering with the content of the written contract is extremely low, and the contract can now be executed automatically. Decentralized applications (DApps) use smart contracts to execute complex processes.

In XRP, approval of transactions is performed according to a predetermined algorithm of a "validator" at a Ripple-designated institution. Approval is faster, but there is concern that it may become more centralized.

10.1.2 Ripple's Crypto Asset XRP

XRP, the crypto asset issued by Ripple Labs Inc., is managed through the approval process in the transaction system operated using Ripple Transaction Protocol, RTXP, where only transaction records are agreed upon by specific approvers, called validators, can be added. The validators, who play an essential role in this process, are chosen from organizations and individuals deemed trustworthy by Ripple. The transaction is recorded to the blockchain if 80% or more of the validators agree to approve it. This is analogous to a type of majority voting allowing quick approval. This approval process is known as Proof of Consensus, or PoC, and it has the advantage of not requiring a large investment or electricity like BTC mining. However, it does require a guarantee that Ripple, the operator of RTXP, will not rewrite the ledger at will. Another advantage is that it does not require much electricity, which corresponds to our society's decarbonization goal. For this less environmental impact property, XRP is often called "green crypto." The payment system RTXP has been updated from time to time, such as the protocol as of 2014 [3], the protocol as of 2018 [4], and the latest Cobalt protocol [5].

When the distributed ledger of XRP started to work in 2005, the entire amount of XRP was already issued, and there will be no more. Therefore, the more people who need XRP, the higher the price of XRP will naturally be. If the price of XRP fluctuates significantly before and after a money transfer by a financial institution or an individual, users will find it difficult to use. As a result, as the price of XRP rises, Ripple will normalize it by releasing its XRP holdings into the market. Ripple, however, has taken steps to lock up its XRP holdings so that it cannot freely sell them.

Currently, money transferred overseas via a financial institution has to go through a correspondent bank before it reaches the bank account in the destination country. XRP can be exchanged for fiat currency such as Japanese Yen, US dollars, Euros, and crypto assets such as Bitcoin. This is referred to as the XRP bridge function. This function allows for international remittances such as Japanese Yen to XRP to USD, Japanese Yen to XRP to Thai Baht, and so on. In such transactions, the gateway's IOU (I owe you.) or the right to receive a certain amount of XRP is used [6]. For example, when sending a certain amount of XRP from A to B, it is not XRP but IOU that is sent to B's account. The gateway guarantees the trustworthiness of the IOU transaction and is selected under a strict screening process by Ripple. The gateway functions as a bank's interceptor, accepting XRP deposits from users and rewriting the balance in the RTXP payment system. This enables international money transfers using XRP to be processed directly between financial institutions in the two countries, bypassing correspondent banks and thus speeding up and lowering the cost of international money transfers.

10.1.3 Using Blockchain Technology to Solve Global Issues

Blockchain can be a fundamental technology to provide solutions to various global issues: international remittance for migrants with low cost, fast, and reliable, digital ID for medical services for refugees, trade-in decarbonized energy, such as renewable energy, nuclear power, and hydrogen, financial inclusion to provide everyone nondiscriminatory financial services, commerce management, such as supply chain and commodity market, economic support for financing, and talent matching [7].

Among these issues, the problem of international remittances, including those of immigrants and developing countries, has been attracting particular attention in recent years. International remittances from migrants and others to their home countries are the most important source of funds for developing countries, and the demand for international remittances is growing as economic globalization progresses. In 2018, the global value of international remittances reached $529 billion; except for China, the value of international remittances exceeded the sum of official development assistance (ODA) and foreign direct investment (FDI) [8]. Unlike ODA, which may not reach its intended recipients due to corruption, and FDI, which may not benefit as many people other than the invested industry sectors, international remittances can be sent directly to households in need of funds. As a result, international remittances have emerged as a viable means of raising funds for education, health, employment, and entrepreneurship. Furthermore, international remittances are a consistent source of income for many households. International remittances are considered counter-cyclical because they are directed to households in the destination country that require more money during periods of low economic activity [9].

To use blockchain and crypto assets for international remittances and other global issues, the price of the crypto assets must be stable, the amount of energy required for transactions must be appropriate, the cost of the transaction must be low, the speed of the transaction must be high, and the occurrence of anomaly events such as money laundering and fraud must be prevented.

We focus on energy trading systems using the blockchain technology, among global issues. In this energy trading, using distributed energy sources, represented by renewable energy, is essential. Blockchain technology is often utilized in energy-related use cases because it enables peer-to-peer transactions. One prime example is Brooklyn Microgrid [10], a distributed electricity marketplace powered by blockchain technology, which allows consumers and prosumers to share electricity demand and supply information and transact electricity directly without an intermediary. Power Ledger, an energy technology company, has also been developing a trustless, transparent, and interoperable energy trading platform [11] powered by a token to align participants' incentives.

Based on these precedents, we implemented an energy trading system using blockchain technology in a dormitory of our school, where approximately 80 students reside. The current system for collecting electricity charges is complicated and burdens the students and staff in charge of accounting. To solve the problems of complexity and transparency in collecting electricity charges, we develop an energy trading system, EDISON-X, that uses blockchain technology to manage the buying (bidding) and selling (asking) of electricity usage rights, i.e., tokens, for students residing in the dormitory of our school.

We developed a blockchain-distributed energy trading system and conducted a small-scale demonstration experiment in our dormitory. Based on the results of this experiment, we discuss the energy trading system using blockchain technology for the practical usage of renewable energy. The goal is to develop methodologies based on topology and network science to understand energy trading characteristics and predict market changes in advance.

10.2 Energy Trading System EDISON-X

10.2.1 System Architecture

We developed an energy trading system, EDISON-X, that uses blockchain technology to manage the buying (bidding) and selling (asking) of electricity usage rights (tokens). Here EDISON-X stands for energy distribution and integration systems on either "campus," "small office," "home," "remote area," "island region," or "rural areas in developing countries." EDISON-X is built on various technologies and frameworks developed by Google, Ripple Labs, and others as open source. EDISON-X uses a cloud-based database called Firestore [12], built on Google Cloud, and has high stability. The school dormitory's electricity usage and the electricity generated by the Solar photovoltaic (PV) system are monitored and controlled by Panasonic's energy server, WeLBA, which sends the data to the Firestore.

Figure 10.7 shows a system configuration of EDISON-X. We use an XRP ledger (XRPL) that consumes little electric energy for validation because it uses an algorithm instead of mining [3, 13]. The XRPL is a public, permissionless blockchain that has been operating since 2012 and is one of the largest blockchain networks in the world. A key difference between XRPL and other major blockchain networks such as Bitcoin and Ethereum is that XRPL uses the Federated Byzantine Agreement (FBA) as its consensus mechanism, which makes its settlement extremely fast (3–4 seconds per settlement), low cost ($\leq$\$0.001 per transaction), and scalable (1,500 transactions per second). Because of these properties, XRPL is ideal for payments and settlement use cases. XRPL also enables a wide range of other use cases, including non-fungible tokens (NFTs) and Decentralized Finance (DeFi), and became the first major blockchain to go carbon neutral in 2020.

Any process on the Google Cloud that changes the token balance is sent to the XRPL as a transaction. Many transactions occur on EDISON-X in a short period. Therefore, the XRPL, which has a superior processing time, is adopted to solve the problem of fast transaction time. Usually, transactions are processed within seconds. These functions and the web application described below run on Firebase [14], a platform on the Google cloud. Firebase has the advantage of ease of management and high security through an authentication system developed by Google.

The buying and selling of electricity usage rights are tokens. UPX and SPX tokens purchase electricity from the utility company's distribution lines and the photovoltaic panels. UPX/SPX tokens guarantee the right to use electricity from utility companies or photovoltaic panels. Users can buy tokens from other users who want to sell their tokens. Users can also sell their tokens to others who want to purchase them. In the case of a shortage of tokens in the entire dormitory, the system

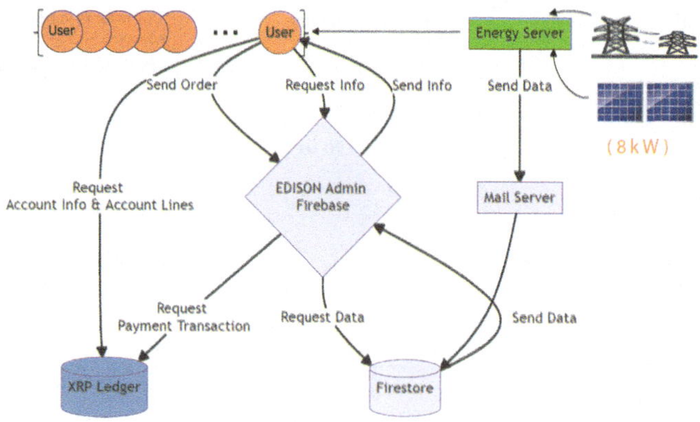

Fig. 10.7 EDISON-X: system configuration, EDISON-X is built on various technologies and frameworks developed by Google, Ripple Labs, and others as open source

will issue additional tokens at a higher price than those issued at the beginning of the month.

10.2.2 Web Application

EDISON-X has a web application for users to access data and place orders. This application is developed in Angular. Angular is a TypeScript-based open-source web application framework led by Google. The application has three main functions. These are creating orders, retrieving transaction data, and checking balances. Users access Firestore through the application to check their token balance and create orders. It is also possible to see some of the information on the XRPL using the XRPL Application Programming Interface.

EDISON-X's Graphical User Interface (GUI) uses Material Design [15]. Material Design was introduced by Google in 2014 and featured simple designs inspired by materials such as paper and ink. The EDISON-X web application uses a white background and indigo as a primary color, with pink as an accent. In addition, because it is a single-page application with no page transitions on the browser, it can run as fast as a native application.

Figure 10.8a–c shows GUI Screens for viewing important EDISON information in one place. The essential information includes the "Current account balance," "Last traded price," "This month's electricity usage," "Ranking of electricity usage," "Percentage of tokens held," and "Daily change of electricity usage." Figure 10.8d shows GUI Screens for viewing Buy/Sell Order. Participants can see the remaining available capacity (UPX, SPX, or at capacity, if any) and the latest transaction price, decide on the type, price, and quantity of tokens, and issue a buy/sell order.

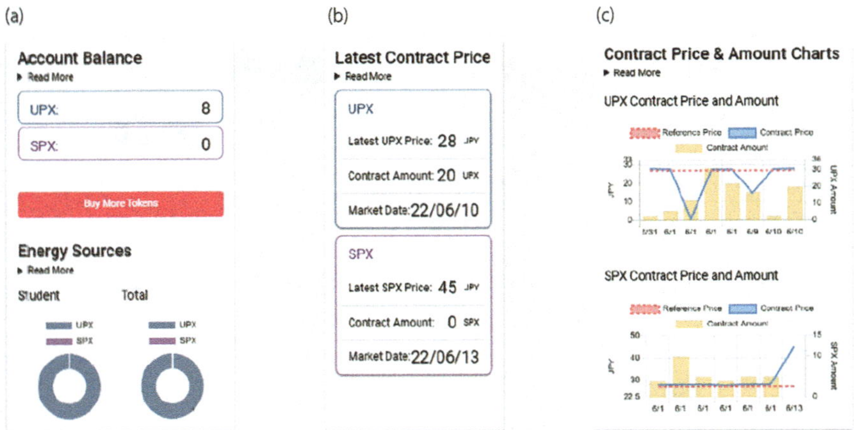

Fig. 10.8 EDISON-X: various GUI, Panels (**a**)–(**c**) show GUI Screens for viewing important EDISON-X information in one place. Panel (**d**) shows GUI Screens for viewing Buy/Sell Order

10.2.3 *Monthly Process*

10.2.3.1 Beginning of the Month

The system calculates the estimated usage for this month by dividing the usage of the same month last year by the number of students. UPX/SPX tokens will be issued to guarantee the right to use electricity in the school dormitory. The price for renewable energy from solar PV will be higher, and the tokens will be sold to students who wish to use them.

10.2.3.2 During the Month

Students can buy tokens from other students who want to sell their tokens. Students can sell their tokens to other students who want to buy tokens. In the case of a shortage of tokens in the entire dormitory, the system will issue additional tokens at a higher price than those issued at the beginning of the month.

Every day, the transaction price and volume are determined at the intersection of buy order and sell order curves, as shown in Fig. 10.9. Buy orders higher than the transaction price and sell orders lower than the transaction price will be accepted. All orders that do not satisfy this condition are rejected.

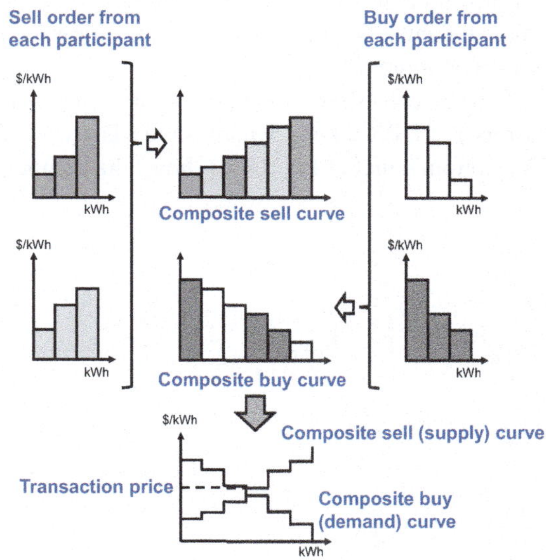

Fig. 10.9 Token market: single-price auction, The transaction price and volume are determined at the intersection of buy order and sell order curves

10.2.3.3 End of the Month

The system buys extra tokens at a discount from students. The system will sell additional tokens to students who need more at a higher price. The purchase and sale price will be determined so that the monthly income and expenses will be zero.

10.2.4 Demonstration Experiment

Approximately 80 students reside in the dormitory of our school. The current system for collecting electricity charges is complicated and burdens the students and staff in charge of accounting. On July 1, 2022, 17 students in the dormitory of our school participated in an experiment to confirm the operation of the EDISON-X system. UPX and SPX tokens are used to purchase electricity supplied from the utility company's distribution lines and the photovoltaic panels (8 kW) installed on the roof of the building of our school, respectively. We recorded the usage data of the previous month in a database. We used this data as a reference and assumed the following electricity usage for the students. The trading was carried out in a single-price auction, with the previous day's bid and ask orders being processed at midnight each day. In addition to the participants' orders, the system issued sell orders for UPX and SPX tokens when the participant's aggregated remaining token was insufficient. At the end of the month, the system buys extra tokens at a discount from students. The system will sell additional tokens at a higher price to students who need more tokens.

Figure 10.10 shows the daily number of orders for buying (bidding) and selling (asking) of UPX tokens, which is the right to use electricity from utility companies. The bottom panel of Fig. 10.10 shows the number of contracted buy and sell orders.

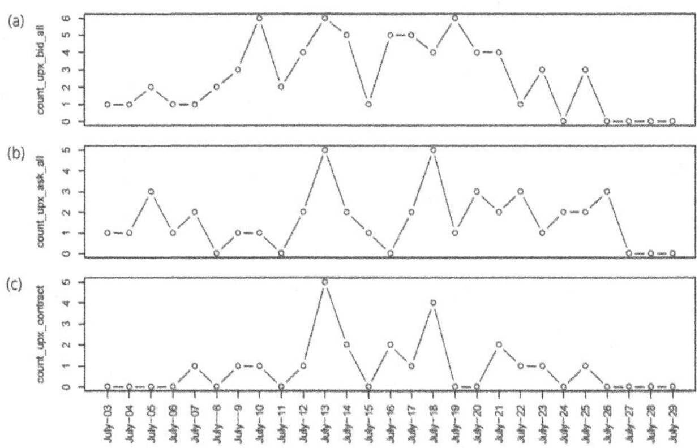

Fig. 10.10 Bid and ask for UPX token. The daily number of orders for (**a**) buying (bidding), (**b**) selling (asking), and (**c**) contracted to buy and sell orders

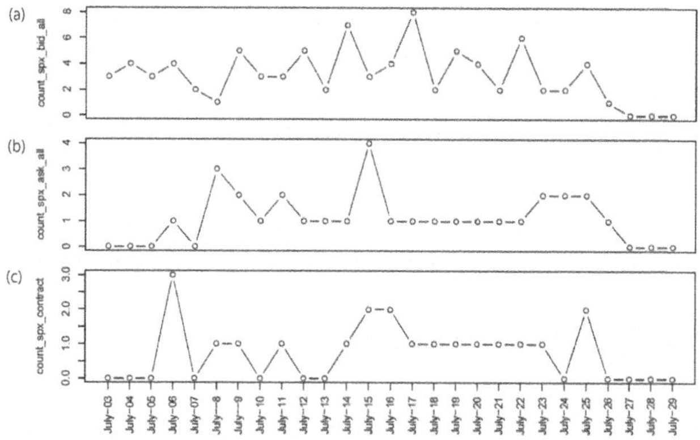

Fig. 10.11 **Bid and ask for SPX token.** The daily number of orders for (**a**) buying (bidding), (**b**) selling (asking), and (**c**) contracted to buy and sell orders

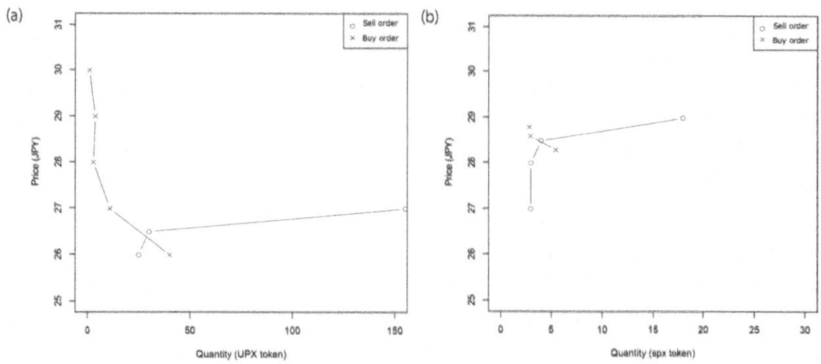

Fig. 10.12 **Typical demand and supply curves**, (**a**) UPX token (**b**) UPX token. Transaction prices are higher on the SPX than on the UPX, and transaction volume is lower on the SPX than on the UPX

At the beginning and end of the month, there are fewer bids for buying and selling orders. On average, the number of buying orders is smaller than the number of selling orders during the month.

Figure 10.11 shows the daily number of orders for buying (bidding) and selling (asking) of SPX tokens, which is the right to use electricity from photovoltaic panels. The bottom panel of Fig. 10.11 shows the number of contracted buying and selling orders. Almost all selling (asking) orders in this experiment were placed from the EDISON-X system. The temporal variation of the number of SPX buying and selling orders shows a similar trend to that of UPX.

Figure 10.12 shows typical composite demand and supply curves. Figure 10.12a is for the UPX Token. Figure 10.12b is for the SPX Token. Here, the reward is given to the top SPX user for the least CO_2 emission. The intersection of the demand and supply curves determines trading price and volume. We can see that the trading price is higher and the volume is lower on the SPX than on the UPX, reflecting that the higher trading price on the SPX is because the system sets the sell order price higher than the market price on the UPX. The high SPX transaction price reflects that the system sets the sell order price higher than the UPX market price. The low trading volume reflects the small installed capacity of solar PV.

10.3 Topological Characterization of Acquired Data

10.3.1 Hypergraph of Transactions

In an ordinary network [16], each edge (link) connects two vertices, but in the case of the EDISON-X energy market, we see an edge (link) must connect any number of vertices. The EDISON-X energy market is traded daily, meaning that a contract to buy or sell tokens is not concluded between two specific users but between market participants on that day. If the number of contracted orders each day equals n, the number of users involved in the contracts equals $n + 1$. The characteristics of the EDISON-X energy market can be regarded as a higher-order network. The higher-order network is represented by a concept of topology, i.e., a simplicial complex formed by a set of simplices that is closed under the inclusion of the faces of each simplex. In network science, we often use hypergraph [17–19], which is an alternative representation of higher-order networks that can be used instead of simplicial complexes.

In the EDISON-X energy market, transaction relationships involving more than two users can be represented by a hypergraph instead of the simplicial complexes. Figure 10.13 is a hypergraph showing the transaction relationship for UPX tokens in July 2022. Figure 10.13a shows Daily Transactions for selected days. The orange circles are users, and the area surrounding two or more users represents the transaction relationship. Transactions were concluded for 4 days among two users, 1 day among three users, and 1 day among six users. Figure 10.13b shows a Hypergraph of Transactions contracted on July 07, 09–10, 12–14, 16–18, 21–23, and 25, 2022. The closer the user's position is to the central part of the graph, the more important role it can be interpreted as playing in the market.

Figure 10.14 is a hypergraph showing the transaction relationship for SPX tokens in July 2022. The orange circles are users, and the area surrounding two or more users represents the transaction relationship. Figure 10.14a shows Daily Transactions for selected days. Transactions were concluded for 3 days between two users and 3 days between three users. Figure 10.14b shows a Hypergraph of Transactions contracted on July 08–09, 11, 14–23, and 25, 2022. Since the EDISON-X system is responsible for selling new SPX tokens, the node in the center of the graph is written "admin," which stands for the system administrator.

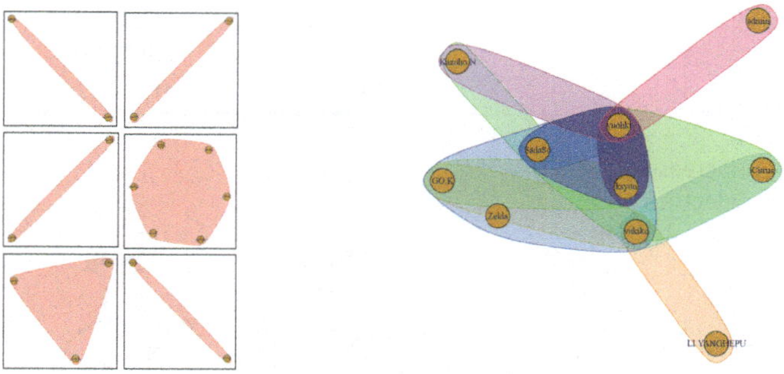

Fig. 10.13 Hypergraph of transactions in July 2022 for UPX token. The orange circles are users, and the area surrounding two or more users represents the transaction relationship. The closer the user's position is to the central part of the graph, the more important role it can be interpreted as playing in the market

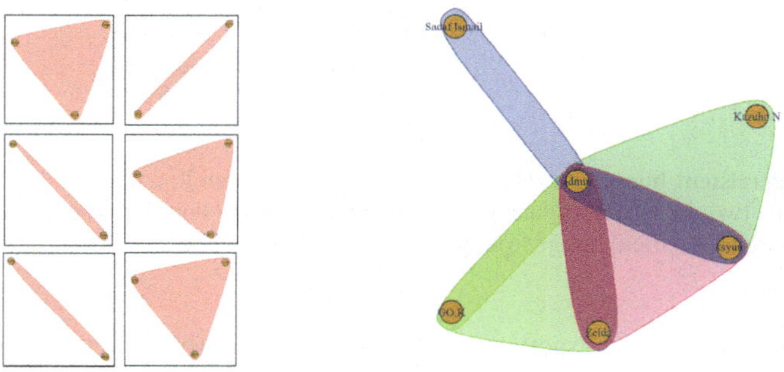

Fig. 10.14 Hypergraph of transactions in July 2022 for SPX token. Since the EDISON-X system is responsible for selling new SPX tokens, the node in the center of the graph is written "admin," which stands for the system administrator

10.3.2 Cavity Detection Using Persistent Homology

How can we characterize the "data shape," a feature of the distribution of many data points? It can be characterized not by "clusters" of data points but by the existence of "cavities (rings, holes)." Topology is useful for examining such characteristics of the shape of data. In particular, cavities can be detected using homology by drawing a sphere of appropriate radius centered on each data point. Furthermore, by using persistent homology [20, 21], which extends the concept of homology, it is possible to obtain not only the existence of a cavity but also detailed geometric characteristics such as its size and stability. Such a methodology is called topological data analysis (TDA) and has recently attracted much attention [22, 23].

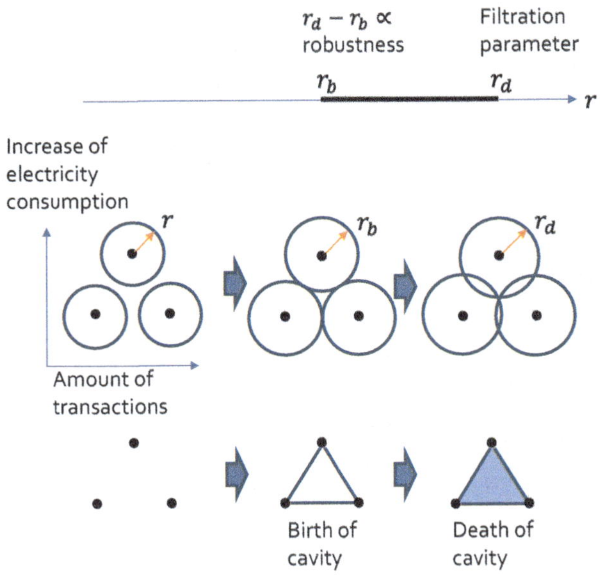

Fig. 10.15 Persistent homology. The characteristics of the overall market are represented as a data point on a two-dimensional plane with the volume of transactions by each user on the x-axis and the increase in electricity usage from the previous day on the y-axis

We expect market transactions to become more active when electricity consumption increases from the previous day. A disturbance in the correlation between electricity consumption and market transactions is observed as a "cavity." Thus, we hypothesize that market transactions become less active when "cavities" appear. To test this hypothesis, we quantify disturbances in the correlation between electricity consumption and market transactions by detecting one-dimensional "cavities."

Figure 10.15 explained the concept of persistent homology. Persistent homology has been recognized as one of the essential tools of TDA. The characteristics of the overall market are represented as a data point on a two-dimensional plane with the volume of transactions by each user on the x-axis and the increase in electricity usage from the previous day on the y-axis. Draw a circle of radius r from each user's data point. The circles do not overlap if the radius r is small, as in the middle left panel. As the radius r increases, three circles come into contact with each other at $r = r_b$, as shown in the middle panel. A cavity appears at this radius's center of the three touching circles. As the radius r is further increased, the cavity disappears at $r = r_d$ as shown in the middle right panel. The radius r is called the filtration parameter, and the larger $r_d - r_b$ is, the more robust the cavity is.

Figure 10.16a shows the overall market characteristics; no market transactions were executed on July 20, 2022. Figure 10.16b plots r on the x-axis when the cavity appears and r on the y-axis when it disappears; the further away from the 45° line, the more robust the cavity is. From this figure, we can see that one cavity is robust.

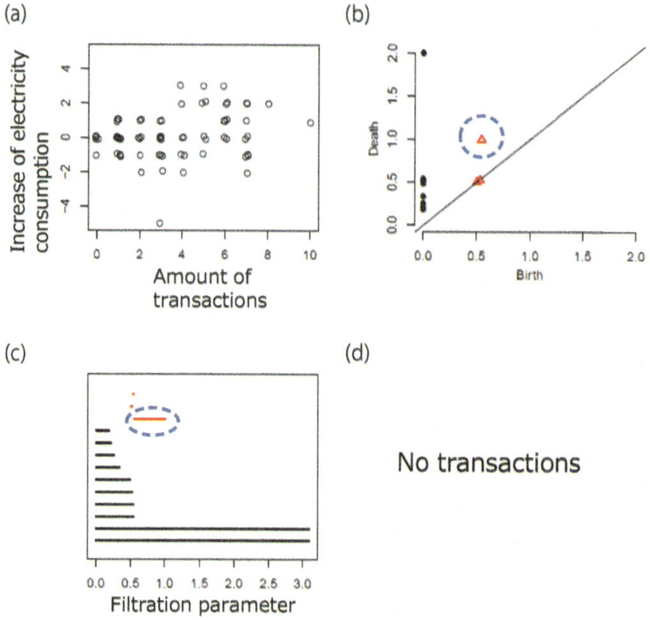

Fig. 10.16 Market characteristics on July 20, 2022. The black circles in panel (**a**) indicate the position of each user, and the red triangles in panel (**b**) indicate robust cavities. No market transactions were executed when one robust cavity was observed

Figure 10.16c shows the filtration parameter from the onset to the disappearance of a cavity, indicated by the red line. The longer this line is, the more robust the cavity is. On July 20, robust cavities were observed. Figure 10.17a shows the overall market characteristics; on July 21, 2022, market transactions were executed. From Fig. 10.17b, c, we can see that one cavity is robust. Figure 10.17d shows three users' transactions. On July 21, a market transaction occurred, although robust cavities were observed.

Table 10.1 shows the relation of Market transactions and "cavities." When market transactions were executed, the number of days robust cavities were observed was twice (=8/4) as many as the number of days they were not observed. However, in the case of no market transactions, the number of days with robust cavities was 2.5 times (=5/2) the number without robust cavities. The results mean market transactions become less active when "cavities" appear. This result implies that the hypothesis that market transactions become less active when "cavities" appear could be adopted. We, however, need more data samples.

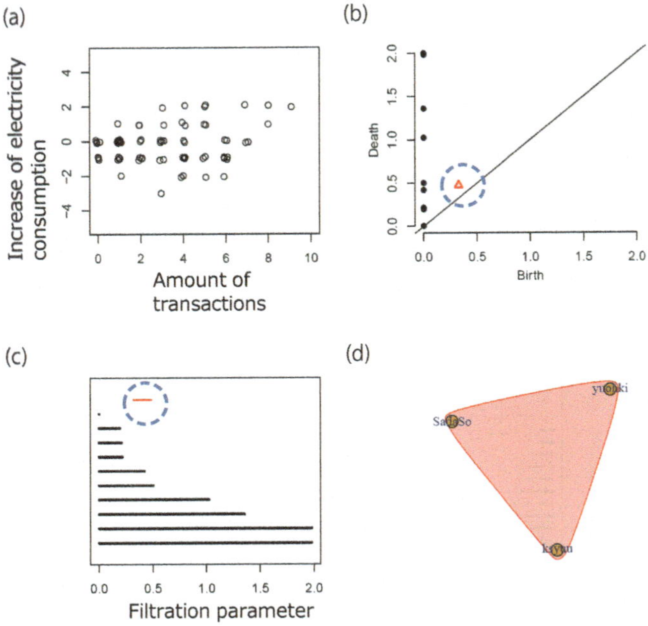

Fig. 10.17 Market characteristics on July 21, 2022. The black circles in panel (**a**) indicate the position of each user, and the red triangles in panel (**b**) indicate robust cavities. Market transactions were executed among three users, although robust cavities were observed

Table 10.1 Market transactions and "cavities"

	With transactions	No transactions
No cavity	4	2
With cavity	8	5

10.4 Summary

To solve the problems of complexity and transparency in collecting electricity charges, we developed the EDISON-X, which uses blockchain technology to manage the buying and selling of electricity usage rights for students residing in our school dormitory. We developed a blockchain-distributed energy trading system in this study and conducted a small-scale demonstration experiment in our school dormitory. From July 1, 2022, 17 students in our school joined the demonstration experiment to confirm the operation of the EDISON-X system. We confirmed that the energy trading system using blockchain technology could contribute to the effective deployment of renewable energy.

We developed topology and network science methodologies to understand the characteristics of energy trading. We show that the developed methodologies help understand energy trading characteristics and predict market changes in advance. We test the hypothesis that market transactions become less active when "cavities" appear using persistent homology. The preliminary analysis showed that market transactions become less active when "cavities" appear. This result implies that the hypothesis could be adopted. Currently, data acquisition and analysis are underway to get more convincing results.

References

1. Peck, M. E. (2017) "Blockchains: How They Work and Why They'll Change the World", p. 26, IEEE Spectrum, October 2017.
2. Nakamoto, S. (2009) "Bitcoin: A Peer-to-Peer Electronic Cash System", 24 May 2009.
3. D. Schwartz, N. Youngs, and A. Britto, "The Ripple Protocol Consensus Algorithm", Ripple Labs Inc, 2014.
4. Chase, B. and MacBrough, E. (2018) "Analysis of the XRP Ledger Consensus Protocol", arXiv:1802.07242v1, 20 Feb, 2018.
5. MacBrough, E. (2018) "Cobalt: BFT Governance in Open Networks", arXiv:1802.07240v1, 20 Feb, 2018.
6. Fugger, R. (2004) "Money as IOUs in Social Trust Networks & A Proposal for a Decentralized Currency Network Protocol", 18 April, 2004.
7. D. J. Galen, et al., "2019 Blockchain for Social Impact", Stanford GSB Center for Social Innovation, September 2019.
8. Barne, D. and Pirlea, F. (2019) "Money sent home by workers now largest source of external financing in low- and middle-income countries (excluding China). World Bank Data Blog." https://blogs.worldbank.org/opendata/money-sent-home-workers-now-largest-source-external-financing-low-and-middle-income. Cited 02 July 2019.
9. Frankel, J. A. et al (2011) "Are bilateral remittances countercyclical?", Open Economies Review, 22(1), 1–16. 2011.
10. Khalil, S., Disfani, V., Ahmadi, D., Rollins, G., "Impact of Blockchain Technology on Electric Power Grids - A case study in LO3 Energy", arXiv:2106.05395, 2021.
11. Power Ledger White Paper, https://www.powerledger.io/company/power-ledger-whitepaper.
12. Cloud Firestore, https://firebase.google.com/docs/firestore, accessed 2022-11-09.
13. The XRP Ledger Foundation, https://xrpl.org/xrp-ledger-overview.html.
14. Developer documentation for Firebase, https://firebase.google.com/docs, accessed 2022-11-09.
15. Material Design, https://m3.material.io/Material.io, accessed 2022-11-09.
16. A.-L. Barabasi, M. Pósfai, "Network science", Cambridge: Cambridge University Press. (2016). ISBN: 9781107076266 1107076269
17. A. Bretto, "Hypergraphs: Basic Concepts". In: Hypergraph Theory. Mathematical Engineering, Springer, Heidelberg (2013).
18. F. Battiston, E. Amico, A. Barrat, et al. "The physics of higher-order interactions in complex systems", Nat. Phys. 17, 1093–1098 (2021).
19. G. Bianconi. "Higher-Order Networks, An introduction to simplicial complexes", (Elements in Structure and Dynamics of Complex Networks), Cambridge, Cambridge University Press.
20. H. Edelsbrunner, D. Letscher, and A. Zomorodian, "Topological persistence and simplification", Discrete & Computational Geometry, 28(4):511–533, Nov 2002.
21. A. Zomorodian and G. Carlsson, "Computing persistent homology", Discrete & Computational Geometry, 33(2):249–274, Feb 2005.
22. H. Edelsbrunner and J. Harer, "Computational topology: an introduction", American Mathematical Soc., 2010.
23. G. Carlsson, "Topology and data", Bull. Amer. Math. Soc., 46:255–308, January 2009.

Chapter 11
V2G: Vehicle to Grid

11.1 System Dynamics

"The limits to growth" is a report on the exponential economic and population growth with a finite supply of resources [1]. This publication predicted that if humankind continues to increase its population and destroy the environment at the current rate, resources may be depleted in about 20 years and that human growth will eventually reach its limits due to environmental degradation and resource constraints. "The limits to growth" was received with shock by the world at a time when the postwar free economic order was under pressure to be rebuilt. The authors developed "the world model," which is formulated using 12 variable simultaneous linear differential equations. They made various simulations with different parameter sets. They found that the simulation results strongly depend on the feedback relationships in the simultaneous differential equations rather than on the values of the parameters. We note that in the world model, no considerations were made to the emerging countries, the role of financial mechanisms on resource price, and the economic and social losses caused by climate change. This research was later generalized and developed into system dynamics, the study of the characteristics of real-world dynamical systems [2].

11.1.1 Predator–Prey Model

In the following, we learn mathematics to understand the system dynamics step by step. The simplest model of the food chain is the predator–prey model, described

Y. Ikeda, *Data Science of Renewable Energy Integration*,
Evolutionary Economics and Social Complexity Science 30,
https://doi.org/10.1007/978-981-99-8779-5_11

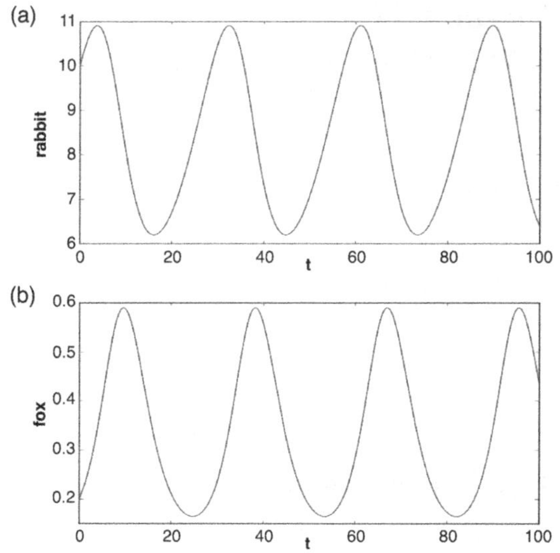

Fig. 11.1 Predator–Prey Model. The change in the number of foxes lags slightly behind the change in the number of rabbits

by the Lotka–Volterra equation with the number of rabbits $R(t)$ and the number of foxes $F(t)$:

$$\begin{cases} \frac{dR(t)}{dt} = R_{in} - R_{out} = rR(t) - hF(t)R(t) \\ \frac{dF(t)}{dt} = F_{in} - F_{out} = ehF(t)R(t) - qF(t). \end{cases} \tag{11.1}$$

Here, r, h, e, and q are the fecundity rate of the rabbit, which means the probability of increasing the rabbit, the rate of predation, which means the probability of a fox catching and eating a rabbit, the fecundity rate of fox, which means the probability of a fox eating a rabbit and increasing, and the rate of natural death of fox, respectively. The Lotka–Volterra equation is a set of rate equations describing a dynamical system. The first equation is nonlinear, but if we want to interpret it as a linear equation for $R(t)$, we can also say that the coefficient in the second term varies in time as $hF(t)$. The second $F(t)$ equation holds the same interpretation.

Figure 11.1 shows a simulation of the predator–prey model with parameters $r = 0.1$, $h = 0.3$, $e = 0.2$, and $q = 0.5$ and the initial conditions $R(0) = 10.0$ and $F(0) = 0.2$. Population cycles of rabbits and foxes are observed. Note that the change in the number of foxes lags slightly behind the number of rabbits.

11.1.2 Stability of Dynamical System

Next, we discuss the stability of the dynamical system. The system is described by variables $x_i (i = 1, 2, \cdots, N)$ and evolves in time by the following simultaneous differential equations. The function $F_i (i = 1, 2, \cdots, N)$ contains parameters

$a_k(k = 1, 2, \cdots, M)$ and is a nonlinear function with respect to the variables x_i:

$$\begin{cases} \frac{dx_1(t)}{dt} = F_1(x_1(t), \cdots, x_N(t); a_1. \cdots, a_M), \\ \cdots \\ \frac{dx_N(t)}{dt} = F_N(x_1(t), \cdots, x_N(t); a_1. \cdots, a_M). \end{cases} \tag{11.2}$$

This system satisfies the following equation in a steady state:

$$\frac{dx_1(t)}{dt} = \cdots = \frac{dx_N(t)}{dt} = 0. \tag{11.3}$$

Thus, the steady-state solution $x_{i0}(i = 1, 2, \cdots, N)$ satisfies the following equation:

$$\begin{cases} F_1(x_{10}, \cdots, x_{N0}; a_1. \cdots, a_M) = 0, \\ \cdots \\ F_N(x_{10}, \cdots, x_{N0}; a_1. \cdots, a_M) = 0. \end{cases} \tag{11.4}$$

Now, assuming that displacements $\Delta x_i(t)(i = 1, 2, \cdots, N)$ from the stationary solution have occurred, we derive the equations that satisfy these displacements. By substituting the following equations into Eq. (11.2) and using Eqs. (11.3) and (11.4)

$$x_i(t) = x_{i0} + \Delta x_i(t), \tag{11.5}$$

we obtain the following differential equations for $\Delta x_i(t)(i = 1, 2, \cdots, N)$:

$$\frac{\Delta \mathbf{x}}{dt} = \mathbf{J}\Delta \mathbf{x}, \tag{11.6}$$

where each element of Jacobi matrix $\mathbf{J}$ which is also called Jacobian is defined by

$$J_{ij} = \left. \frac{\partial F_i(x_1, \cdots, x_N)}{\partial x_j} \right|_{x_{10}, \cdots, x_{N0}}. \tag{11.7}$$

The time evolution of the displacements can be characterized by the eigenvalues $\lambda_\alpha(\alpha = 1, 2, \cdots, N)$ of the Jacobian:

$$\mathbf{J}\mathbf{V}_\alpha = \lambda_\alpha \mathbf{V}_\alpha, \tag{11.8}$$

where $\mathbf{V}_\alpha$ are N-dimensional eigen vectors. In such eigenvalue problems, the eigenvalues are generally complex numbers:

$$\lambda_\alpha = \mathrm{Re}(\lambda_\alpha) + i\mathrm{Im}(\lambda_\alpha) = \lambda_\alpha^{(R)} + i\lambda_\alpha^{(I)}. \tag{11.9}$$

The time evolution of the displacements can be written using the eigenvalues, as follows:

$$\Delta \mathbf{x}(t) \propto e^{\lambda_\alpha t} = e^{\lambda_\alpha^{(R)} t} e^{i\lambda_\alpha^{(I)} t} = e^{\lambda_\alpha^{(R)} t} \left(\cos \lambda_\alpha^{(R)} t + i\lambda_\alpha^{(I)} \sin t \right). \tag{11.10}$$

The temporal change of the displacements can be classified using the real and imaginary parts of the eigenvalues as follows:

- If $\lambda_\alpha^{(R)} \leq 0$ and $\lambda_\alpha^{(I)} = 0$, the system exhibits monotonic convergence.
- If $\lambda_\alpha^{(R)} \geq 0$ and $\lambda_\alpha^{(I)} = 0$, the system exhibits monotonic divergence.
- If $\lambda_\alpha^{(R)} = 0$ and $\lambda_\alpha^{(I)} = 0$, the system is in a steady state.
- If $\lambda_\alpha^{(R)} \leq 0$ and $\lambda_\alpha^{(I)} \neq 0$, the system exhibits convergence with oscillation.
- If $\lambda_\alpha^{(R)} \geq 0$ and $\lambda_\alpha^{(I)} \neq 0$, the system exhibits divergence with oscillation.
- If $\lambda_\alpha^{(R)} = 0$ and $\lambda_\alpha^{(I)} \neq 0$, the system is in steady-state oscillation.

11.1.3 Feedback of Dynamical System

Next, we discuss the feedback of the dynamical system as shown in Fig. 11.2. Variables indicated by squares are linked to other variables by links; Fig. 11.2 shows how the four variables A to D are connected in a loop. Each link has a direction, indicating the sign of the effect of a positive change in the source on the destination. In such a loop, feedback occurs as a whole. The signs of each link shown in Fig. 11.2 are (-), (-), (+), and (-). The product of these signs is negative value; we observe negative feedback, which exhibits balancing. Alternatively, if the signs of each link are (-), (+), (+), and (-), the product of these signs is a positive value. We observe positive feedback, which exhibits reinforcement.

Variable A is a change that returns to itself via variable B, variable C, and variable D. This is called the direct effect. The total effect $R^{(T)}$ of the feedback loop is equal to the direct effect $R^{(D)}$:

$$R^{(T)} = R^{(D)}. \tag{11.11}$$

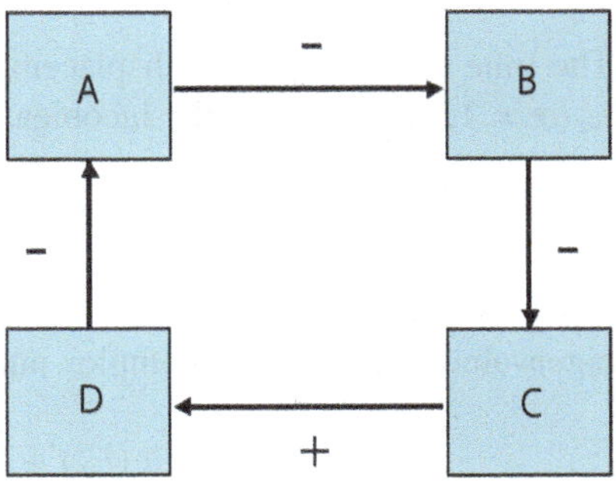

Fig. 11.2 Feedback. (+) means that if the source of the arrow is larger (smaller), the destination will also be larger (smaller). (-) means the opposite

However, the effect returned to variable A is transmitted to variable B. In the same way, the second effect returns to variable A through variable C and variable D. This second effect is smaller than the direct effect and is $g(< 1)$ times the direct effect.

$$R^{(T)} = R^{(D)} + g \times R^{(D)}. \tag{11.12}$$

The second effect generates a third effect, and so on. The same feedback is repeated an infinite number of times. In this way, the effect of the feedback converges to a particular value.

$$R^{(T)} = R^{(D)} \times (1 + g + g^2 + g^3 + \cdots) = \frac{R^{(D)}}{1-g}. \tag{11.13}$$

In the subsequent sections of this chapter, we develop the system dynamics model to evaluate the impact of electric vehicles on the power grid.

11.2 Electric Vehicle

The evolution of peak load can be considered a vital grid design metric [3, 4]. Overall electricity demand and the daily schedules of electricity usage are currently major influences on peak load. In the future, peak load will be influenced by new or increased demands, such as the penetration of clean energy technologies, such as electric vehicles (EVs).

11.2.1 G2V and V2G

A high penetration of EVs will affect the peak load significantly if the charging is not controlled. Many EV users might charge their EVs simultaneously in the early evening when peak system demand already exists and further stress the grid capacity. Scheduled charging, however, can shift some load from the evening to the early morning and thus flatten the load curve. EVs are powered by an electric battery that must be charged from the grid, called Grid-to-Vehicle (G2V). EVs will increase peak load if connected during peak hours and access to the grid is not managed appropriately. The number of EVs determines the total stress on the grid resulting from all individual vehicle loads.

In a more advanced deployment stage, EVs could provide vehicle-to-grid (V2G) services. They could be used as electric storage to reduce peak generation capacity by discharging electricity from the battery to the grid. Smart-grid technology, through advanced monitoring and control equipment, could reduce peak demand and thus prolong and optimize the use of the existing infrastructure. Investments in generation, transmission, and distribution might be deferred by adopting smart

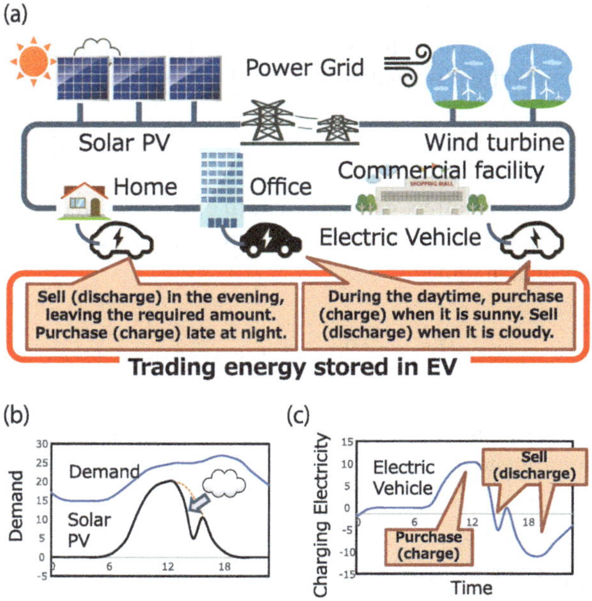

Fig. 11.3 Concept of V2X. EVs can serve as a balancing power to achieve a balance between supply and demand

grids, and smart-grid investment might be partially facilitated by using synergies between grid renewal and modernization. EVs, however, also have the potential to act as grid storage devices and feed electricity back into the grid at peak hours. V2G relies on a smart grid that enables greater demand side participation for balancing peak reduction of contingency actions [5–7].

If enough power is stored in the EV, the car can be used as a power source. This is Vehicle-to-Load (V2L), which can supply external power, allowing the use of electric appliances such as microwave ovens and electric kettles. Some EVs are equipped with a power outlet in advance, while others can be used as a power outlet by attaching a special adapter to the charging port.

In recent years, an increasing number of people have purchased EVs as "emergency preparedness" for disasters or as storage batteries to store surplus electricity from solar power generation installed in their homes and use it. The method of supplying power from an EV to a home via a dedicated device is called Vehicle-to-Home (V2H). V2L, V2H, and V2G are sometimes collectively called Vehicle-to-Everything (V2X). The concept of V2X is shown in Fig. 11.3. In this figure, a running EV is not depicted for a simple explanation. Figure 11.3a assumes that the EV is parked in a private home, office, or commercial facility and connected to the power distribution grid through charging and discharging equipment. As shown in Fig. 11.3b, electricity demand is mainly during the daytime, with peak demand during the evening and nighttime hours. Electricity generated by solar PV panels peaks during the daytime and is zero during the nighttime. Even during the daytime, the power generated from solar PV panels may decrease due to rain clouds and other factors. As shown in Fig. 11.3c, when the weather is clear and the power generated from the solar PV panels is large, EVs parked in private homes, offices, and commercial facilities can store enough electricity during the daytime. When the power generated from solar panels decreases due to rain clouds or when the power demand increases in the evening or later and the power generated from solar PV panels decreases, the EVs sell the electricity stored in the EVs. In this way, EVs can balance supply and demand as a balancing power. Thus, if EVs can be used as a new balancing power, more renewable energy can be integrated into the power grid.

There are two major advantages of V2H. The first advantage is that the battery capacity is larger than that of general household storage batteries, enabling electricity to be used for longer periods. The electrical energy of a typical household storage battery is about 5 to 12 kWh, while a small EV can hold 20 kWh, and a large EV can hold 100 kWh. Even if such large capacities are not necessary for daily use, they can provide power during a disaster or other power outage. The second advantage is the possibility of shifting or shaving electricity peaks. EVs can be charged late at night when electricity rates are less expensive, and the EVs can be used as a power source to run household appliances during the daytime. In addition, people with solar PV generation systems installed in their homes can store the electricity generated during the day and use it when necessary. In this way, V2H has the great merit of lowering electricity bills.

V2G is an attempt to pursue the second benefit of V2H on a larger scale, as well as shifting/shaving electricity peaks not only to households. V2G enables the power charged in the batteries of EVs to be supplied to the power grid, making it possible to use EVs as infrastructure for energy supply. More renewable energies are installed shortly, the supplied power will likely become unstable due to weather, and V2G will be required to function to enhance the balancing power. Currently, demonstration experiments are being conducted worldwide, and large-scale efforts are underway in Japan to verify various issues, such as whether charging from the power grid and discharging to the power grid go smoothly.

11.2.2 Charging Power

A typical household plug has between 1.8 kW in Northern America (110 V at 16 A) and 3.7 kW (230 V at 16 A) in Europe. Many existing battery vehicle chargers use the National Electrical Code (NEC) "Level 2" standard of 6.6 kW. The first automotive power electronics unit designed for G2V (in production by AC Propulsion) provides 80 A in either direction. Thus, by Joule's law (Power [W] = potential difference [V] × electric current [I]), it provides 19.2 kW at a residence (240 V) or 16.6 kW at a commercial building (208 V) [8]. In addition, industrial chargers (or so-called fast-charging stations) have capacities of 43 kW to 200 kW [9]. The French utility EDF uses the following categorization levels for chargers: 3 kW, 6 kW, 24 kW, 43 kW, and 150 kW [10]. The US Department of Energy uses three classification levels: 1.44 kW, 3.3 kW, and 60 kW to 150 kW [11]. The last level is for fast charging.

Fast charging can reduce battery life, and it is not assumed to become the preferred charging mode. An alternative fast-charging mode will potentially be battery swapping, where a fully charged one replaces the discharged battery within a few minutes. The battery swapping station would instead use slow charging to optimize battery life.

France expects that 90% of vehicles will be charged using slow charging and 10% using quick charging [12]. By using this distribution and assuming 3.7 kW for slow charging and 40 kW for fast, an average charging load PCHG of 7.3 kW is assumed.

11.2.3 Discharging Power

The charging power is governed by the charger, which determines the load of the charging process on the grid (G2V). If a vehicle provides V2G services to the grid, it is uncertain whether the charger or the battery limits the discharging power. The power that a vehicle can deliver can be approximated with the following formula [8]:

$$P_{VEH} = \frac{(E_S - \eta_{EV}\,(d_D + d_{RB}))\,\eta_{INV}}{t_{DISP}}, \tag{11.14}$$

where P_{VEH} is the maximum power from V2G (kW), E_S the stored energy available as direct current (kWh) to the inverter, d_D the distance driven (km) since battery was full, d_{RB} the distance (km) of the range buffer required by the driver, η_{EV} the vehicle electric driving efficiency (kWh/km), η_{INV} the electrical conversion efficiency of the direct to alternating current inverter (dimensionless), and t_{DISP} the time the vehicle's stored energy is dispatched (hours).

The IEA's EV/PHEV Technology Roadmap assumes that EVs will have a battery storage capacity ES of about 30–80 kWh. It also assumes an average efficiency η_{EV} of 0.15 kWh/km to 0.2 kWh/km, with some additional reserve battery capacity [3].

The distance driven since the battery was full, d_D, depends on various characteristics: the driving pattern, the vehicle type, and the driver's strategies for being prepared to provide power. The average daily vehicle distance traveled per driver in the United States is 51.2 km. Kempton [8] assumes that half the average daily vehicle distance has been traveled when the driver parks the vehicle and plugs it into the grid, $d_D = 26$ km. The distance of the range buffer d_{RB} is the minimum remaining range required by the driver as a reserve. The range buffer is determined by the V2G service company or the driver and does not depend on the vehicle characteristics. A study with Californian drivers found that 32 km was sufficient for most drivers [13]. This paper uses a range buffer d_{RB} of 32 km for EVs. An inverter efficiency η_{INV} is 0.93.

The share of vehicles connected to the grid during peak load will strongly influence the load of the total vehicle fleet on the grid. In the United States, vehicles are parked 96% of the time. The fraction of vehicles connected during peak hours is expected to be lower since the vehicle use is also higher during peak hours.

The smart grid, however, can actively control the load by partially disconnecting some vehicles and thus throttle peak demand. Managed charging is expected to penetrate the market soon, as it is a low-cost and mature technology available today. It only requires a communication device in the charger that transmits relevant battery information to the grid operator, who, by remote control, can partially disconnect a certain number of charging vehicles if the grid capacity is saturated. Passenger car electrification is underway, and the additional load must be managed to a certain degree.

11.2.4 Simple Grid-EV Model

Temporal evolution of the stored electricity in EVs, $E(t)$ [Wh], and the kinetic energy of generator rotor $G(t)$ [Wh] are described by the following system dynamics equations:

$$\begin{cases} \frac{dE(t)}{dt} = E_{in} - E_{out} = \text{G2V} - (\text{Driving} + \text{V2G})\,, \\ \frac{dG(t)}{dt} = G_{in} - G_{out} = \text{EnergyInput} - (\text{Generation} + \text{G2V} + \text{V2G})\,, \end{cases} \tag{11.15}$$

where G2V, V2G, Driving, Generation, and demand are electric power from the grid to EVs [W], electric power from EV to the grid [W], electric power consumed in driving EVs [W], generated electricity [W], and electricity demand [W], respectively.

We note the negative sign of V2G in the last term on the right side of the equation $G(t)$. V2G allows electricity to be returned from the storage battery of the electric vehicle to the power grid. This has the effect of reducing the effective electricity demand. Therefore, the generator needs to generate less power for the V2G power. This is why V2G in the last term has a negative sign.

The system is illustrated in Fig. 11.4. The two stock variables $E(t)$ and $G(t)$ are indicated by rectangles, and the relationship between the stock variables is represented by arrows.

In the simple model of this chapter, we assume that the various quantities can be written in the following form. Such an expression may not be realistic, but more is needed to investigate the quantitative behavior of the system. In the following, $d(t)$, $1 - d(t)$, V_{max}, and $D(t)$ are the percentage of EVs on the road, the percentage of EVs parked, storage capacity, and electricity demand [W], respectively.

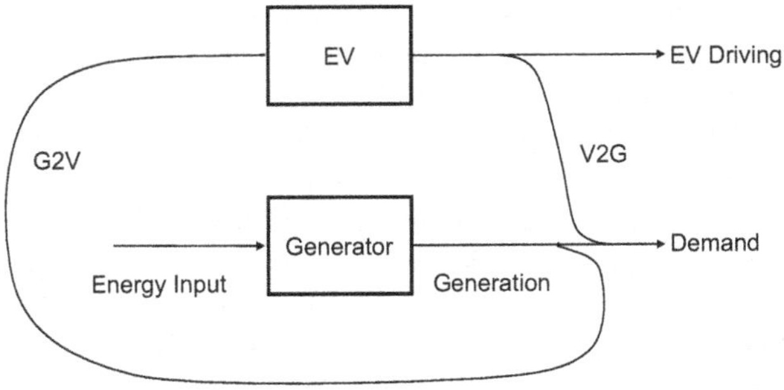

Fig. 11.4 G2V and V2G. Image of system dynamics equations for a simple grid-EV model is drawn

Energy Input
EnergyInput [W] is proportional to the difference between demand $D(t)$ and the kinetic energy of the generator rotor $G(t)$ [Wh]:

$$\text{EnergyInput} = D(t) - \mu G(t), \tag{11.16}$$

where constant μ has a dimension [h^{-1}].

Generated Electricity
Generation [W] is proportional to demand $D(t)$ [W]:

$$\text{Generation} = \varphi D(t), \tag{11.17}$$

where constant φ has no dimension.

Electric Power Consumed in Driving EVs
Driving [W] is proportional to stored electricity in EVs, $E(t)$ [Wh], and the percentage of EVs on the road $d(t)$:

$$\text{Driving} = \varepsilon E(t)d(t), \tag{11.18}$$

where constant ε has a dimension [h^{-1}].

Electric Power from Grid to EVs
G2V [W] is proportional to the difference between the storage capacity V_{max} and the stored electricity in EVs, $E(t)$ [Wh], and the percentage of EVs parked $1 - d(t)$:

$$\text{G2V} = \xi\,(V_{max} - E(t))\,(1 - d(t)), \tag{11.19}$$

where constant ξ has a dimension [h^{-1}].

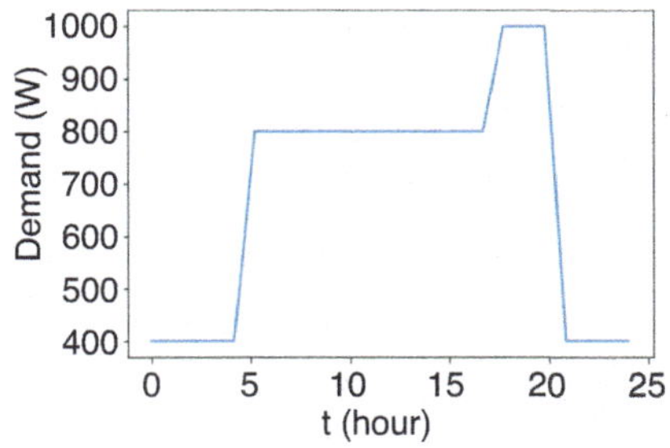

Fig. 11.5 Effective Demand. $D(t)$: Evening demand is high, but the output of solar PV power is lost, so effective demand is the highest level

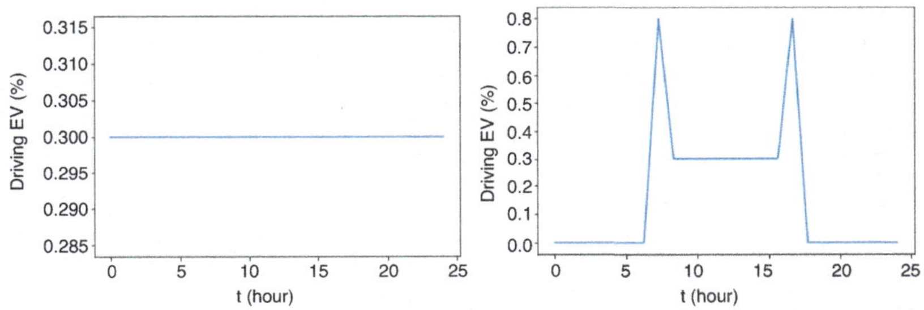

Fig. 11.6 Two scenarios for the percentage of EVs on the road. $d(t)$: The left panel shows a constant percentage of EVs running throughout the day, while the right panel shows a higher percentage of EVs running during morning and evening commuting

Electric Power from EVs to Grid

V2G [W] is proportional to the difference between the demand $D(t)$ and generated electricity $\mu G(t)$ [Wh] and the percentage of EVs parked $1 - d(t)$:

$$\mathrm{V2G} = \eta\,(D(t) - \mu G(t))\,(1 - d(t)), \tag{11.20}$$

where constant η has no dimension.

In the following simulations, we assumed two levels of daytime and nighttime demand. We also assumed that a certain amount of electricity is generated by solar PV power during the daytime. As a result, the effective demand is three levels, as shown in Fig. 11.5. Evening demand is high, but the output of solar PV power is lost, so it is the highest demand level. Furthermore, we assumed two scenarios for the percentage of EVs on the road $d(t)$, as shown in Fig. 11.6. In the first scenario, a certain percentage of EVs continue to run throughout the day. In the second scenario, many EVs run during the morning and evening commute, and all EVs stop at night.

Simulation 1

First, we explain the results for Simulation 1, where the system consists of a generator only; no EVs are used. The parameters for the generator are $\mu = 1.0$

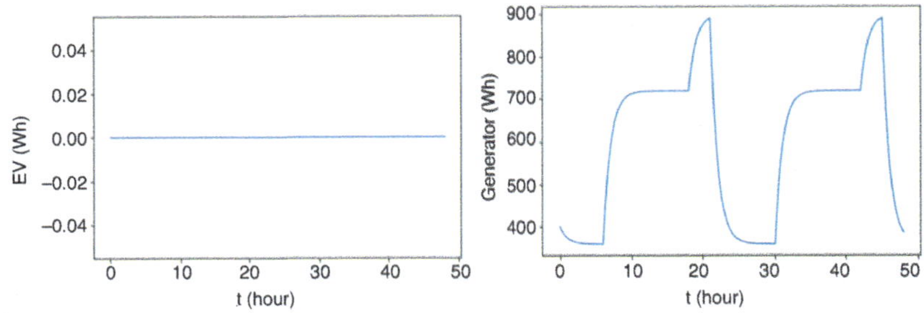

Fig. 11.7 **Simulation 1:** The system consists of a generator only; no EVs are used

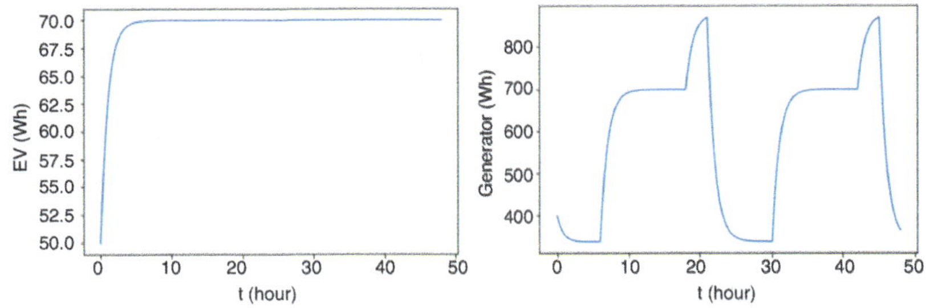

Fig. 11.8 **Simulation 2:** The system includes both generators and EVs and only performs G2V; EV driving is scenario 1

and $\phi = 1 \times 10^{-1}$. The value of the parameter ϕ was determined to reproduce the effective demand $D(t)$. The results for the stored electricity in EVs, $E(t)$ [Wh], and the kinetic energy of generator rotor $G(t)$ [Wh] are shown in Fig. 11.7. All simulations were performed over 2 days. The electric energy of the EVs $E(t)$ remained zero throughout the 2 days. The kinetic energy of the generator rotor $G(t)$ shows the same shape as the electricity demand $D(t)$.

Simulation 2
Next, we explain the results for Simulation 2, where the system includes both generators and EVs and only performs G2V; EV driving is scenario 1. The parameters for G2V are $\xi = 1.0$ and $\epsilon = 1.0$. For all subsequent simulations, the EV storage capacities shall be the following values: $V_{max} = 100$. The results for the stored electricity in EVs, $E(t)$ [Wh], and the kinetic energy of generator rotor $G(t)$ [Wh] are shown in Fig. 11.8. Since the running EVs is 30% as shown in the left panel of Fig. 11.6, the amount of stored power $E(t)$ in the left panel of Fig. 11.8 shows a constant value of 30%. The kinetic energy of the generator rotor $D(t)$ shown in the right panel is the same as in Simulation 1.

Simulation 3
Then, we explain the results for Simulation 3, where the system includes both generators and EVs and only performs G2V; EV driving is scenario 2. The

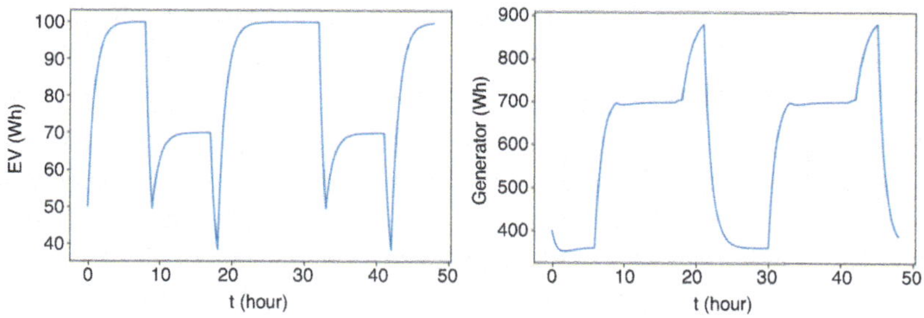

Fig. 11.9 **Simulation 3:** The system includes both generators and EVs and only performs G2V; EV driving is scenario 2

parameters for G2V are the same as in Simulation 2: $\xi = 1.0$ and $\epsilon = 1.0$. The results for the stored electricity in EVs, $E(t)$ [Wh], and the kinetic energy of generator rotor $G(t)$ [Wh] are shown in Fig. 11.9. The amount of electricity stored in EVs $E(t)$ is higher at night and lower during the day. Furthermore, $E(t)$ is considerably lower in the morning and evening, reflecting the driving scenario shown in the right panel of Fig. 11.6. The kinetic energy of the generator rotor $D(t)$ shown in the right panel is similar in Simulation 1 and Simulation 2. Due to the peak of EV driving in the morning and evening, the maximum values of $D(t)$ are much higher than in the previous two simulations.

Simulation 4

Finally, we explain the results for Simulation 4, where the system includes both generators and EVs and performs both G2V and V2G. EV driving is scenario 2. The parameters for G2V are the same as in Simulation 2 and Simulation 3: $\xi = 1.0$ and $\epsilon = 1.0$. The parameters for V2G is $\eta = 2 \times 10^{-1}$. The results for the stored electricity in EVs, $E(t)$ [Wh], and the kinetic energy of generator rotor $G(t)$ [Wh] are shown in Fig. 11.10. The amount of stored electric energy for EVs $E(t)$ is similar to that in Simulation 3, but with somewhat erratic changes. The portion of $E(t)$ that

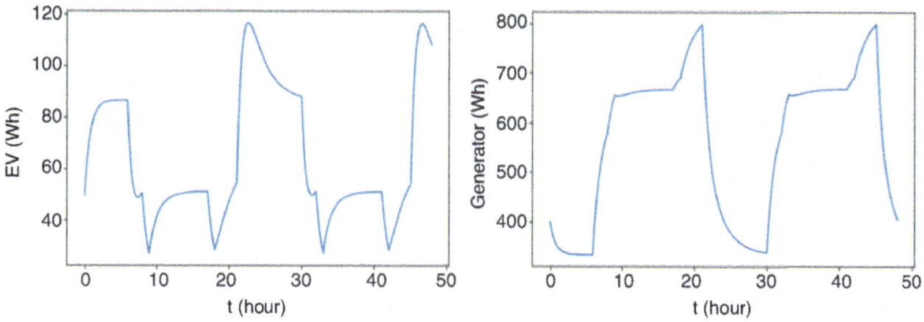

Fig. 11.10 **Simulation 4:** The system includes both generators and EVs and performs both G2V and V2G. EV driving is scenario 2. The right panel shows that $G(t)$ is smaller by performing V2G compared to Simulation 3

exceeds its capacity V_{max} is an artifact of using a simple model. The right panel shows that the kinetic energy of the generator rotor $G(t)$ is smaller compared to Simulation 3. This change is the effect of V2G.

We developed the system dynamics model to evaluate the impact of EVs on the power grid. In particular, the effects of V2G are very interesting. As suggested by the previous studies, V2G can be expected to reduce peak demand. Although very simple, the system dynamics model developed in this chapter can reproduce the effect of V2G in reducing peak demand. Further development of a more realistic model based on this model is expected to lead to a quantitative evaluation of the effects of V2G and to promote innovation toward decarbonization through concrete policy recommendations.

11.3 Summary

In this chapter, the basics of system dynamics are explained, and stability and feedback are discussed. The impact of electric vehicles on the electric grid, i.e., G2V and V2G, is discussed, and a system dynamics model is developed to evaluate their effects. Although the model is simple, it confirms the effectiveness of V2G in reducing peak demand.

References

1. Meadows, Donella H; Meadows, Dennis L; Randers, Jørgen; Behrens III, William W (1972). "The Limits to Growth; A Report for the Club of Rome's Project on the Predicament of Mankind". New York: Universe Books. ISBN 0876631650. Retrieved 26 November 2017.
2. John Sterman, "Business Dynamics: Systems Thinking and Modeling for a Complex World", Irwin Professional Pub, ISBN-13: 978-0072389159, February 2000.
3. International Energy Agency, "EV/PHEV Technology Roadmap", OECD/IEA, Paris, 2009.
4. Steve Heinen, David Elzinga, Seul-Ki Kim, and Yuichi Ikeda, "Impact of Smart Grid Technologies on Peak Load to 2050", International Energy Agency, WORKING PAPER, 2011
5. K. Rahimi and M. Davoudi, "Electric vehicles for improving resilience of distribution systems", Sustain. Cities Soc., vol. 36, no. November 2017, pp. 246–256, 2018
6. G. Merhy, A. Nait-Sidi-Moh, and N. Moubayed, "Control, regulation and optimization of bidirectional energy flows for electric vehicles charging and discharging", Sustain. Cities Soc., vol. 57, no. March, p. 102129, 2020
7. T. Kobashi, K. Say, J. Wang, M. Yarime, D. Wang, T. Yoshida, Y. Yamagata, "Techno-economic assessment of photovoltaics plus electric vehicles towards household-sector decarbonization in Kyoto and Shenzhen by the year 2030", J. Clean. Prod., vol. 253, p. 119933, 2020
8. Kempton, W., "Vehicle-to-Grid Power Fundamentals: Calculating Capacity and Net Revenue", Journal of Power Sources, Elsevier Inc., Vol. 144, pp. 268–279, 2005.
9. ABB, "Electric Vehicle Infrastructure: DC Fast Charge Station", ABB, Turgi, Switzerland, 2010.

10. Legrand, J.-L., "Groupe de travail sur les infrastructures de recharge pour les véhicules électriques ou hybrides rechargeables", workshop presentation, 2009.
11. US Department of Energy, "Plug-in Hybrid Electric Vehicle Charging Infrastructure Review", DOE, Idaho, 2008.
12. Rheinisch-Westfälische Technische Hochschule Aachen, "Grid for Vehicles", 2010.
13. Kurani, K., T. Turrentine and D. Sperling, "Demand for Electric Vehicles in Hybrid Households: An Exploratory Analysis", Transportation Policy, Vol. 1, pp. 244–256, 1994.

Chapter 12
Conclusion: Prospects for the Future

Significant changes have begun to take place all over the world and in every part of the society. In this chapter, we will consider the future of the decarbonized society in developed countries, taking Japan as an example, where natural resources are scarce, and the society is aging rapidly.

12.1 Society with Distributed Autonomous Organizations

In the future, Distributed Autonomous Organizations (DAOs) will become widespread in all parts of the society. DAOs are a new organizational structure involving multiple participants online and relying on software systems such as blockchain and smart contracts. DAOs are increasingly important due to the crypto asset and decentralized financial ecosystems, in contrast to traditional organizational forms that are private, closed, and centralized. The term DAO does not represent a specific type of organizational structure; the legal treatment of organizations classified as DAOs depends on how their organizational arrangements are structured. Many DAOs use smart contracts to automate or program some elements of their internal activities.

Examples of DAOs include social structures or organizations that include multiple participants setup for investment purposes, including investment or trading in crypto tokens or non-fungible tokens (NFTs) and fundraising, crowdsourcing, or charitable purposes. Many DAOs are involved in software engineering, developing, modifying, and maintaining open-source software infrastructure (e.g., blockchain systems and distributed financial applications). There is ambiguity about what constitutes a DAO, how they can be structured, and the possibility of multiple types of organizational forms or legal characterizations.

Y. Ikeda, *Data Science of Renewable Energy Integration*,
Evolutionary Economics and Social Complexity Science 30,
https://doi.org/10.1007/978-981-99-8779-5_12

Today, there are numerous DAOs, but only some are structured according to national laws. A tremendous amount of value flows is created, used, and sometimes lost by DAOs, raising questions about the legal status of DAOs, the responsibilities of their participants, and the rules applicable to DAOs. Therefore, appropriate legislation is urgently needed to promote and develop DAOs.

In Japan, DAOs promoting "digital citizens" have begun to take off. In particular, local governments suffering from depopulation have been active in this area. For example, "e-Kaga Citizen" in Ishikawa Prefecture and "Yamakoshi DAO" in Niigata Prefecture are well known. The "e-Kaga Citizen" does not include any foreign participants, while the "Yamakoshi DAO" includes Chinese and American participants. DAOs are open to foreign countries; however, it will be necessary in the future to set "borders" in consideration of security.

In recent years, there has been an increase in various start-up companies using blockchain technology. Under the current taxation system, start-ups are required to pay taxes in the Japanese yen when they make unrealized profits on crypto assets. In order to avoid paying the tax, there is no end to the number of start-ups that are forced to leave Japan for Singapore and other countries. For new industries based on blockchain technology to take root in Japan, it is essential to establish appropriate and urgent legislation.

In the first half of 2023, Japan is in the process of developing stablecoin legislation. Megabanks and regional banks will begin to offer services using stablecoins. This trend will enable regional banks to find investment targets other than government bonds, improve their business performance, and contribute to revitalizing local economies, the original raison d'être of regional banks. Stablecoins will likely become increasingly popular in the future. However, are stable coins stable? How stable are they? This question may be a subject for new research. From another perspective, it is interesting to ask whether the role of cryptographic assets such as Bitcoin will change when stablecoins become popular. Presumably, Bitcoin will survive as a speculative destination even if stablecoin becomes popular.

12.2 Decarbonized Society

As lifestyles change and innovations become more widespread, as described in the previous section, society will become increasingly decentralized. The new society that will emerge as a result will be called a "Decarbonized Society." Today, employees of large companies living in urban areas are fed up with long commuting and working hours. Long commuting and working hours are the working environment or lifestyle of the twentieth century. Many city dwellers want to move to regional cities. Reflecting such needs, some large companies have begun to encourage remote work by reducing the number of times people come to the office because of the COVID-19 pandemic. On the other hand, local governments suffering from depopulation want to promote a new work style in which remote employees are encouraged to move to rural areas as an effective countermeasure against depopulation. The new work style, which involves the movement of people, is thought to have a certain degree of effectiveness, but it is unlikely to be a powerful countermeasure.

Will "digital citizens" cope with the decline in population due to depopulation and aging? In a DAO, anyone can work from anywhere (even abroad). As a result, they can flexibly respond to a shortage of workers in a particular country. However, "digital citizens" do not increase the demand within a country or a specific municipality. Therefore, "digital citizens" still cannot cope with population decline. "Digital citizens" will not be able to fill the labor force in the production departments of manufacturing plants. The manufacturing industry in the high-tech sector is capital-intensive, and workers are expected to be concentrated in R&D and design engineers. At the same time, the production of products will be automated, with robots taking over in the production departments.

In addition, in recent years, in the logistics industry, which is responsible for the supply chain, there have been concerns for the first time about the emergence of a shortage of truck drivers and the need to avoid working long hours. More than drone transportation is needed in response to the shortage of truck drivers. Therefore, a method has been proposed in which packets are carried by truck drivers engaged in short-distance transportation and general drivers when they travel short distances and then transported over long distances using a bucket relay method. If the packet transportation is recorded on a blockchain, traceability of the bucket relay can be ensured. This method may be a promising solution to the shortage of truck drivers in the logistics industry. As a result of these innovations, labor shortages in the manufacturing and logistics industries are not expected to become a severe problem, and the most significant challenge facing Japan will not be labor shortage.

In a decarbonized society, how energy is produced and consumed will change as we move away from fossil fuels. In order to achieve large-scale grid integration of renewable energy with fluctuating output, wide-area operation of power grids, demand side management, and the use of energy storage devices, including hydrogen, will be necessary. In addition, using electric vehicles in the logistics sector is essential to reduce greenhouse gas emissions. The social background that will promote the spread of these vehicles is the depopulation and aging of rural areas. In particular, rural communities with depopulation and aging populations will require systems to watch over the daily lives of their elderly citizens. The consistent introduction of renewable energy + electric vehicles + automatic driving + storage batteries + life monitoring will make it possible to realize the following. This will be a countermeasure against depopulation and aging of rural areas. In addition, industries need to respond to carbon pricing, and large companies can do so. However, small- and medium-sized enterprises (SMEs), many located around urban areas or in regional cities, may need help understanding the carbon pricing system and may have a hard time dealing with it. Therefore, a service that records information on greenhouse gas emissions of SMEs on a blockchain and enables mutual surveillance is needed.

In this way, in a decarbonized society, various innovations will be made to solve problems such as changes in people's lifestyles, the aging of society, local economies facing depopulation, and changes in scarce factors of production due to information technology and technological innovation. For such innovations to succeed, it is necessary to change how people think.

12.3 Alternative to Economic Growth

DAOs change the way people work. DAOs compensate for the shortcomings of joint-stock companies. Not only will the industrial form change, but behind this change will be a significant shift in people's thinking and values as well. The era of pursuing economic growth through continued mass production and consumption will end soon. Economic growth in the new decarbonized society will use less energy and materials to improve the quality of life. At the same time, people in both developed and developing countries will pursue social fairness and well-being more than ever before.

Productivity will increase, wages will increase, supply will increase, well-being will increase, demand will increase, and the economy will grow. Here, productivity is not only labor and capital but also materials and energy productivity. The labor shortage will diminish due to artificial intelligence, robots, and automation of labor and services. In advanced economies such as Japan, capital will not be a scarce factor of production either due to limited future demand growth or due to aging, which will influence limited investment needs. Many have pointed out that the high consumption of materials and energy is a cause of climate change and a significant impediment to sustainability. Therefore, the productivity in terms of materials and energy is considered to be the most important indicator.

Today, the welfare that people pursue will not be economic wealth as represented by GDP but social fairness and well-being. The most important thing at this point is to express social fairness and well-being with quantitative measures or proxy variables. Once this quantitative measure or proxy variable is established, it will be possible to formulate a new capitalism. In order to consider measures of social fairness and wellbeing, the "old values and social norms" that we discarded in the process of industrialization in the twentieth century may provide a hint. It seems that the following three values that have existed in Kyoto for a long time are essential:

Value1: "KIWAME" To sharpen everything to the ultimate limit
Value2: "MOTENASHI" A warm welcome to new commers
Value3: "SHIMATSU" An attitude of moderation and frugality in life

When these three values are restored, "economic growth" will no longer mean an increase in GDP but a concept that means an increase in social fairness and well-being. Thus, based on a criterion that differs from a value system that is simply focused on economic growth, the "new system" referred to in Chap. 1 will be built to enhance the sustainability of human society.

Appendix A
Time Series of Wind Speed

Hourly change of wind speed at 50 m from the surface in London for each month of 2020 is shown in Figs. A.1, A.2, A.3, A.4, A.5, A.6, A.7, A.8, A.9, A.10, A.11, and A.12. The top panel shows wind speed s_t in m/s, and the bottom panel shows its change rate defined as log-difference of the wind speed: $r_t = \log s_t - \log s_{t-1}$. All data were from NASA's "The POWER Project."

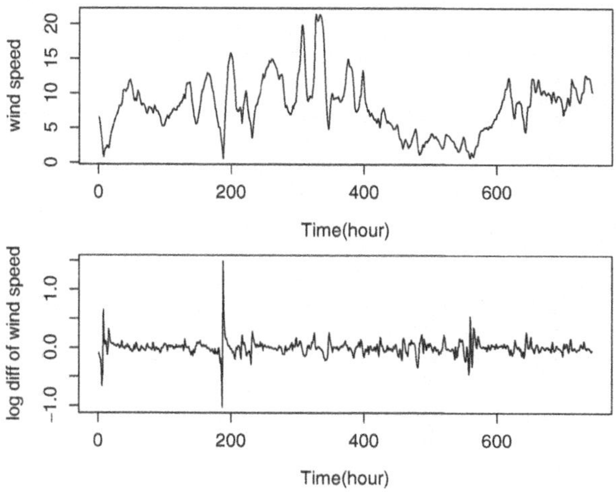

Fig. A.1 Wind speed (m/s) and its change rate at 50 m from the surface in London: January 2020

Y. Ikeda, *Data Science of Renewable Energy Integration*, Evolutionary Economics and Social Complexity Science 30,
https://doi.org/10.1007/978-981-99-8779-5

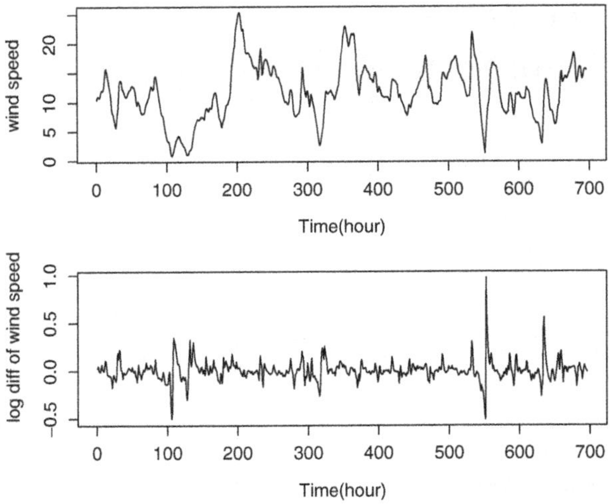

Fig. A.2 Wind speed (m/s) and its change rate at 50 m from the surface in London: February 2020

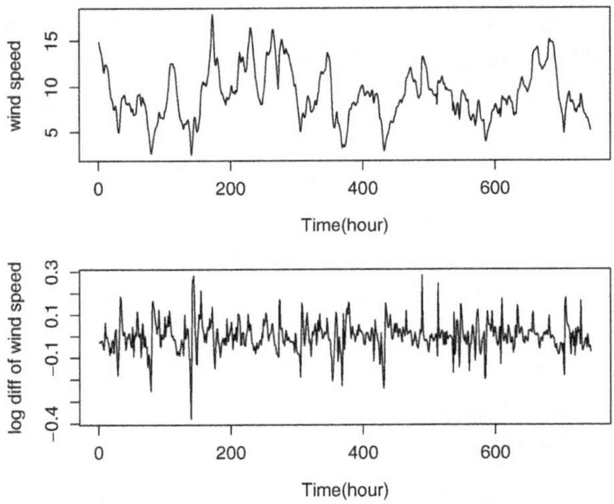

Fig. A.3 Wind speed (m/s) and its change rate at 50 m from the surface in London: March 2020

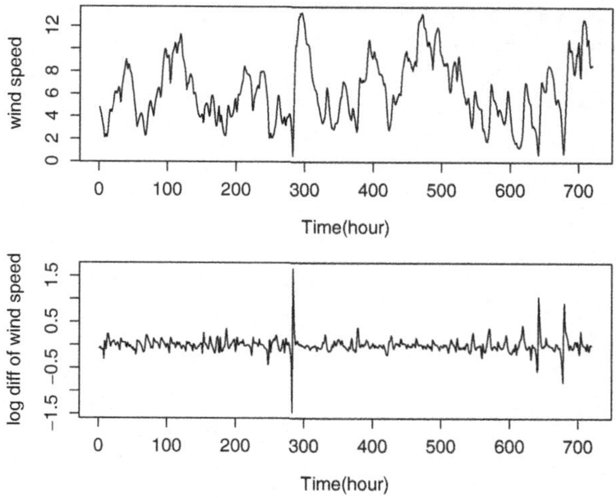

Fig. A.4 Wind speed (m/s) and its change rate at 50 m from the surface in London: April 2020

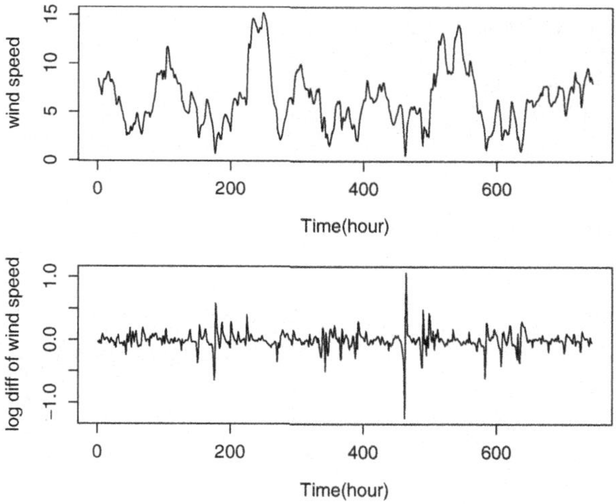

Fig. A.5 Wind speed (m/s) and its change rate at 50 m from the surface in London: May 2020

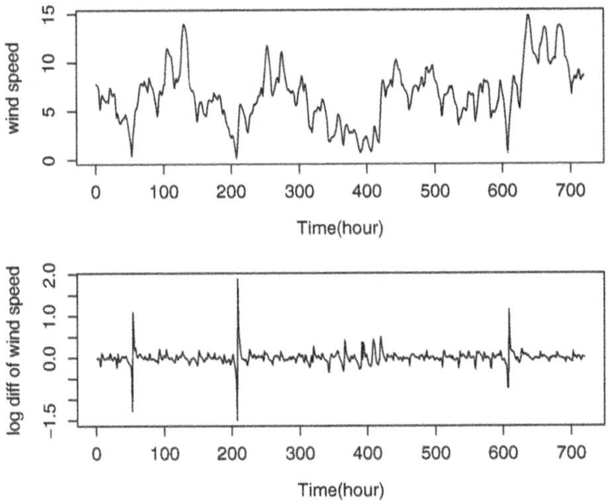

Fig. A.6 Wind speed (m/s) and its change rate at 50 m from the surface in London: June 2020

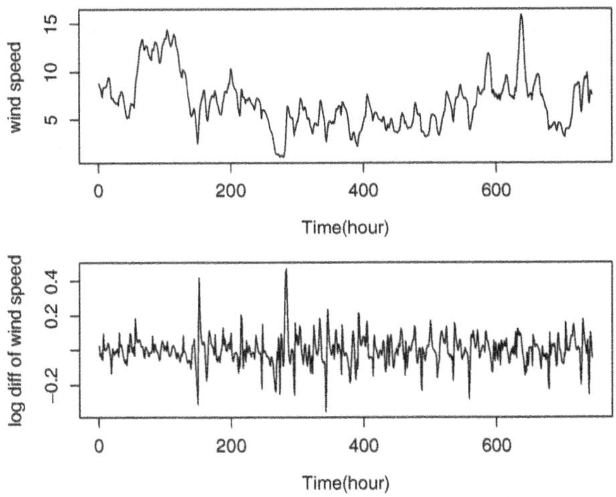

Fig. A.7 Wind speed (m/s) and its change rate at 50 m from the surface in London: July 2020

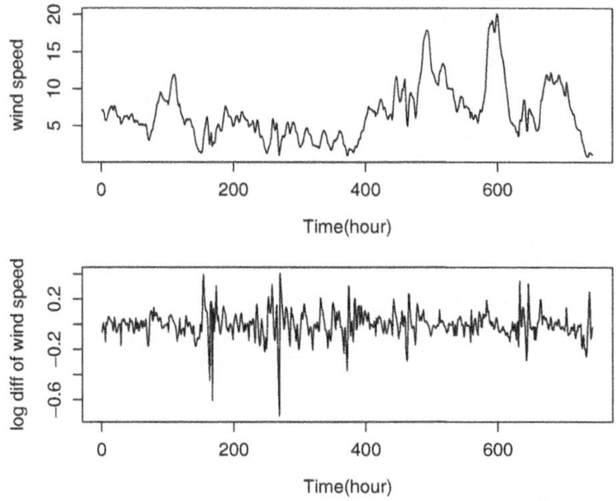

Fig. A.8 Wind speed (m/s) and its change rate at 50 m from the surface in London: August 2020

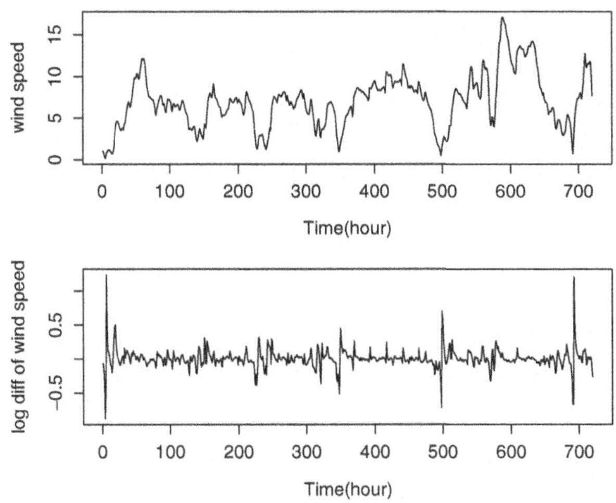

Fig. A.9 Wind speed (m/s) and its change rate at 50 m from the surface in London: September 2020

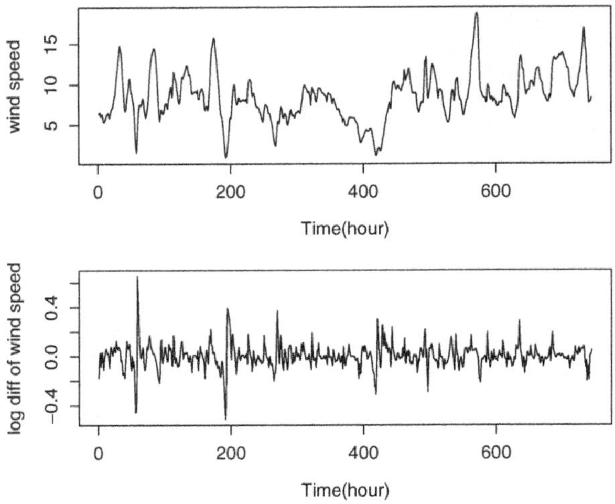

Fig. A.10 Wind speed (m/s) and its change rate at 50 m from the surface in London: October 2020

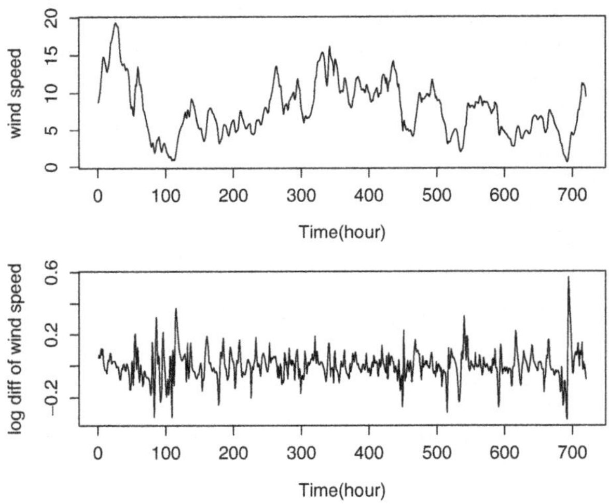

Fig. A.11 Wind speed (m/s) and its change rate at 50 m from the surface in London: November 2020

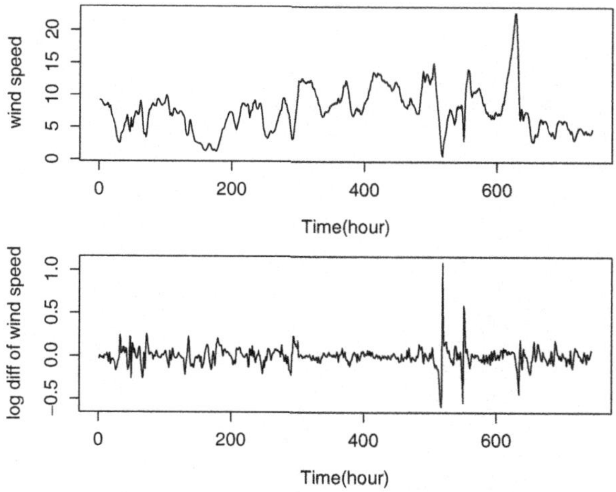

Fig. A.12 Wind speed (m/s) and its change rate at 50 m from the surface in London: December 2020

Appendix B
Random Walk Model for Wind Speed

We examine the random walk model for wind speed time series. The random walk model is expressed as

$$x_{t+1} = a_0 x_t + a_1 + u_t, \tag{B.1}$$

with $a_0 = 1$. Here u_t is the residual in regression analysis or interpreted as noise in the random walk model. Parameters of fitting wind speed (m/s) at 50 m from the surface in London for each month of 2020 are shown in Tables B.1, B.2, B.3, B.4, B.5, B.6, B.7, B.8, B.9, B.10, B.11, and B.12. QQ plot for residuals of fitting wind speed (m/s) at 50 m from the surface in London for each month of 2020 is shown in Figs. B.1, B.2, B.3, B.4, B.5, B.6, B.7, B.8, B.9, B.10, B.11, and B.12. All data were from NASA's "The POWER Project."

Table B.1 Random walk model: January 2020

	Estimate	Std. error	t value	Pr(>|t|)
a_1	0.113497	0.054065	2.099	0.0361 *
a_0	0.986541	0.006027	163.693	<2e-16 ***

Residual standard error: 0.644 on 741 degrees of freedom
Multiple R-squared: 0.9731, Adjusted R-squared: 0.9731
F-statistic: 2.68e+04 on 1 and 741 DF, p-value: < 2.2e-16

Table B.2 Random walk model: February 2020

	Estimate	Std. error	t value	Pr(>|t|)
a_1	0.190154	0.085133	2.234	0.0258 *
a_0	0.984323	0.006784	145.100	<2e-16 ***

Residual standard error: 0.8218 on 693 degrees of freedom
Multiple R-squared: 0.9681, Adjusted R-squared: 0.9681
F-statistic: 2.105e+04 on 1 and 693 DF, p-value: < 2.2e-16

Y. Ikeda, *Data Science of Renewable Energy Integration*,
Evolutionary Economics and Social Complexity Science 30,
https://doi.org/10.1007/978-981-99-8779-5

Table B.3 Random walk model: March 2020

	Estimate	Std. error	t value	Pr(>|t|)
a_1	0.170764	0.067530	2.529	0.0117 *
a_0	0.979902	0.007054	138.913	<2e-16 ***

Residual standard error: 0.5453 on 741 degrees of freedom
Multiple R-squared: 0.963, Adjusted R-squared: 0.963
F-statistic: 1.93e+04 on 1 and 741 DF, p-value: < 2.2e-16

Table B.4 Random walk model: April 2020

	Estimate	Std. error	t value	Pr(>|t|)
a_1	0.142192	0.053669	2.649	0.00824 **
a_0	0.978346	0.007775	125.825	< 2e-16 ***

Residual standard error: 0.5771 on 717 degrees of freedom
Multiple R-squared: 0.9567, Adjusted R-squared: 0.9566
F-statistic: 1.583e+04 on 1 and 717 DF, p-value: < 2.2e-16

Table B.5 Random walk model: May 2020

	Estimate	Std. error	t value	Pr(>|t|)
a_1	0.093312	0.043909	2.125	0.0339 *
a_0	0.985236	0.006266	157.246	<2e-16 ***

Residual standard error: 0.5045 on 741 degrees of freedom
Multiple R-squared: 0.9709, Adjusted R-squared: 0.9709
F-statistic: 2.473e+04 on 1 and 741 DF, p-value: < 2.2e-16

Table B.6 Random walk model: June 2020

	Estimate	Std. error	t value	Pr(>|t|)
a_1	0.10040	0.04697	2.138	0.0329 *
a_0	0.98505	0.00648	152.012	<2e-16 ***

Residual standard error: 0.5055 on 717 degrees of freedom
Multiple R-squared: 0.9699, Adjusted R-squared: 0.9699
F-statistic: 2.311e+04 on 1 and 717 DF, p-value: < 2.2e-16

Table B.7 Random walk model: July 2020

	Estimate	Std. error	t value	Pr(>|t|)
a_1	0.093125	0.044403	2.097	0.0363 *
a_0	0.986066	0.006043	163.175	<2e-16 ***

Residual standard error: 0.4578 on 741 degrees of freedom
Multiple R-squared: 0.9729, Adjusted R-squared: 0.9729
F-statistic: 2.663e+04 on 1 and 741 DF, p-value: < 2.2e-16

Table B.8 Random walk model: August 2020

	Estimate	Std. error	t value	Pr(>|t|)
a_1	0.045607	0.038904	1.172	0.241
a_0	0.992022	0.005051	196.401	<2e-16 ***

Residual standard error: 0.5148 on 741 degrees of freedom
Multiple R-squared: 0.9812, Adjusted R-squared: 0.9811
F-statistic: 3.857e+04 on 1 and 741 DF, p-value: < 2.2e-16

Table B.9 Random walk model: September 2020

	Estimate	Std. error	t value	Pr(>|t|)
a_1	0.121618	0.047304	2.571	0.0103 *
a_0	0.983957	0.006144	160.161	<2e-16 ***

Residual standard error: 0.5267 on 717 degrees of freedom
Multiple R-squared: 0.9728, Adjusted R-squared: 0.9728
F-statistic: 2.565e+04 on 1 and 717 DF, p-value: < 2.2e-16

Table B.10 Random walk model: October 2020

	Estimate	Std. error	t value	Pr(>|t|)
a_1	0.216631	0.073039	2.966	0.00311 **
a_0	0.974686	0.008161	119.427	< 2e-16 ***

Residual standard error: 0.6512 on 741 degrees of freedom
Multiple R-squared: 0.9506, Adjusted R-squared: 0.9505
F-statistic: 1.426e+04 on 1 and 741 DF, p-value: < 2.2e-16

Table B.11 Random walk model: November 2020

	Estimate	Std. error	t value	Pr(>|t|)
a_1	0.086369	0.047773	1.808	0.071 .
a_0	0.989114	0.005519	179.210	<2e-16 ***

Residual standard error: 0.542 on 717 degrees of freedom
Multiple R-squared: 0.9782, Adjusted R-squared: 0.9781
F-statistic: 3.212e+04 on 1 and 717 DF, p-value: < 2.2e-16

Table B.12 Random walk model: December 2020

	Estimate	Std. error	t value	Pr(>|t|)
a_1	0.107236	0.053235	2.014	0.0443 *
a_0	0.985211	0.006366	154.750	<2e-16 ***

Residual standard error: 0.5871 on 741 degrees of freedom
Multiple R-squared: 0.97, Adjusted R-squared: 0.9699
F-statistic: 2.395e+04 on 1 and 741 DF, p-value: < 2.2e-16

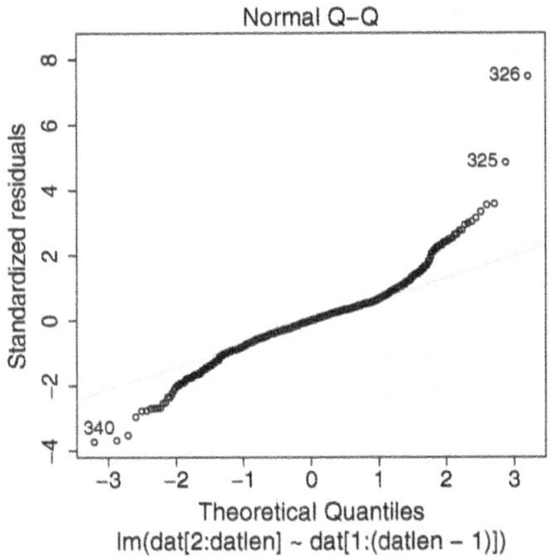

Fig. B.1 QQ plot for residuals of fitting wind speed (m/s) at 50 m from the surface in London: January 2020

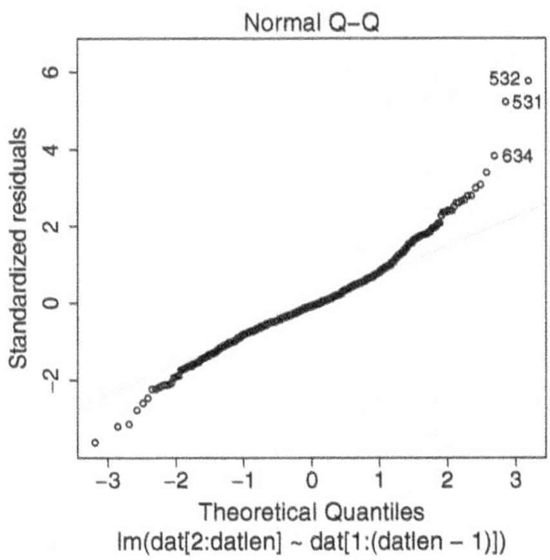

Fig. B.2 QQ plot for residuals of fitting wind speed (m/s) at 50 m from the surface in London: February 2020

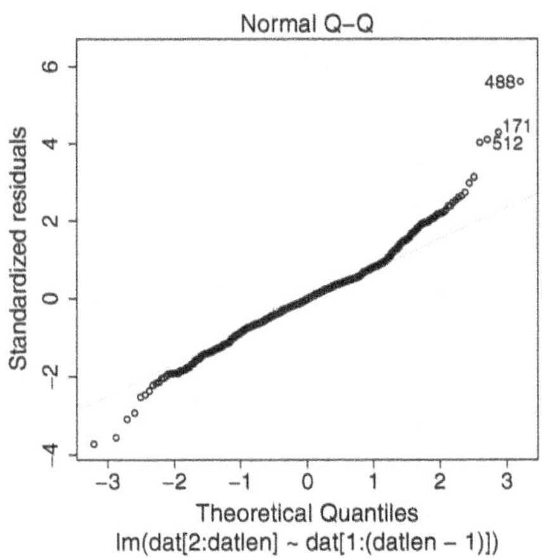

Fig. B.3 QQ plot for residuals of fitting wind speed (m/s) at 50 m from the surface in London: March 2020

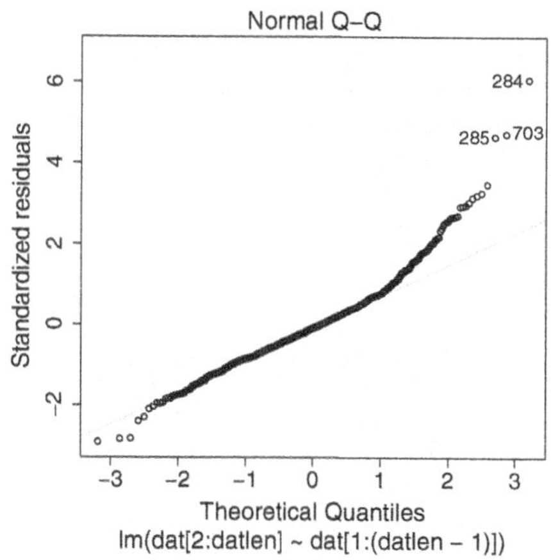

Fig. B.4 QQ plot for residuals of fitting wind speed (m/s) at 50 m from the surface in London: April 2020

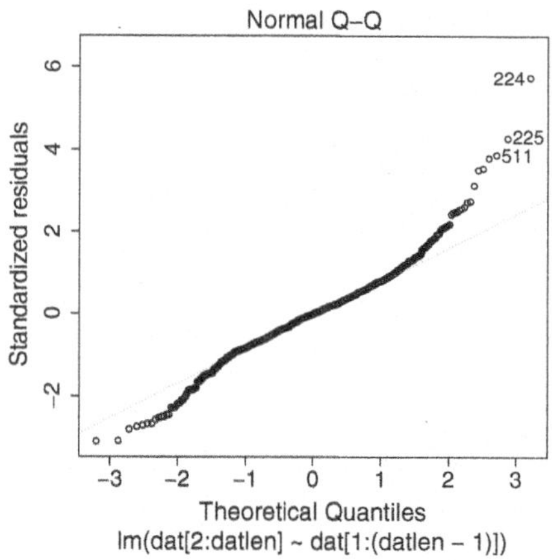

Fig. B.5 QQ plot for residuals of fitting wind speed (m/s) at 50 m from the surface in London: May 2020

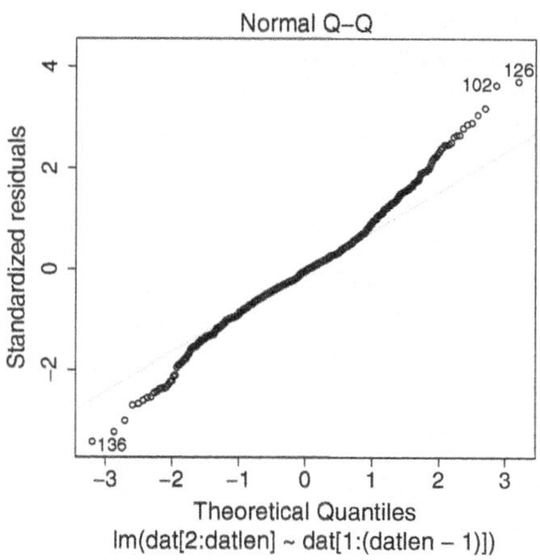

Fig. B.6 QQ plot for residuals of fitting wind speed (m/s) at 50 m from the surface in London: June 2020

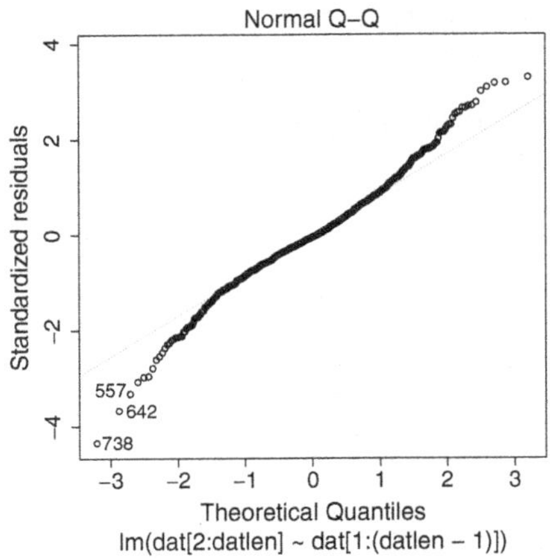

Fig. B.7 QQ plot for residuals of fitting wind speed (m/s) at 50 m from the surface in London: July 2020

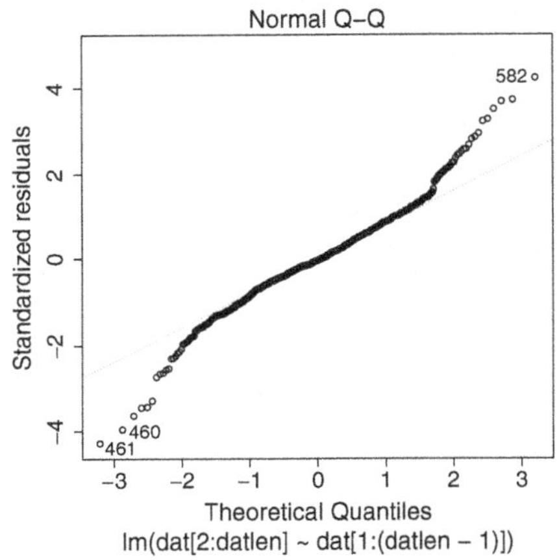

Fig. B.8 QQ plot for residuals of fitting wind speed (m/s) at 50 m from the surface in London: August 2020

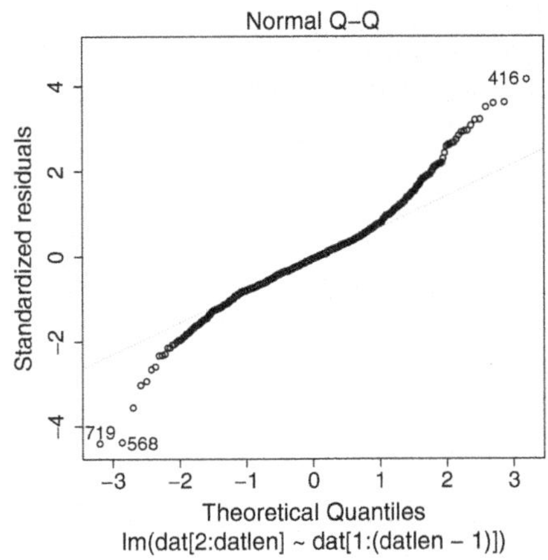

Fig. B.9 QQ plot for residuals of fitting wind speed (m/s) at 50 m from the surface in London: September 2020

Fig. B.10 QQ plot for residuals of fitting wind speed (m/s) at 50 m from the surface in London: October 2020

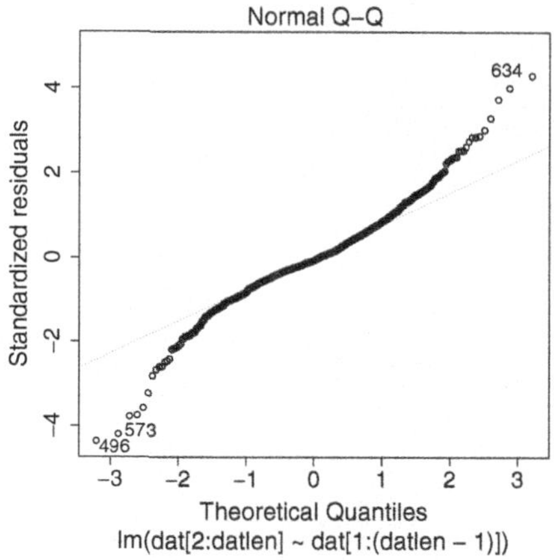

Fig. B.11 QQ plot for residuals of fitting wind speed (m/s) at 50 m from the surface in London: November 2020

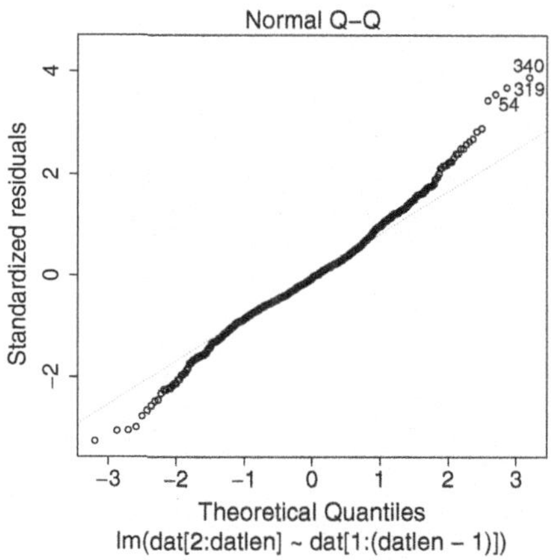

Fig. B.12 QQ plot for residuals of fitting wind speed (m/s) at 50 m from the surface in London: December 2020

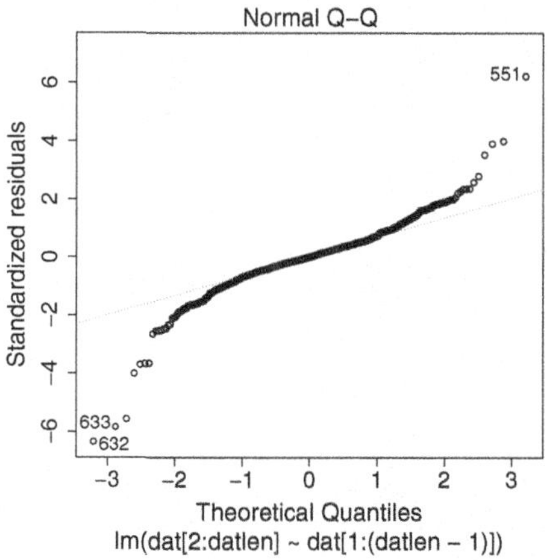

Appendix C
Correlation of Wind Speed

Figures C.1, C.2, C.3, C.4, C.5, C.6, C.7, C.8, C.9, C.10, C.11, and C.12 show the correlations of wind speeds at 50 m above the surface for each of these capitals from January 2020 to December 2020, respectively. These figures show that there is a positive correlation among relatively neighboring countries throughout the year and a negative correlation among geographically distant countries during the winter season. All data were from NASA's "The POWER Project."

Y. Ikeda, *Data Science of Renewable Energy Integration*,
Evolutionary Economics and Social Complexity Science 30,
https://doi.org/10.1007/978-981-99-8779-5

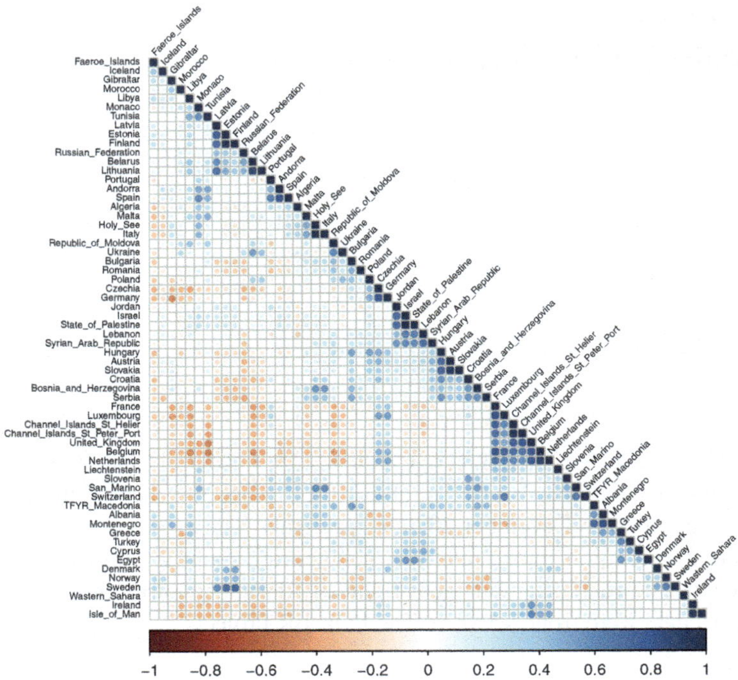

Fig. C.1 Wind speed correlation among European countries at 50 m from the surface: January 2020

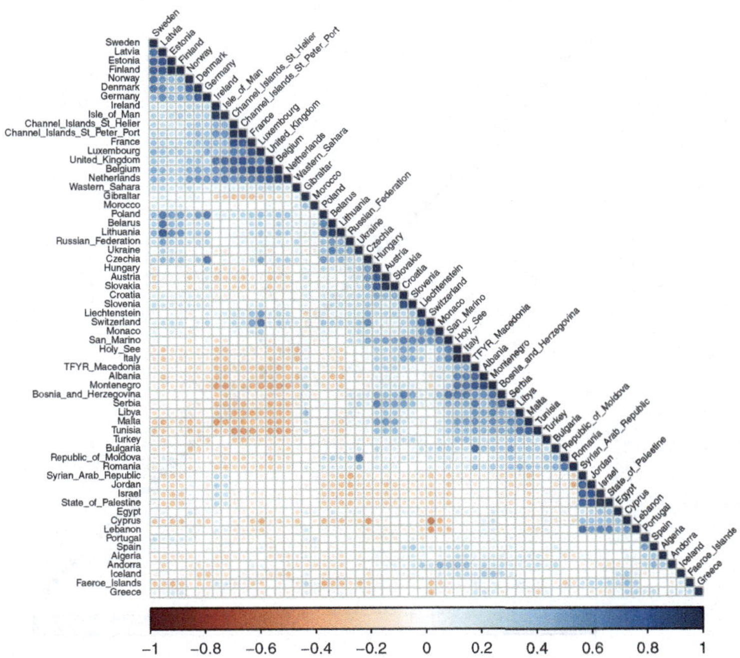

Fig. C.2 Wind speed correlation among European countries at 50 m from the surface: February 2020

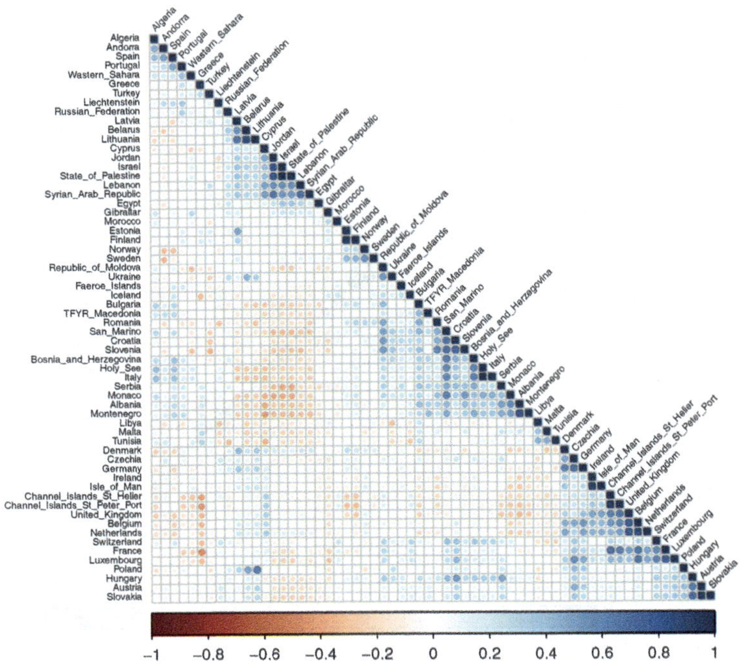

Fig. C.3 Wind speed correlation among European countries at 50 m from the surface: March 2020

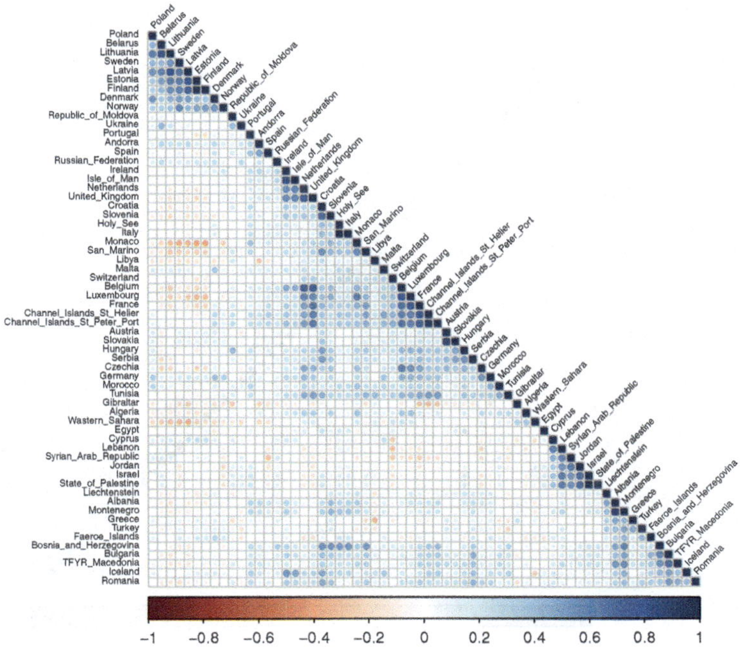

Fig. C.4 Wind speed correlation among European countries at 50 m from the surface: April 2020

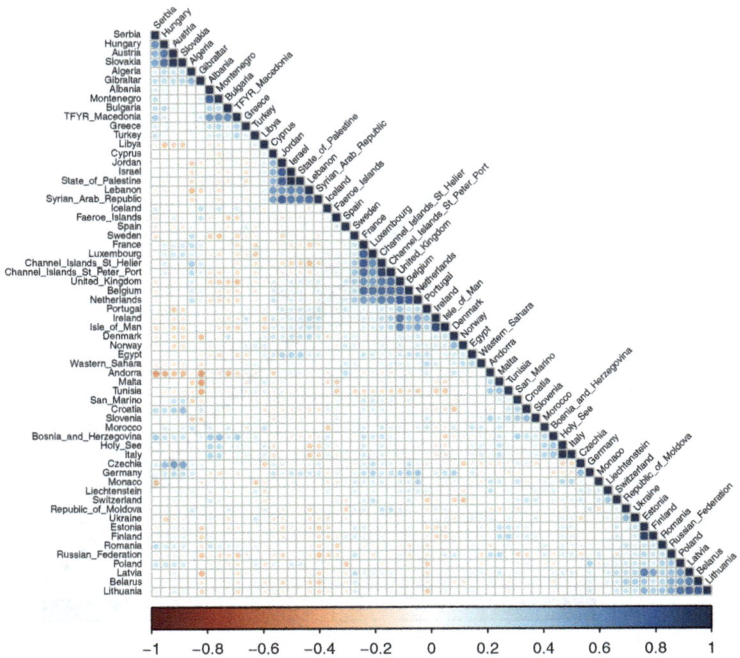

Fig. C.5 Wind speed correlation among European countries at 50 m from the surface: May 2020

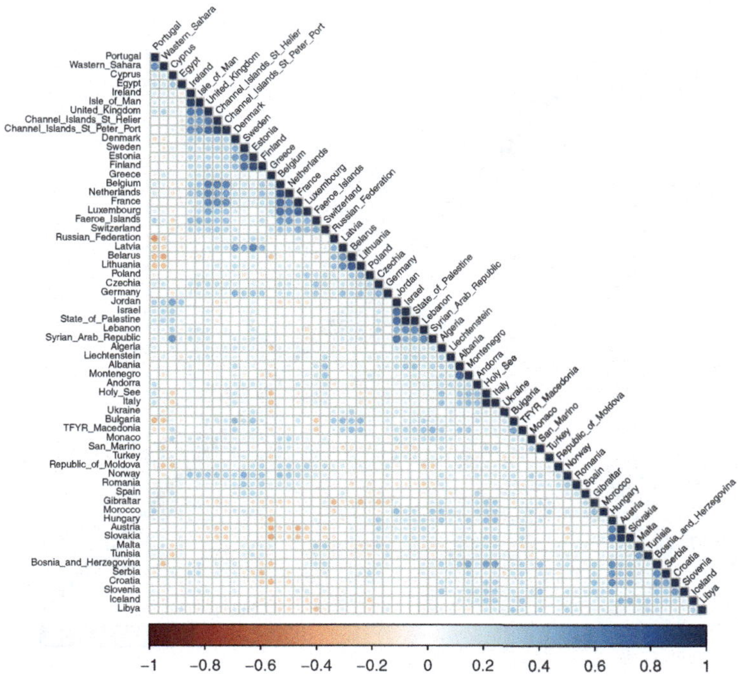

Fig. C.6 Wind speed correlation among European countries at 50 m from the surface: June 2020

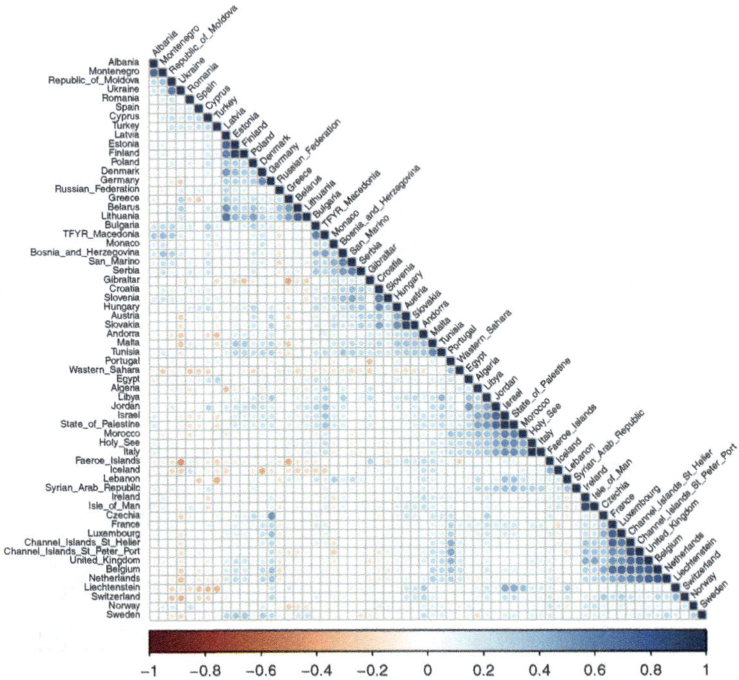

Fig. C.7 Wind speed correlation among European countries at 50 m from the surface: July 2020

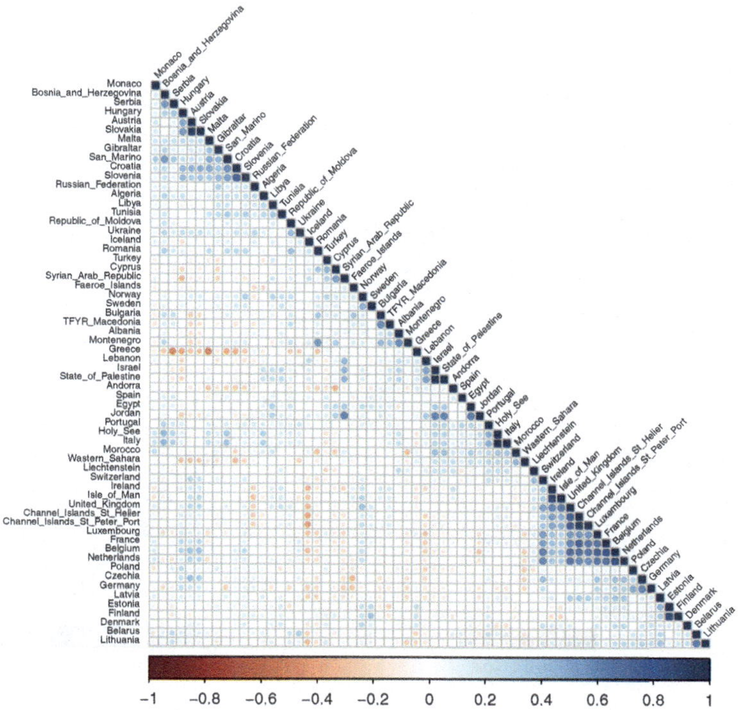

Fig. C.8 Wind speed correlation among European countries at 50 m from the surface: August 2020

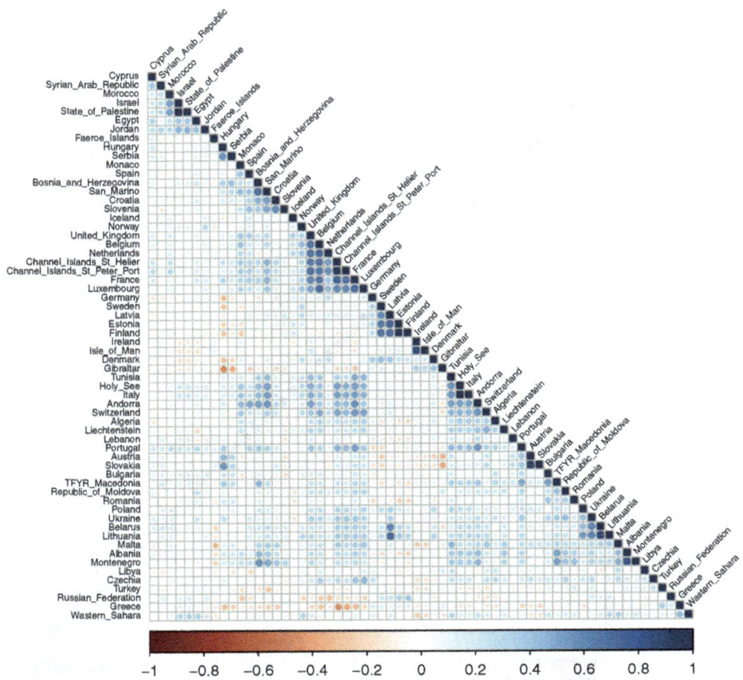

Fig. C.9 Wind speed correlation among European countries at 50 m from the surface: September 2020

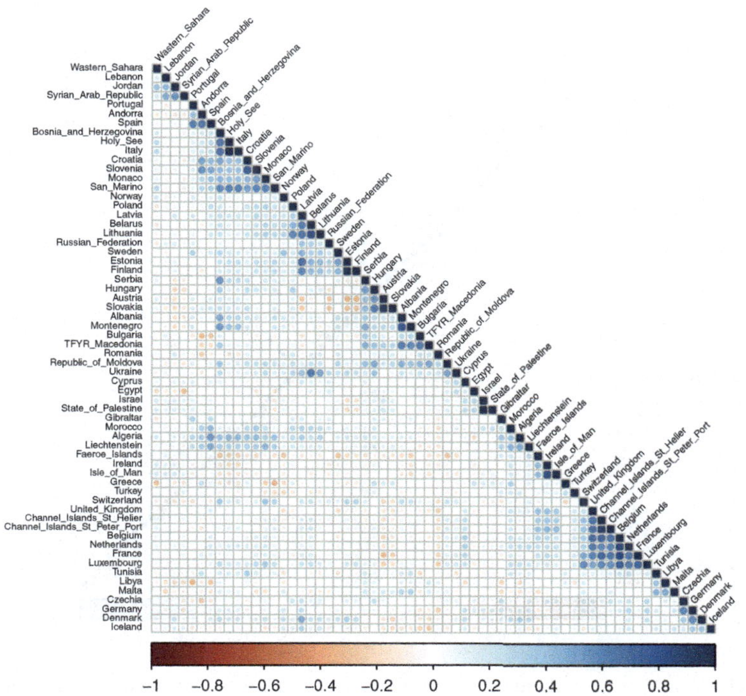

Fig. C.10 Wind speed correlation among European countries at 50 m from the surface: October 2020

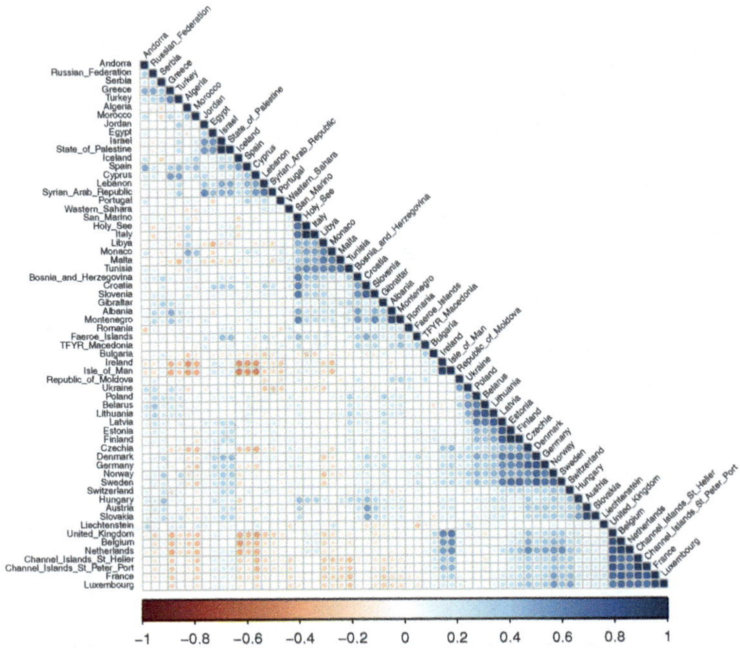

Fig. C.11 Wind speed correlation among European countries at 50 m from the surface: November 2020

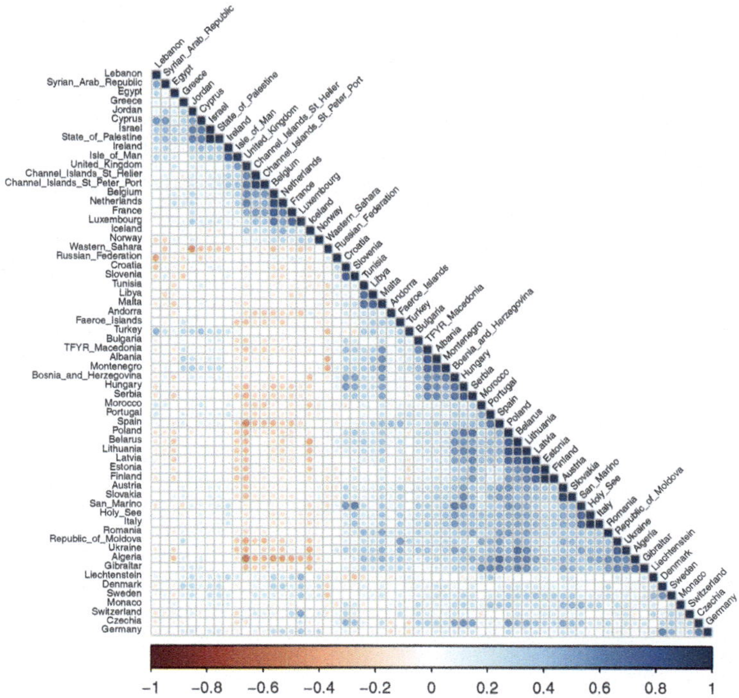

Fig. C.12 Wind speed correlation among European countries at 50 m from the surface: December 2020

Index

Y. Ikeda, *Data Science of Renewable Energy Integration*,
Evolutionary Economics and Social Complexity Science 30,
https://doi.org/10.1007/978-981-99-8779-5

The manufacturer's authorised representative in the EU is Springer Nature Customer Service Centre GmbH, Europaplatz 3, 69115 Heidelberg, Germany. If you have any concerns regarding our products, please contact ProductSafety@springernature.com

Printed and bound by CPI Group (UK) Ltd, Croydon, CR0 4YY
30/09/2026
02225318-0015